William James Lloyd Wharton

Hydrographical Surveying

A description of the means and methods employed in constructing marine charts.

Second Edition

William James Lloyd Wharton

Hydrographical Surveying
A description of the means and methods employed in constructing marine charts. Second Edition

ISBN/EAN: 9783337418694

Printed in Europe, USA, Canada, Australia, Japan

Cover: Foto ©berggeist007 / pixelio.de

More available books at **www.hansebooks.com**

HYDROGRAPHICAL SURVEYING.

A DESCRIPTION OF THE MEANS AND METHODS
EMPLOYED IN CONSTRUCTING

MARINE CHARTS.

By REAR-ADMIRAL

SIR WILLIAM J. L. WHARTON, K.C.B.,
HYDROGRAPHER TO THE ADMIRALTY.

SECOND AND REVISED EDITION

LONDON:
JOHN MURRAY, ALBEMARLE STREET.
1898.

LONDON:
PRINTED BY WILLIAM CLOWES AND SONS, Limited,
STAMFORD STREET AND CHARING CROSS.

PREFACE TO FIRST EDITION.

I HAVE endeavoured in the following pages to collect together information which has for the most part existed, but in a traditionary form, for many years.

Circumstances have led to a partial break in this tradition, and it is no disparagement to the more modern treatises on nautical surveying to say that the young surveyor has no book to which he can refer for information on all details, as they, confessedly, do not enter into them.

Belcher's, the former standard work, is out of print, and in many ways out of date, as though the main principles of chart-making must remain the same, the lapse of time and introduction of steam, &c., have placed many additional means at the disposal of the marine surveyor of to-day.

Knowing that there are many older and more experienced hydrographical surveyors than myself, it has been with considerable diffidence that I have applied myself during a period of leisure to stop the gap, and I hope my brother surveyors will remember that I write not for them, but for those young officers who wish to become acquainted with the practical portion of our branch of the naval profession.

The Appendix, largely composed of reprints of tables either falling out of print, or scattered about in different works, will, it is hoped, be found to save labour and time, and be useful to the Surveying Service at large.

W. J. L. WHARTON.

H. M. S. SYLVIA,
  March 29th, 1882.

PREFACE TO SECOND EDITION.

In preparing a second edition of Hydrographical Surveying I have endeavoured to alter as little as possible, confining myself mainly to bringing the book up to date in matters connected with instruments and fittings which have changed since the first edition was published.

In deep-sea sounding there has been an entire revolution from the substitution of wire for hemp, and as my own experience with the former is very slight, I am greatly indebted to Captain A. M. Field and Captain W. U. Moore for information on the subject. I have also received from these officers many notes on other subjects treated of, and wish to express my acknowledgements for the valuable aid they have thus afforded me.

W. J. L. WHARTON.

April, 1898.

CONTENTS.

——◆◇◆——

	PAGE
PRELIMINARY	1

CHAPTER I.

INSTRUMENTS AND FITTINGS .. .. — 4

Sextants and Stands—Horizon—Theodolite—Station Pointer—
Scales — Straight-edges — Chains — Protractors — Pocket
Aneroids—Heliostat — Ten-foot Pole—Drawing-Boards—
Weights—Transfer Paper—Paper—Books—Chronometers
—Marks—Boat's Fittings—Lead-lines—Beacons.

CHAPTER II.

A MARINE SURVEY IN GENERAL 57

CHAPTER III.

BASES 63

By Chain—By difference of Latitude—By Angle subtended by
known length—By Measured Rope—By Sound.

CHAPTER IV.

THE MAIN TRIANGULATION 73

General—Making a Main Station—False Station—Sketch—
Convergency—Calculation.

CHAPTER V.

PLOTTING 101

CHAPTER VI.

PAGE

RUNNING SURVEY 122

CHAPTER VII.

COAST-LINING 133

CHAPTER VIII.

SOUNDING 142
Boat Sounding—Ship Sounding—Searching for Vigias.

CHAPTER IX.

TIDES 159

CHAPTER X.

TOPOGRAPHY 178

CHAPTER XI.

HEIGHTS 183
By Theodolite—By Sextant—Obtaining Distance from Elevation of a known Height—Levelling.

CHAPTER XII.

OBSERVATIONS FOR LATITUDE 197
By Circum-meridian Altitudes of Stars—By Circum-meridian
Altitudes of Sun.

CHAPTER XIII.

OBSERVATIONS FOR ERROR OF CHRONOMETER 218
General remarks on obtaining Longitude—Error by Equal
Altitudes—Errors by two Stars at Equal Altitude.

CHAPTER XIV.

PAGE

MERIDIAN DISTANCES 248
 Telegraphic—Chronometric.

CHAPTER XV.

TRUE BEARING 272
 By Theodolite—By Sextant—Variation.

CHAPTER XVI.

SEA OBSERVATIONS 286
 Double Altitude—Sumner's Method—New Navigation—Short
  Equal Altitude—Circum-meridian Altitudes of Sun.

CHAPTER XVII.

THE COMPLETED CHART 295
 Fair Chart—Reducing Plans—Delineation—Symbols—Colouring—Graduation.

CHAPTER XVIII.

DEEP-SEA SOUNDINGS 308
 Wire Sounding—Dredging.

CHAPTER XIX.

MISCELLANEOUS 321
 Distortion of Printed Charts—Observations on Under-Currents
  — Exploring a River—Swinging Ship.

CONTENTS OF APPENDIX.

APP.		PAGE

A.—To prove that Tan Convergency = Tan Dep . Tan Mid Lat .. 331

B.—In Graduating a Chart on the Gnomonic Projection 332

C.—To prove Chord $= 2 \text{ rad} \left\{ \text{Vers} \left(90 + \frac{\theta}{2} \right) - 1 \right\}$ 334

D.—To prove Reduction to the Meridian $= \dfrac{\text{Cos } l \,.\, \text{Cos } d}{\text{Sin } z} \,.\, \dfrac{\text{Vers } h}{\text{Sin } 1''}$ .. 335

E.—To show that the Distance of Horizon in English Miles

$$= \sqrt{\frac{3}{2} \text{ height in feet}}$$ 336

F.—Base by Sound 337

G.—Form for Deck Book 338

H.—   ,,   Chronometer Comparison Book 339

J.—Table of Chords of Arcs 340

L.—   ,,   Lengths of Degrees, Minutes and Seconds 352

M.—   ,,   Reduction to the Meridian 366

N.—   ,,   Dip for Calculating Heights 368

O.—   ,,   Angles subtended by various lengths at different distances 369

P.—   ,,   Distance of Visible Horizon 370

Q.—   ,,   Distance of Sea Horizon 371

R.—   ,, for Ten-foot pole 372

S.—   ,, of Time in decimals of a day 373

T.—   ,,   Metrical and English Barometers 374

U.—   ,,   Corresponding Thermometers 375

V.—   ,,   Foreign Measures of Depth 376

HYDROGRAPHICAL SURVEYING.

PRELIMINARY.

THERE is nothing mysteriously difficult in the art of Hydrographical or Marine Surveying. For the ordinary details, no deep theoretical or mathematical knowledge is needed ; on the contrary, it is an eminently practical branch of the Naval Profession.

An aspirant to its acquirement should have a quick eye, should possess the ordinary good common-sense that is necessary to secure success in all walks of life, but above all he must have a boundless capacity for taking pains in details at all times and seasons.

The advice, "Surtout, point de zèle," does not apply to surveying. Without zeal, and the utmost keenness for the progress of the work, the attention and interest will soon fail ; and the necessity for constant application throughout long days, often extended into the night, will soon seem monotonous, and become a bore to one whose heart is not thoroughly in it.

Happily, it is a profession of volunteers, and the author's experience is, that in no branch of the public service can the juniors be more anxious to do their duty, not only to the letter, but to the utmost of the spirit, and to such as these no day seems long enough. To them, the interest is constantly kept up. Every day has its incidents. The accuracy of the

B

work of each assistant, when proved, is an infinite gratification
to him, and he has also the continual satisfaction of feeling
that of all he does a permanent record will remain, in the
chart which is to guide hundreds of his fellow-seamen on
their way.

For any naval officer, then, who is really anxious to learn,
the practical part of surveying will soon be mastered. It
will quickly become a labour of love, and the constant attention and trouble necessary will be merged in the interest
taken in the work. Thorough honesty must always guide
him, so that nothing may appear that is not known to be
correct. Omissions there must always be, but let there be
no sins of commission, that pains and care will prevent.

It is not of course suggested that all can become thorough
good surveyors in all branches. One man will have a particular aptitude for astronomical observing; another will
have a natural talent as a draughtsman, that no efforts on
the part of another can compete with, and so on; but to
become a good practical hand is within reach of all who
seriously are desirous of being so, and will take the trouble
to gain the necessary experience, without which all theory
and book-teaching will be useless.

One crucial test of a surveyor's capability is his power of
so planning and carrying out his work as to economise time.
Even in a plan of an ordinary bay or harbour, which every
naval officer should be able to make, the trained surveyor, by
his experience of how to set about it, will accomplish it in a
fraction of the time required by another. Nothing is more
important than the knowledge of how to suit means to the
end, and many hours are wasted by the anxious tyro in
endeavouring to attain an accuracy in detail which cannot be
utilised in the finished plan.

Inside of the broad principles of map making, marine
surveying is made up of numerous dodges and details, for
which there is nothing like practical exposition on the ground,
and those who can get others to show them will need but
little other help, but as in many cases this instructor will

not be at hand, it is hoped that the following pages may sometimes supply the information required.

It may seem to many that some points remarked on are too insignificant to be heeded, but those who are acquainted with the work will know how much time is lost by inattention to, or ignorance of, these little things, and a young surveyor will be a very few days at work before he finds this out.

We assume our reader to have the ordinary knowledge of the sextant that all naval officers are taught, and that he is not entirely ignorant of the first principles of making a plan from a base by means of angles.

We write mainly for those who join the Surveying Service, and shall speak throughout as though we had the resources of an ordinarily fitted surveying ship at command.

We have endeavoured to take things in the order that they will generally come in the prosecution of a survey.

In work of the nature of Hydrographical Surveying, it is impossible to give directions as to how to undertake every detail. Ordinary means fail now and again from exceptional, local, or other circumstances, and ready resource in overcoming difficulties is one of the most important requisites in a nautical surveyor. To invent or improvise a method of doing a particular piece of work is a most satisfactory achievement when successful, but it is scarcely necessary to say that this can only come to the most naturally talented with experience.

The following pages will then not be found to provide for every occasion, but will only describe the ordinary and accepted modes of setting about work.

CHAPTER I.

INSTRUMENTS AND FITTINGS.

Sextants and Stands—Horizon—Theodolite—Station Pointer—Scales—
Straight-edges—Chains—Protractors—Pocket Aneroids—Heliostat—
Ten-foot Pole—Drawing-Boards—Weights—Transfer Paper—Paper
—Books—Chronometers—Marks—Boat's Fittings—Lead-lines—
Beacons.

Errors of instruments to be ascertained. IN preparing for any surveying work, whether in a regularly fitted surveying ship or not, the first thing is to test all instruments and ascertain their errors. To do the former well, it is necessary to have an intimate knowledge of the points on which each instrument is liable to go wrong, which is only thoroughly to be learnt by experience; but a few hints will assist the beginner.

A thorough acquaintance with the construction of instruments will save many an hour, lost by one whose instrument has gone wrong while in the middle of his work, and spent in fruitless efforts to make out where the fault lies.

No instrument perfect. No instrument, not even engine-divided protractors, can be assumed to be without error, and are seldom found so, and though those errors may be small, in some cases they are of importance, and no work can be deemed satisfactory without the knowledge of how much correction should be applied, in such instances as it may be necessary to do so.

Contents of Chapter. We shall therefore commence by some observations on instruments, and on all materials and fittings required for conducting a regular marine survey, embodying in these such hints on using each instrument in general, as are likely to be useful, and also some on choosing them that are not mentioned by Heather in his work on Instruments.*

This useful work, which should be in the hands of every Heather's Work on Instruments. surveyor, goes so fully into the construction of instruments, and in most cases into the methods of ascertaining, and, as far as may be, correcting their errors, that we shall refer the reader to it on most points, adding only certain practical suggestions that are not therein mentioned.

HADLEY'S SEXTANT.

It is not, perhaps, necessary to say much about the sextant, as so many works have already treated the subject; but there are several practical points not generally mentioned, which may be of value in selecting a sextant with a view to the work of a nautical surveyor.

Besides those noted by Heather, then,—

1. One of the eye-pieces of the inverting telescope should have a high magnifying power, about 15 diameters, as contacts of the sun's limbs in observations with the artificial horizon are far easier made the larger the suns.

2. Several dark eye-pieces should be provided, with neutral Dark eye-pieces. tint glass in them of different intensities. These should be fitted, not to *screw* on to the eye-piece, but ground conical, to slip on to a similar conically ground surface on the telescope eye-piece. These will be found very useful on cloudy days, as a little practice will soon enable the observer to substitute one shade for another in a fraction of a second, as clouds sweep on or off the sun, and many sights will thereby be saved. It is very important to have the suns in artificial horizon observations of the same brilliancy, and for this reason the hinged shades on the sextant should never be used for the purpose; as, in the first place, they introduce error, and also, if the shades have to be altered to suit the varying

---

* "Mathematical Instruments." J. F. Heather, M.A. Lockwood and Co., London.

brightness of the sun during the observation, the suns will
be of different brilliancies, as these shades are never of the
same tints.

By using the dark eye-pieces, the up-and-down piece,*
when adjusted to equalise the suns, will bring the axis of the
telescope nearly exactly in line with the edge of the silvered
surface of the horizon-glass, which is the best position for
observing, and from which it must never be moved until the
equal altitudes or other observations are complete. No
matter what depth of shade is then used by shifting the
dark eye-pieces, the two images will be of the same tint.

The darker the shade used the better. Beginners are very
apt to use too bright suns.

If in observing with the sun the observer can accustom
himself to use one eye for taking the observation, and the
other for reading and setting the vernier, he will find it very
convenient, and it will tend to keep both his eyes in good
order.

Position of
up-and-
down
piece.

3. It is very convenient for picking up the images in
the artificial horizon, if the up-and-down piece is so placed
as to enable the observer to look over it into the horizon-
glass.

In many sextants the up-and-down piece is placed so close
to the index glass that this is not possible, and regard should
be had to this point.

4. An interrupted thread, to screw the telescope into the
collar of the up-and-down piece, is a great convenience.

5. An extended vernier, *i.e.* a vernier whose divisions are
twice the distance apart of those on the arc, will be found
convenient for accurate observing.

6. A steel tangent screw will be found to last longer and
work more evenly than a brass one.

The methods of ascertaining the index and other errors of
Hadley's sextant, and correcting them, are so fully entered

---

* The up-and-down piece of a sextant is the portion that bears the
collar for the telescope.

into by Heather, that they are here omitted, with the exception of the following remarks on the centring error :—

This very important error of the sextant cannot be corrected **Centring** in the instrument, and it requires a considerable amount of **Error.** labour to settle its quantity, which in an indifferent instrument may be quite sufficient to vitiate the result of any observations on one side only of the zenith.

The centring error, pure and simple, arises from the non-coincidence of the centres of the index arm and of the graduated arc, so that the vernier does not move truly along the arc, and the angle read off will not be correct. This error varies with the angle, and is generally greater as the angle increases, but the same result of error appears from the index arm becoming bent ; from any part of the frame receiving a blow which alters its shape ; from the flexure of the instrument from varying temperature ; and from defective graduation ; but, as it is generally impossible to disentangle the errors arising from these different sources, they are all included in the one correction for centring.

Centring error is to be obtained by comparing the angle measured by the sextant with the true angle.

It is to be found roughly by measuring a series of angles carefully, by repetition, with a large theodolite, between well-defined objects on the horizontal plane at different angular distances, and then measuring the same with the sextant placed on a stand. The difference will be the centring error at each angle, index error being first applied.

The most accurate method, because it employs a large number of observations for the same, or nearly the same, angle, is by observation of pairs of circum-meridian stars in the artificial horizon, at various altitudes. Double the difference between the resulting latitude by each star, and the mean latitude, will be the centring error for an angle equal to the double altitude of that star, that is the angle actually measured by the sextant, index error being carefully determined and applied before working out.

The sign of the correction is easy to determine from a

consideration of whether the altitude is too little or too great. Thus in north latitude, if stars south of the zenith give a latitude too great, their altitudes have been too little, and the correction for centring will be plus.

It is hardly necessary to say that every precaution must be taken to eliminate other errors, such as choosing stars of a closely similar altitude, unless the latitude is already accurately known; determining the roof error of the horizon, or eliminating it by reversion; carefully correcting the refraction for temperature, etc., and that it requires considerable accuracy of observation, and many sets to arrive at a good result. The agreement, or otherwise, of the mean latitude by each pair will form an excellent test of the general accuracy of observation, and the agreement of the resulting centring errors by different observations at the same altitudes will enable the observer to judge of the truth of his final errors. Thus every careful set of observations for latitude affords a means of testing this error.

Centring error may also be obtained by careful measurement of the angles between stars. The correct apparent distances must be found in the same manner as in clearing a lunar distance; the true distance being first calculated from their declinations and right ascensions, but if stars in the same vertical plane can be chosen, the apparent distance can be arrived at by simple application of the refractions.

There are other methods, involving more calculation, which need not be described.

The centring error is determined at **Kew Observatory** for certain angles by fixed collimators, and is given on every Kew certificate, but it must be remembered that in any case it can never be considered as determined for good. Including, as it does, errors from so many causes, it does not remain perfectly steady, but its amount should be ascertained from time to time for any sextant which is to be employed for accurate determination of positions, for circumstances often prevent the use of methods whereby it as well as other errors are eliminated. For instance, a latitude may have to be obtained by altitude of the sun only, when, without knowledge of the centring error, it may easily be incorrect to as much as a minute, or even more.

As an example, the author's Troughton sextant had at 120° a centring error of – 20″. After a fall and repair by the maker, it was + 50″.

To find the error caused by the refraction, through non-parallelism of the sides, of the coloured shades. **Errors of hinged shades.**

Measure the diameter of the sun, with different combinations of the shades. Take out the pin which supports one set of the coloured shades, and replace the shades *reversed*, so that the face before next the index glass is now away from it. Remeasure the diameter of the sun with same combinations as before, and half the difference of the measurements of each set will be the error due to the shade reversed.

These errors can be neglected in sea observations, and if coloured eye-pieces are fitted as recommended above, the shades are not required when the artificial horizon is made use of.

SOUNDING SEXTANT.

This useful form of sextant is made of various sizes. It chiefly differs from the observing sextant in being generally lighter and handier, in having the arc cut only to minutes, and having a tube of a bell shape so as to include a larger field in the telescope.

All angles in the frame of the instrument should be rounded off, especially that at the zero end of the arc. Considerable injuries may result to the face of the observer when using the sextant in a boat in a lively sea, if this is not done.

The graduation of the arc should be plain enough to read without a magnifying-glass.

The measurable angle should be as large as possible, *i.e.* about 140°.

The index glass should be large, so as easily to pick up objects.

Good tube invalu-able. The telescope should be of a high magnifying power and clear definition.

These sextants are now supplied by the Hydrographic Office with two telescopes—one for ordinary use, and another, of aluminium, with a larger object glass for occasions when faint objects are required to be seen. The collimation of these large telescopes is however a delicate matter, and when accuracy is required, should be tested.

When in good adjustment, a sounding sextant so fitted is invaluable for star observations with a faint sea horizon.

RESILVERING MIRRORS.

On service, the mirrors of sextants, especially sounding sextants, frequently get dimmed by damp, and the surveyor must be able to resilver them himself.

A supply of tinfoil, of good quality, for this purpose, is one of the necessary stores. Mercury is always to be had. The operation has been frequently described, but it is perhaps better to repeat it.

Take a piece of tinfoil, a little larger than the glass to be silvered, and smooth it out on a perfectly flat surface, as a sheet of plate glass, or a thick smooth book-cover. This smoothing can be well done by a little pad of chamois leather, which can be kept for the purpose, or by the finger.

Drop a small bubble of mercury on to the foil, and by gentle rubbing with the pad, spread it over the former so that it shows a bright surface. Pour mercury on until the piece of foil is quite fluid, and brush any large spots of dross lightly off. Lay a piece of clean paper, long enough to handle easily, on the mercury, and the glass, previously well cleaned by means of spirits of wine, on the paper. Pressing on the glass with one hand, withdraw the paper with the other, slowly and steadily, and a pure surface

will appear under the glass, the dross all coming away with the paper.

Incline the book, or whatever surface we have been working on, so as to let superfluous mercury run off, placing strips of tinfoil at the lower edge to assist in sopping this up.

After from twelve to twenty-four hours, the amalgam will be dry, and firmly adhering to the glass. Cut the edges carefully round with a sharp knife, and varnish lightly over, either with the clear stuff used by the instrument makers, or with varnish that can be made on board, by dissolving sealing-wax in spirits of wine.

The glasses of some sextants seem fitted on purpose to invite the damp to penetrate between glass and silvered surface. These will want protection by sticking thin strips of paper along the edges exposed, and well varnishing. In some cases a stopping of thick amalgam, placed between the glass and the frame at the back, where there is one, will answer well, and prevent any damp getting at the back of the glass at all.

The mercury which remains will contain tinfoil in amal- **Amalgamated mercury.** gam, and should be preserved in a bottle by itself, draining off the thick of the amalgam by a sharply twisted paper funnel. It can then be used again for resilvering. Care must be taken not to allow any of this to get into the artificial horizon bottles, as the smallest quantity of it will spoil a whole bottle of pure mercury, and the amalgam can only be removed by evaporation.

Notwithstanding, mercury containing tin in amalgam can be used for artificial horizon work, by carefully sweeping the surface after it is poured out, with a piece of paper. Some observers have gone so far as to prefer amalgamated mercury for this purpose, but we do not agree, except when used in connection with the amalgamated trough described on page 13.

In resilvering an horizon glass, only the portion required **Horizon Glass.** should be operated on, leaving one half clear. The edge of the foil must be sharply and smoothly cut before applying the mercury, and not the smallest nick or cut permitted to remain in it.

SEXTANT STAND.

Though a practised observer will get good observations in an artificial horizon, with a sextant without a stand, he will get them far better *with* one, and in all work where accuracy is aimed at, a stand should be used.

Unsteadiness of hand, to which all are so liable, from previous exertion, indisposition, and many other causes, is put out of the question by using a stand.

With star observations this is especially the case, as it is extremely difficult to hold the instrument in the hand firmly enough to prevent a little vibration of the images.

Sextant stands should be lacquered, not bright, and should have large heads to the foot screws, so as to be grasped easily while observing.

The bearing which carries the sextant should be accurately fitted into the socket in the handle, and should be very slightly conical. If too much so, it is liable to jam.

The counterbalances are usually too heavy for an ordinary sextant. They should be of such a weight as to balance the sextant without the screws at the ends of the pivot being set up too taut. Sometimes one weight is enough, or as much lead can be taken out of each as is necessary to reduce the weights to balance. The weights are now sometimes fitted to slide in and out, thus allowing of adjustment.

Stools for stand. Small three-legged stools about 14 inches high, on which to place the sextant stand, should be made, and it will be found convenient to sink hollows in the top to correspond with the three foot screws to prevent slipping.

Other little hollows sunk in the top for the spare dark eye-pieces to lie in, will also prevent these falling off, and by placing them in regular order, any one can be at once picked up without delay, when it is requisite to change them.

Another similar stool for the observer will make him comfortable, a great point for good observing.

ARTIFICIAL HORIZON.

The glass in the roof should be of the best quality, and the faces of each pane accurately parallel.

A wooden trough to place inside the iron one is a convenience, as it raises the level of the mercury up to the height of the lower edge of the glasses in the horizon roof, a consideration where low altitudes have to be observed. The reduced area of mercury will not matter when observing the sun. When taking stars, the iron trough only should be used, as stars are more difficult to pick up, and its larger area will facilitate operations.

Three short wooden legs or buttons, fitted to the iron trough, will enable it to stand steadier on uneven ground than the four projections usually cast on the under side.

In connection with this, an artificial horizon stand is very useful. This consists of two iron plates; the lower one has three short legs on which it stands firmly; the upper one is pierced by three long large-headed screws, which serve as legs and fit into slight hollows on the lower plate. By adjusting these, the horizon laid on the upper plate can be levelled, when we have uneven ground. Four iron battens, screwed on to the upper plate so as just to permit the horizon roof to fit inside them, will prevent any wind getting to the mercury. Horizon stand.

The horizon cover should be marked at one end, or side, and this mark should in most cases be in the same position with regard to the observer. Of this more is said under "Observations." Mark on cover.

A new form of horizon is now being introduced, with the object of diminishing the waves set up in mercury by vibrations. Amalgamated Trough.

It consists of a circular shallow trough, of metal gilt. This is amalgamated, after getting the surface absolutely clean and free from grease, by wetting it with a few drops of dilute sulphuric acid, and then rubbing into it a drop of mercury until the whole surface is bright, when a very small quantity of mercury added will flow evenly and form a horizontal surface.   The dross is wiped off with a broad camel's hair brush.

In this shallow trough waves are killed almost instantaneously.   The trough should be thoroughly washed on each occasion before being used.

THEODOLITE.

The less a theodolite is tampered with by unpractised hands the better, but they must be adjusted from time to time, and little things are constantly wanting attention.

The adjustments are well described by Heather, but as it is very important to know them, they are here given, in case the former work should not be at hand.   The adjustments are—

1. Adjustments of the telescope, viz., for parallax and for collimation.

2. Adjustment of horizontal limb, viz., to set the levels on the horizontal limb to indicate the verticality of the azimuthal axis.

3. Adjustment of the vertical limb, viz., to set the level beneath the telescope to indicate the horizontality of the line of collimation.

Commence operations by setting up the theodolite as level as you can by eye, by moving the legs.   See that the legs are firm, and everything tight.   Set all levels as true as you can, by the parallel plate screws and vertical arc tangent screw.

Adjustment for Parallax.   Parallax is occasioned by the image formed by the object-glass not falling exactly on the cross-wires.

First adjust the moveable eye-piece until the cross-wires are sharply defined.   Then obtain the proper focus for the object by moving the milled head on the telescope.

This will throw out the image of the cross-wires, and the eye-piece must be again adjusted, until cross-wires and object are both truly in focus.

This has to be done each time the theodolite is set up, and is therefore only a temporary adjustment. The others are more permanent.

Collimation is effected by directing the telescope on some well-defined point, and bringing it to coincide with the intersection of the wires, with the level downwards. **Adjustment for Collimation.**

Turn the telescope in the Y's, until the level is uppermost. If the object is still at the intersection of the wires, the collimation in altitude is correct.

If not, bring the wires half-way towards the object by turning the screws holding the diaphragm. Then re-set the telescope by the tangent screw for the object, and bring the telescope round in the Y's to its former position, when any displacement still existing must be corrected in the same way, half by the diaphragm screws, and half by the tangent screw. After a few trials the error should be corrected.

Do the same with the telescope with level right, and level left, at right angles to its former positions in the Y's, for azimuth error. When this is done, the cross-wires, while the telescope is slowly revolved, should remain over the object.

* The collar being tightened by its clamping screw, unclamp the vernier plate, and turn it round till the telescope is over two of the parallel plate screws. Bring the bubble of the level beneath the telescope to the centre of its run by turning the tangent screw of the vertical arc. Turn the vernier plate half round, bringing the telescope again over the same pair of the parallel plate screws; and, if the bubble of the level be not still in the centre of its run, bring it back to the centre, half way, by turning the parallel plate screws over which it is placed, and half way by turning the tangent screw of the vertical arc. Repeat this operation till the bubble remains accurately in the centre of its run in both **Adjustment of Horizontal Limb.**

---

* From Heather.

positions of the telescope; and then turning the vernier plate round till the telescope is over the other pair of parallel plate screws, bring the bubble again to the centre of its run by turning these screws. The bubble will now retain its position, while the vernier plate is turned completely round, showing that the internal azimuthal axis, about which it turns, is truly vertical.

If the bubbles of the levels on the vernier plate are now brought to the centre of their tubes, by means of the screws fitted for the purpose, they will be adjusted to show the verticality of the internal azimuthal axis.

Now, having clamped the vernier plate, loosen the collar, by turning back the screw, and move the whole instrument slowly round upon the external azimuthal axis, and, if the bubble of the level beneath the telescope maintains its position during a complete revolution, the external azimuthal axis is truly parallel with the internal, and both are vertical at the same time; but, if the bubble does not maintain its position, it shows that the two parts of the axis have been inaccurately ground, and the fault can only be remedied by the instrument maker.

Adjustment for vertical limb.

To adjust for the vertical limb, the bubble of the level being in the centre of its run, reverse the telescope, end for end, in the Y's, and if the bubble does not remain in the same position, correct for one half the error by the capstan-headed adjusting screw at one end of the level, and for the other half, by the vertical tangent screw. Repeat the operation till the result is perfectly satisfactory. Next turn the telescope round a little, both to the right and to the left, and if the bubble does not still remain in the centre of its run, the level must be adjusted laterally by means of the screw at its other end. This adjustment will probably disturb the first, and the whole operation must then be carefully repeated. By means of a small screw, fastening the vernier of the vertical limb to the vernier plate over the compass box, the zero of this vernier may now be set to the zero of the limb, and the vertical limb will be adjusted for horizontality.

The vertical limb should move in a truly vertical plane. **Adjustment for vertical plane.** Any error can only be adjusted in the larger instruments, but every theodolite must be tested for it, as, if much error exists, the instrument requires alteration by the maker. It will introduce error into all angles to objects much elevated or depressed, and it is especially important for observations for true bearing to know that this adjustment is perfect.

To test it, direct the theodolite when horizontal to either the edge of a well-built wall, or still better, a steady plumbline. The cross-wires, when the instrument is elevated and depressed, should still intersect the line.

If they do not, in 6-inch theodolites, the adjustment can generally be made by means of screws on one of the Y frames. In smaller theodolites we must accept the error, and take care not to use them for true bearings.

These adjustments completed, the instrument will be ready for work.

There are, however, a variety of small points on which a **Points liable to derangement.** theodolite may go wrong while away in the field, and a knowledge of the general causes of these temporary derangements is very useful, and may prevent loss of a day's work, and much aggravation to all concerned.

The parts of a theodolite, especially in an old instrument, that most frequently get out of order, are the small screws which hold the milled heads of the tangent screws in their places. A young observer is often much bothered and puzzled by his instrument not coming back to zero, which may result from many things, but most frequently from one of the small screws above mentioned being loose. Screwing it up tight enough to prevent any play when the instrument is clamped, but not so tight as to make the tangent screw work hard, will often remove the difficulty.

Other causes of not coming back to zero are:—

1. Looseness of the sockets through which the tangent screws work, and which can be easily tightened by their screws.

2. Looseness of the fittings of the brass stand on the theodolite legs. There are many working parts here, and any

of them are liable to get loose.   The leverage on the brass
plate that fits on each leg head is enormous, if the leg should
be allowed to swing out in taking off the rings; and as the
screws that hold them on are small, looseness may easily
take place here.

3. In an old instrument, the faces of the clamping plates
may screw close together without clamping the instrument
tightly.   This is from the part that holds the instrument, and
which gets all the friction, being much worn.   The parts into
which the clamping screw fits must be smoothly filed on
their inner faces, so as to ensure the other parts coming into
contact with the body of the instrument, before the faces of
the clamping plates meet.

4. The upper plate will sometimes not revolve freely, but
catches every now and then.   This is from the piece of metal
which clamps the two plates together either not fitting very
well, or being dirty inside, or perhaps bent.   Placing the
finger underneath so as to press it up to the lower plate,
whenever the plates are to be revolved, will ensure its working smoothly, and is a better thing for a young hand to do
than to attempt to take it off.   The same thing will happen
to the reading-glass plate; but here it is often the little screw
underneath which is loose, and simply screwing it up will
relieve it.   Lifting it with the finger will always assist it to
run round easily.

Putting in new webs.   An operation the nautical surveyor has frequently to per-
form is replacing the wires, or rather cobwebs, of his theodo-
lite telescope.

For this purpose catch a garden spider, as a house spider
does not spin his rope taut enough.   Having cut some holes,
say two inches square, in a strip of cardboard about three
inches wide, place the spider on it, and shake him off.   As
he throws out his web in falling, twist it up on the cardboard
so as to cross the holes, and lay it on one side.

Having taken the diaphragm from the telescope, and
scraped off the old balsam, lay it on the table and place the
smallest drop of Canada balsam on its edges.   With the aid

of a magnifying glass, place the cardboard across it, in such a manner that the web will lie in the notches cut in the diaphragm, when it will adhere to the balsam.

Heather gives a good description of measuring angles with the theodolite, to which we will add, that, Measuring angles with the theodolite.

Regard must be had to the purpose for which the angles are to be taken, in settling how many times, and in what manner, the angles shall be repeated. An error of two minutes will make no perceptible difference when plotted, unless the line be very long, say five feet. All objects, therefore, that are simply to be plotted, and do not come into the triangulation, can be taken round once with zero at 360°, and a second round taken after with another zero, say 100° for convenience' sake, simply for the purpose of making sure that there are no gross errors, as no theodolite in adjustment should give an angle in error as much as two minutes.

Angles to main stations, however, will be very likely required to enter into the calculation, and the correctness of the plotting will any way depend on them. These must therefore be repeated, the number of times varying according to the degree of accuracy required. Repeating angles.

One method of repeating angles is thus given in Heather, somewhat altered.

Having taken the first measurement, loosen the clamp of the lower plate, turn the theodolite bodily round until the telescope is directed upon the zero-object, and again clamping the instrument, perfect the bisection of the zero by the cross-wires by means of the slow-motion screw on the neck of the instrument. The index of the vernier, together with the co-incident division of the limb, will thus have been brought from the position in which it was when the telescope pointed at the object to be measured, round to the previous position of the 360°. First method.

Now release the upper or vernier plate (looking again at the vernier first to see it has not been moved), turn it until the telescope is again directed towards the object, clamp and perfect the bisection by the tangent screw moving the upper plate. The reading now on the vernier will be twice that formerly read off, or nearly so, and will be entered in the book under the former observation. This process can be repeated as often as required. The mean angle can be obtained by dividing the last reading (increased by as many three hundred and sixty degrees as the plate has revolved) by the number of observations, but it is better for our purposes to put down each individual reading. The difference between every two consecutive readings will give a value for the angle, and we can then see how they agree with one another An example of this kind of repeating is given on page 78.

The above method is perhaps the most accurate ; but, when many angles are to be taken, requires much time, and we shall arrive at a conclusion quite near enough for any hydrographi-
Second method. cal triangulation by taking all the angles in succession with the vernier set to 360°, and then, changing the degree of the zero to some even submultiple of 360°, as 90°, 180°, etc., take all objects again, repeating thus as often as necessary, which will be found much quicker. 360°, 120°, and 240° divide the arc equally and give three readings, which is often sufficient. The other method can be reserved for taking single angles, as, for example, a flash from a distant station.

Reading both Verniers. If both verniers are read, any error arising from bad centring should be eliminated for any given position of the plates.

For practical hydrographical purposes if one vernier is read, with the zero in several positions, submultiples of 360°, it is as a rule sufficient.

Different to the sextant, the theodolite has no index error to apply to horizontal angles, but to the vertical arc there is a correction to be found and applied, which will be mentioned in discussing the method of ascertaining heights.

Coloured shades to eye-pieces. A theodolite for hydrographical purposes should be fitted with coloured shades to the eye-piece of the telescope for observing the sun for true bearing.

STATION POINTER.

This useful instrument is of hourly service in nautical surveying.

Either in sounding, coast-lining, or topographical plotting, the position of the observer depends mainly on it.

The station pointer is used to plot a position on the chart, by means of angles taken *at* it, to other objects already fixed.

Its construction depends upon the fact that the angles subtended by the chord of the segment of a circle measured from any point in the circumference, are equal. (Euclid III. 21.)

Theory of Station Pointer.

Fig. 1.

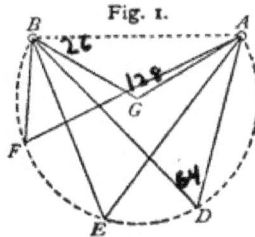

Thus, in the figure, the angles A D B, A E B, A F B are all equal, so that if we have observed the angle subtended by A B, we know at any rate that we are somewhere on the circumference of a circle, the size of which depends on the angle observed.

To draw this circle, we take advantage of the fact that the angle at the centre of any segment is double the angle at the circumference. (Euclid III. 20.) We lay off, therefore, from either end of the line whose subtended angle we have observed, the complement of the angle. The point where these lines meet is the centre of the circle, which we describe with the distance from this centre to either end of the line, as a radius.

Thus if our observed angle is 64°, we lay off A G, B G each making an angle of 26° with A B, and describing the

circle with centre G and radius A G or G B, we get the circle
we want, for A G B = 180° − (B A G + G B A)

$$= 180° − 52°.$$
$$= 128°.$$

And as A G B = 2 A E B,
the angle **A E B** and **all other** angles on the circumference
will be 64°.

If the angle observed is more than 90°, we describe the
circle by laying off the number of degrees over 90°, on the
opposite side of the line to that on which we know we are,
and proceed as before.

If we can obtain, besides the angle subtended by A B, the

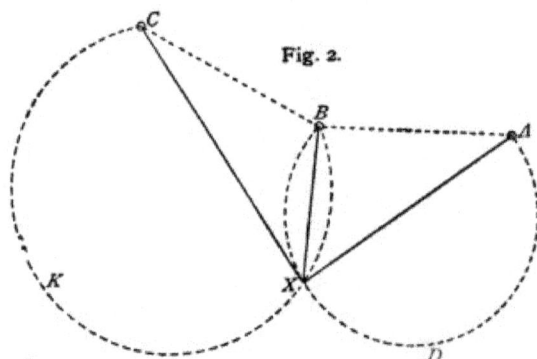

Fig. 2.

one subtended by B C, another line, one of whose ends is
identical with A B, we can draw another circle on whose circumference we must also be, and the intersection of these
two circles must be our exact position X, as it is the only one
from which we could have obtained these two angles at the
same time. See Fig. 2.

The station pointer obtains us this position X without the
trouble of drawing the circles, as it is manifest that, if we
have the angle A X B on one leg of the station pointer and
B X C on the other, the only spot at which we can get the
three legs to coincide with the points A, B, and C, will
be X.

We place the station pointer, therefore, on the paper, bringing the chamfered edges of the three legs of the instrument to pass over the three points observed, and make a prick with a needle in the nick in the centre, which will then mark the spot.

A piece of tracing-paper on which the three angles are protracted will answer the same purpose, but, of course, this will entail more time, and in the open air will give trouble, is liable to be blown about by the wind. Nevertheless, this has often to be used, as when points are close together on a small scale, the central part of the station pointer will hide them, and prevent the use of the instrument.

A very useful instrument has been devised by Commander Cust for such occasions, and consists in a plate of transparent zylonite on which a graduated arc is engraved. The requisite angles are drawn on this with pencil, with the angles reversed, and the plate being turned over, so as to bring the pencil lines in contact with the paper to obviate parallax, is used as a station pointer.

This method of fixing is generally known as the "two circle" method, but it is really the "three circle" method, for the circle drawn through the two outer points and the observer's position is also involved. A comprehension of this is of value. *Three Circle Method.*

The chance of error in a "fix" varies greatly with the position of the three points with regard to one another and the observer. *Position of "Points."*

It is in general sufficient to realise that the more rectangular the intersection of the two circles, the less chance there is of any error in the resulting fix, but there are cases where the fix is admirable though these circles are almost tangential, because the third and larger circle produces a rectangular cut.

With points and the observer's position placed as in Fig. 2, the two circles give a good intersection, and the fix is good. Let us, however, take the same points with the observer's position close to the centre object B, as in Fig. 3. We there see that the two circles are nearly tangential, but the third circle through the outer points and the observer, which the station pointer also gives us, cuts at a right angle, and as the position X cannot be off it, the fix is one of the best.

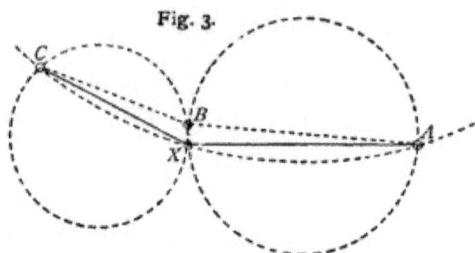

Fig. 3.

In such a case the whole angle between A and C should be observed, if not too large (as in our figure), as the accuracy of the fix depends entirely on this whole angle, and when so near to B a little movement may make considerable difference in B X A, and B X C are separately measured.

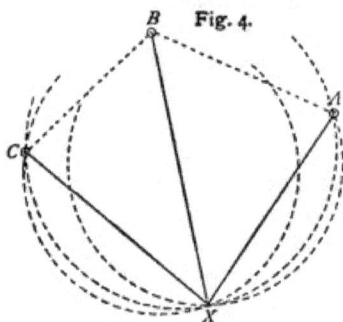

Fig. 4.

Let us now take three points and the observer's position as in Fig. 4, using the same letters.

The angles we have observed give us X as the point of intersection. It is evident that it is difficult to localise this point exactly, as all three circles so nearly coincide as make it impossible to say where the precise point is at which they intersect, and, with the station pointer, we should find that we could move the centre of the instrument considerably, without materially affecting the coincidence of the legs with the three points.

When X is so placed as to fall on a circle passing through the three points A B C, there will be no intersection whatever, as the two circles will coincide; and we cannot tell where we are on the circumference of this practically single circle. *On the Circle.*

The nearer, therefore, we are to being on a circle, whose circumference will include all the three points and our own position, the worse will be what is technically called the "fix," and this must always be guarded against in selecting objects to observe.

When one object is farther from us than the central one, we shall, as a rule, have a good fix; but when the central object is the farthest, the two circles will begin to make a bad intersection.

There is nothing in the whole range of surveying that requires so much attention and knowledge as the "fix," and many are the errors which have crept into surveys from disregard of its conditions. *General Rules.*

When moving along, as when sounding, and fixing from time to time, if both angles change slowly, the fix will be bad, for we must be moving nearly along the circumferences of both circles, and they must therefore nearly coincide.

In plotting the angles with the station pointer, the fix will be good if a very slight movement of the centre of the instrument throws one or more of the points away from the leg; but if this can be done without disturbing the coincidence of the legs and all three points appreciably, the fix is bad.

This is perhaps the most important thing for a beginner to remember and to practise, as it is a practical test involving no theory nor complications.

Theoretically, one of the best positions is inside the triangle formed by the objects, but in practice it is often impossible to observe the large angles incidental to this position.

Practically the beginner will find the following rules safe :—

    1. Never observe objects of which the central is the farthest.

    2. Choose objects disposed as follows :—

        (*a*) One outside object distant and the other two near, the angle between the two near objects being not less than 30° or more than 140°. The amount of the angle between the middle and distant objects is immaterial.

        (*b*) The three objects nearly in a straight line, the angle between any two being not less than 30°.

        (*c*) As before remarked, that the observer is inside the triangle formed by the objects.

There are, nevertheless, cases where the middle object is very distant, when the fix will be good enough for many purposes, but such cases require a thorough grasp of the subject, and should not be adopted by the beginner unless forced to it.

Size of angles admissible. The size of angles admissible in a good fix depends on the position of the three objects. If two objects are equidistant, the angle must not be small, for a slight error in the angle will make a great difference in the position ; but if one object be much farther off than the other, a very small angle between these will suffice, so long as the third object is so placed as to make a fairly large angle.

An arrangement of the objects not yet considered, is when two of them are in line from the observer's position.

"Points" in Transit This is technically called "transit," and no transit of known marks is allowed to take place without making use of it.

One angle to a third object is here enough to fix the position, which is one advantage, another being that if two angles are taken and placed on the station pointer, the coincidence of the position, as plotted by these two angles, with the transit line, gives an excellent check.

Here, Fig. 5, A and B are in line of transit (ϕ) ; C is a third object.

It will be evident that when the observed angle is on the

station pointer, and the latter is placed with one leg coinciding with the line A B, that we only have to move it up or down that line, until C coincides with the other leg, which gives us X. Any other position, as X_1, would not allow the leg to pass over H.

It will also be seen that the farther apart A and B are, the truer will be the direction of the transit line. If one object was at B_1, the position pricked through at X might be a little right or left of the true transit line, without the deviation being visible on the leg of the station pointer.

Also, it will be seen that the angle to the third object should

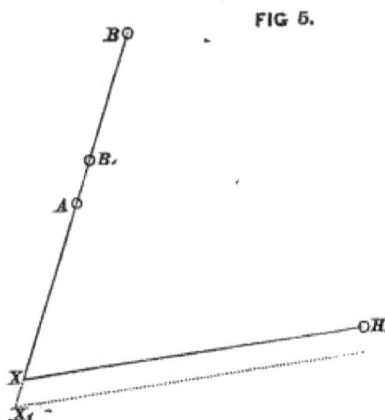

FIG 5.

be as near 90° as possible, anything under 25° being inadmissible, as the angle of intersection at X would permit of a false position without detection.

When using this method, the distance of the third object should also be considered. It must not be too far, or both theoretically and practically, by reason of the imperfection of instruments, the fix may be in error.

In choosing a station pointer, of which instrument Heather Choosing gives but a meagre account, the first important thing to look a Station at is that the smallest angle to be read on the leg which will Pointer. not come to zero, is as small as it should be. A well-planned modern station pointer should allow this leg to return to 3°

or 4°, but old instruments frequently will not read less than 12°.

It is a great nuisance to find that the only angles you can take cannot be plotted by means of your station pointer, and the chance of this should therefore be minimised as much as possible.

Station pointers are generally made to allow the left angle to come to 0°.

When the angle on the right is too small to set, and the left angle is more than 90°, the difficulty can be got over by setting the small angle on the left leg and bringing the right leg round to the left until the required left angle is made between it and the left leg.

Another method when the above cannot be carried out, but care must be taken not to employ it when exact accuracy is required, is to set the angles on the station pointer reversed, *i.e.* the right angle on the left, and *vice versâ*. Place the legs containing the larger angle on its "points," with the centre near the supposed position, and make a prick. Move the centre a little, right and left of the first prick, keeping the points on, and make other pricks. A line drawn joining these pricks will form an arc of the circle for those "points."

Now do the same for the small angle and its "points," and the intersection of the two small arcs will give the position required.

It is, in fact, projecting the circles by means of the station pointer.

Station pointers are made with brass, and with silver arcs; the latter are of course more durable, but for many purposes the brass are to be preferred. When sounding, or doing any work in the open, the reflection from a silver arc is often a bother, and hinders speedy setting of the vernier. The use of a reading-glass is almost a necessity with silver arcs for this reason, and also on account of the fineness of the cutting; whereas, with the brass arcs, a surveyor with good eyes can set his instrument quite correctly without one, a great point in a boat.

For chart-room use the silver arc is to be preferred.

The nick in the centre of the instrument should be small, *i.e.* just deep and wide enough to admit of a needle fairly catching in it. The needle-pricker should always be used for marking the position; not a pencil-point, which soon wears blunt, and will not mark truly in the centre. The prick also remains, and can be seen under the figure with a reading-glass, when inked in.

The prick should be on the continuation of the edge of the bar in which is the nick.

For ordinary soundings and field work, a station pointer of about 5 inches diameter of arc is most convenient. For ship sounding and chart-room work larger ones are supplied.

In testing a station pointer, the first thing is to see that the vernier of the leg which comes back to zero reads exactly 0°, using a magnifying-glass to read off accurately. If it does not, the screws which hold the vernier must be loosened slightly, and the vernier plate moved, until the arrow on the vernier corresponds exactly with 0° of the arc, and the 30′ on the vernier with a division of the arc. **Testing a Station Pointer.**

Take either a large sheet of backed paper, or a white Bristol board, and mark out, by means of chords, lines radiating from a centre, and 10° apart. These lines must be very carefully ruled, and in Indian ink, as this sheet must be kept as the test of all station pointers and protractors, which should be from time to time examined by its means. Screwing on the lengthening legs, and placing the station pointer on this sheet, with the nick in the centre of the arc corresponding exactly with the prick in the centre of your testing circle, and putting weights on the central part of the station pointer, each leg can be in turn moved to correspond with the ruled 10° lines, and the reading of the vernier compared. The error at each 10° should be written on a small piece of paper in the form of a table, and pasted on the inside of the box. **Testing Circle.**

If the legs of the instrument are exactly centred, the readings will either be correct, or the same amount in error all round, for each leg; but as this is a degree of delicacy rarely attained, it will usually be found that the error varies for different positions of the leg. The verniers should be set to minimise the errors between 0° and 90°, which is the amount of angle most used in actual work.

The chamfered edge of the leg and lengthening piece should correspond exactly with the line in all its length ; if it does not, it is also a result of bad centring or bad fitting on of the lengthening piece, but a good instrument should not have this error in any appreciable extent.

It need scarcely be added that if the instrument is found very badly centred, it should be returned to the maker, or not be chosen if buying ; but when an instrument is sent to the other end of the world, you may have to make the best of it, and registering all the errors on the table, be careful to apply them when using the instrument.

Discretion as to applying errors. The necessity for applying a small error depends upon circumstances, as in some cases, the position of the points used will admit of a difference of several minutes in the angle, without any appreciable alteration of the position of the observer ; in others, it is necessary to be exact. As the surveyor gains experience, he will learn when to apply the error, and when not. At the commencement, he must always apply the error.

Caution as to use of Station Pointer. It may here be noted, that, if the points used to fix by are not correctly placed on the chart, the station pointer will not indicate anything wrong, *unless a third, or "check" angle, be taken and plotted.* This must always be remembered in using a station pointer on a published chart, or the adoption of this instrument may have a disastrous result.

In the first place, the chart may be from a rough survey, and there may be absolute errors in the points on it; and secondly, the distortion caused in printing with damp paper always changes the position of points, more or less, and with objects in certain positions, this alone may make an error in a station pointer fix.

In navigating, therefore, with a published chart, of the accuracy of which you are not certain, *always use a bearing,* as well as the sextant angles plotted by station pointers, or *use check angles to each fix.* If the result is to show that the points are not correct relatively to one another, use the compass only, as it is less likely to get you into trouble with a defective chart, for the reason that the non-intersection of three bearings will at once indicate something wrong, and the navigator will choose the points of danger in his course ahead to steer by, rejecting the others whose positions with regard to him are of little moment.

BRASS SCALES.

These must be examined by means of the beam compasses, to see that their divisions are correct, more especially the diagonal portion, as the makers are sometimes not careful enough. If a scale is found to vary, it should be rejected.

A brass scale should never be used for ruling, and *never be taken out of its box.* If it is, some day it will fall from the table, get bent, and its correctness is gone.

STEEL STRAIGHT-EDGE.

This must be examined to see if its edge is exactly straight, by ruling a very fine line, and reversing the straight-edge, when, either ruling another line over the first, or examining the coincidence of the edge with the line already ruled by means of a reading-glass, will prove whether it is perfect.

Placing steel straight-edges edge to edge is another method when there are more than one; but great care must be taken with regard to the light if this is done, as it is difficult to detect a small error if the light falls across. They ought, of course, to touch throughout their whole length.

A steel straight-edge must be kept very clean, and carefully wiped before using, or the paper will soon become very dirty.

If kept bright, care must be taken that no emery is allowed to touch the chamfered edge, or it will get so sharp in time as to cut the pencil, and even the fingers of the operator. When once clean, rubbing daily with a warm dry soft cloth will keep it so, with an occasional rub of emery in damp weather.

Straight-edges are now generally supplied nickelled.

MEASURING CHAINS.

To test measuring chains, which are generally 100 feet long, a hundred feet should be accurately measured along a plank of the upper deck, and marked by nails driven in.

Always to be re-tested when used. Before measuring a base, all the links of the chains should be examined, and bent ones straightened; the chains are then compared with this fixed length and the errors noted. The same reference should be made after measuring, and the mean of these errors applied to the distance measured.

It may be as well to note that, when the chain on comparison proves to be longer than 100 feet, the surplus is to be added to each length measured, and when it is shorter, subtracted.

Points of measurement. The length is to be measured from the *outer* side of one handle, to the *inner* side of the other. This is to allow for the necessity of having a pin to put in the ground at each length.

Each link is a foot long, and every tenth link is marked by a brass label, with as many fingers on it as there are tens of feet from the nearest end.

PROTRACTORS.

Protractors of all kinds must be tested for correctness of division by the same testing-sheet ruled for the station pointers.

This is especially necessary in the case of Bullock's protractors, which have extended arms, generally of very light construction, which a slight blow will bend out of the direct line. These sometimes admit of correction by means of screws, which is easily accomplished by placing the protractor on the testing-sheet, with the opposite verniers exactly coinciding with the same line, and adjusting the extending points until they also prick precisely on the line.

If the divisions of an ordinary protractor are found to be incorrect, there is of course nothing for it but either to return it to the maker to be re-cut, or to mark the errors at each ten degrees, or wherever necessary, on small bits of paper pasted on the protractor.

A boxwood or vulcanite protractor is easily kept clean by rubbing it over with a piece of india-rubber, but a brass or electro-plated one is very apt to dirty the paper in plotting. It is a good plan to carefully paste a piece of tracing-paper on the under-side of these, when a rub with the india-rubber before use will ensure cleanliness. The thinness of the tracing-paper will not interfere with correctness in laying off the angles.

Vulcanite protractors are admirable for field work, as they are light, easily read, and when made thick do not chip like boxwood ones.

Large brass circular protractors of 10 inches or so radius are very useful for laying off the secondary points of a survey, saving the time involved by using chords.

POCKET ANEROID BAROMETERS.

These are very useful when putting in the topography of a country, as they give sufficiently accurate results for minor heights, with but little loss of time; but for the more exact measurement of conspicuous hills, &c., they are but of little use, and the theodolite and sextant must be had recourse to.

In choosing a pocket barometer for the above-named service, therefore, it is not necessary that it should read very low, as it will be but rarely that nautical surveyors have to deal with the intricacies of land over a few thousand feet high. For this purpose, 25 inches is quite low enough, and the five inches of barometric range thus obtained can be so largely marked on the dial as greatly to facilitate the reading to two places of decimals.

An important point for delicate reading is the construction of the index needle. This should be very thin towards the point, and turned with its edge at right angles to the plane of the dial. In this position it should admit of very accurate reading, and moreover assists the observer to hold the instrument at right angles to his line of sight, and thereby to avoid parallactic errors. For this reason the point, though as thin and fine as possible one way, should be tolerably wide in the other, so as to show plainly by its apparent increase of width when it is being looked at from any direction but at right angles to the dial.

The point of the index needle should cover about half the graduation of the arc. If it is too short to reach to it, or so long as to project over it, it is not so easy to read accurately.

In reading, the best position for the aneroid to be held is upright, on a level with the eye, the index being vertical. Read with one eye only, or parallax will creep in. Tap gently each time before reading, and turn the instrument flat, and then vertical again for a second reading, to prevent mistakes, tapping as before. In whatever position, however, the instrument is read the first time, it must be always held for all other readings on that day, the reason being that the weights of the different parts in such a delicately made little instrument have considerable influence on its free movement, and that this influence must be so disposed as to act in the same manner at each reading.

The surveyor will of course never let the pocket aneroid out of his own possession, and will place it about his person in such a manner as to minimise chances of shocks or blows in scrambling through rough country, getting wet in rainy weather, or tumbling out of the pocket. To prevent latter accident, always use a lanyard. The instrument should always be carried in its case.

If a small aneroid is not in regular use, the delicate internal parts, especially the chain, are liable to stick from oxydization, when at length taken up a height. It is therefore convenient, if an air-pump be on board, to place the instrument under the receiver from time to time, so as to keep all working parts in order.

HELIOSTAT.

This instrument, which is simply a mirror mounted in gimbols, so as to turn and reflect the sun in every direction, is of great use. In a survey where many assistants are at work together, it saves an immense amount of time. Smaller beacons or marks can be erected, and the position of a theodolite station that has to be made on the side of a hill, or with dense foliage behind it, is at once made apparent to another observer, who has to take angles to it by the flash, which can be seen a long distance by the naked eye. *Most valuable adjunct for Marine Surveying.*

Some heliostats supplied are mirrors in gimbols, mounted either on stands or in portable cases, with a spike to drive into the ground. *Fitting.*

Neither of these forms is satisfactory, as in many places from which it is desired to use them they cannot be conveniently and firmly placed. Tripod legs of some description on which to place the mirror are best, and a movable arm working round the centre, and carrying an adjustable ring through which to direct the flash, will be found very handy. If the surveyor has to trust to placing some separate object, such as a stick or another tripod, a few feet from the mirror, by which to direct his beam of light, he will soon find himself in some position where there is no standing-ground for such object, as when his theodolite is on the top of a sharp hill, or on a steep coast-line under cliffs at the edge of the sea.

A better instrument is the excellent and convenient Galton's Sun Signal, now also supplied. This is fitted with a telescope, by looking through which and adjusting the mirror, a dim image of the sun is seen covering the object *Galton's Sun Signal.*

required to flash to. Nothing can be better adapted to the
purposes of the nautical surveyor's work than this (when he
is once accustomed to it, as at first it is a little awkward
to manage), and when obtainable they should always be
used. Care must be taken, however, that the instrument is
in adjustment. This can be ascertained as follows : Place
a board, with a sheet of white paper pinned on to it, about
50 yards off. Direct the sun signal flash on to it, and looking
through the telescope, screen and unscreen rapidly with
the hand the direct flash from the mirror. If the circular
image formed by the direct flash on the sheet is not
coincident with the image of the sun as seen through the
telescope, take off the cap at the end of the tube and adjust
with the screw that will be found underneath.

A very workable arrangement can be fitted on board any
ship as follows :—

Instru-
ment
easily im-
provised
on board.

A blacksmith will soon make a frame which will convert
an ordinary looking-glass into a perfect instrument for surveying work, as it must be remembered that we do not
propose to use it for talking, and therefore do not require the
extreme accuracy in directing the beam necessary in the
military heliograph.

The sketch annexed shows a looking-glass fitted in this
manner by a ship's blacksmith. The standard can be made
of any height as convenient; about two feet and a half is a
good length. In soft ground the end of the legs can be
pressed into the earth, and on rocky ground stones placed
against the legs will hold the instrument steady. The arm, *m*,
of light iron, is carried separately, and slips over the shaft
of the standard, clamping where required with a screw.

Into a circular socket in head of standard shaft, the leg of
the frame holding the mirror is shipped; this is also to be
tightened by a retaining screw.

The mirror, which can be of any size from 2 to 6 inches
or more in diameter, revolves on its retaining screws, as an
ordinary toilet-table glass, and can be held in any position by
tightening these screws.

The ring, of flat wood, is made as light as possible, so as to exert less strain in wind. Across it are nailed crossed strips of copper, with a white cardboard disc, about an inch in diameter, fastened to their centre.

The rod that carries this ring slips up and down in a hole at the end of the arm, and is clamped by a retaining screw.

FIC 6.

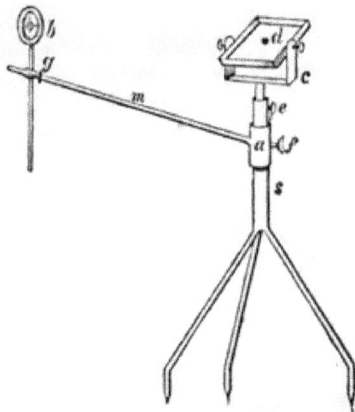

LOOKING-GLASS AS FITTED BY BLACKSMITH FOR HELIOSTAT.

a. Sliding collar carrying arm *m*, revolving round S.
b. Wooden ring, painted black, with cross-wires and white cardboard centre, sliding vertically by means of rod through arm *m*.
c. Iron frame to hold mirror, fitting into socket in top of Standard S.
S. Iron standard with fixed tripod legs.
d. Blind spot in mirror.
e. Screw for clamping mirror frame.
f Screw for clamping arm.
g. Screw for clamping ring rod.

In the centre of the back of the mirror, a hole of about ¾-inch diameter is scraped in the tinfoil, being careful to leave a sharp edge. A similar hole is cut out of the wooden back of the glass frame. This we shall call the "blind spot."

To direct the flash to an object, bring the mirror vertical, Using the and looking through the hole in the centre, revolve the arm Heliostat. until in the direction of the object nearly, clamp it, and adjust the disc rod as nearly as may be, for elevation or depression. Then, slightly loosening the screw clamping the arm, finally adjust the latter, so that the object, as regarded through the hole in the mirror, is obscured by the white cardboard disc in centre of the ring. By turning the mirror so that the dark shade caused by the blind spot is thrown on to the disc, the flash will be truly directed, and must be kept so by slight alterations of the position of the mirror, which should therefore be clamped only sufficiently to hold it steady, and yet admit of gentle movement. The shadow of the blind spot should be slightly smaller than the disc, so as to ensure having it truly in the centre of the latter.

Best glass necessary. The mirror must be of the best glass, with its faces parallel, or the shadow of the blind spot will be very indistinct when the mirror is at a large angle, and also the beam of light will be dispersed before it has traversed many miles.

Size of Mirror. It is well to have the mirror a fair size, say 6 inches square, as in practice it will be found generally necessary, in order to save time, after once adjusting the flash, to leave a bluejacket to keep it on, while the surveyor is taking his angles; and although a man will soon pick up the knack, a larger mirror will allow for eccentricities on his part, and also, on a dull day, a faint flash will be detected from a large mirror, where a small one would not carry any distance.

Power of penetration of flash. On a bright day, a flash from a 3-in. by 2-in. mirror has been seen 55 miles and more.

In hazy weather, angles have been got when the place from which the flash was sent was entirely invisible; and thus whole days have been saved by this simple contrivance. Only those who have spent hours, or even days, in straining their eyes to see a distant mark, can appreciate the value of a heliostat.

TEN-FOOT POLE.

For coast-lining, a pole of measured length is often required, to get distances by observing the angle subtended by it.

A convenient form is as follows —

Two oblong wooden frames, about 18 in. by 2 ft., are made as light as possible, and covered with canvas. These will fit, by means of sockets at the back, on to the ends of a pole, and copper pins passed through socket and pole will keep them at a certain fixed distance apart. Ten feet is a convenient length for transport.

The face of the canvas on the frames is painted white, with a broad vertical black stripe in the centre, and the ten feet will be measured from centre to centre of the black stripes.

In measuring with a sextant the angle subtended by such a pole, the image of one stripe will be brought to cover the other stripe.

A table of distances, corresponding to the angle subtended

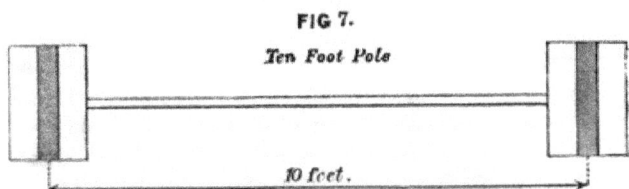

FIG 7.

Ten Foot Pole

10 feet.

by the length of the pole used, should be in each assistant's possession for reference on the spot.*

Fig. 7 represents a ten-foot pole.

DRAWING-BOARDS.

In a surveying vessel it is convenient to have a considerable number of these, and of various sizes, so as to fit all scales. The largest may be about 29 in. by 25 in. The size of which most will be wanted will be about 27 in. to 25 in. by 20 in. to 22 in. There should be some smaller ones, 22 in. by 16 in.

Lightness, combined with sufficient strength not to warp, is the requisite, and seasoned wood is therefore necessary.

---

* Appendix, Table S.

White pine, three-quarters of an inch thick, is as good as
anything, though the smaller boards may be made of thinner
mahogany.

Duck covers to fit the boards are necessary for field work,
when the work is plotted at the time, to carry them in, and
prevent rubbing and wetting from rain.

If it is intended to do most of the detail plotting in the
field and boats, there should be about three or four boards to
every assistant in the survey.

WEIGHTS.

The weights supplied by the Stationery Office are of iron,
flat, oblong, and covered with leather.

Drum-shaped leaden weights to supplement these, covered
with duck or baize, will be found very handy. These can be
of various sizes and weights, to suit all requirements·
Cylinders $2\frac{1}{2}$ in. in diameter and $1\frac{1}{2}$ in. in height are an
average size, and can be cast on board in a wooden mould.

A few others heavier are useful, and some flat weights,
$2\frac{1}{2}$ in. in diameter and $\frac{1}{3}$ of an inch thick, are good for
keeping down small tracings.

TRANSFER PAPER.

This must be made, not bought, as the stuff sold by
stationers always has some oily material in it.

On to a damp sheet of tracing-paper scrape finely some
blacklead, and rub it well in with the hand, a little at
a time, allowing it to dry between each application. Rub
off the loose particles before rubbing in more. The black-
lead is only to be applied on one side of the tracing-paper.
It must be done as evenly as possible, so as to ensure uniformity in the tint, but this is assisted by a good rub with
a soft cloth when the sheet is finished. Two or three applications will be sufficient.

A lump of blacklead for this purpose is supplied to all

surveying vessels, but the lead from a soft pencil will answer as well.

It is a dirty process to the operator, but in a few hours enough can be made to satisfy the requirements of the survey for a long time.

MOUNTING PAPER.

The consideration of what to plot the intending chart on must be undergone before commencing work. Two methods are in use.

To stretch, and firmly paste or glue the paper, on a drawing-board or table, where it must remain until the chart is complete; or to plot on a piece of drawing-paper mounted on holland or calico, and simply flattened out before use.

The advantage of the former is that the paper remains flat, Loose and free from wrinkles or movement, the whole time work is sheets best with being done on it; but for many reasons, the latter is most large staff. convenient for ship work. If many sheets are under weigh at one time, which frequently occurs in an extended survey and large staff, they take up less room, and interfere less with one another, when several persons are working at one table, than when sheets mounted on boards are used, and they are easier put out of the way. If the plotting sheet is very large, and formed of many pieces of paper, which it must often of necessity be, it is very difficult to stretch such a paper, and it would take up the whole of the table, where it would have to be placed, as a board of sufficient size would be very inconvenient in a ship,

The drawback is the constant stretching and taking up of the sheet, with the variation in temperature and dampness of the air, which is undoubtedly a source of annoyance in plotting long lines, as the radii measured the day before, or even a few hours before, will frequently be found so much altered in length as to necessitate remeasurement.

This variation of the sheet may also produce distortion; but Distortion.

with good paper, well mounted, it will be nearly the same for all parts of the chart, and the distortion is so slight as to be of no practical inconvenience, and it is a question whether more distortion is not produced when, in using the other method, the stretched sheet is finally cut off the board when finished.

The fact is, that a material on which to draw charts free from the possibility of stretching and distortion has yet to be discovered, and we must put up with these inconveniences as long as we use the paper of the present day.

Taking one thing with the other, then, the author recommends the use of loose mounted sheets for general ship work.

Care of paper. Paper, whether mounted or not, in damp climates rapidly gets into a useless condition, and this even in hermetically sealed tins.

The stock of paper should be kept in tins in the driest place in the ship, which is probably in or near the engine-room.

The mounting of the backed sheets supplied from the Stationery Office is usually very well done, and it saves time to use these; but as it may be necessary for the surveyor to mount sheets himself, the method will be described.

Mounting loose sheets. The holland or calico on which the paper is to be mounted, and which must be in one piece and larger than the board, must be lightly damped. It is then stretched over the board and tacked to the edges, care being taken to stretch it equally and squarely with the woof and warp. Rub plenty of strong paste into it with the hand, and see there are no lumps left. The sheet of paper must be well damped with a sponge on both sides, taking care to *dab* only, on the side on which the work is to be done, and not *rub* with the sponge. The sheet is then carefully lifted by the four corners, one edge laid on the holland while the rest is kept clear of it, and the paper gently rubbed on to the board with a soft handkerchief, the paper being gradually lowered, so as to allow air bubbles to escape. It will take two people to do this, and it must be done with great care. It must be left to dry by itself, and no hot sun should be allowed to get to it, so that it may dry evenly.

If the plotting sheet is to be formed of more than one piece of paper, the edges of the paper which will overlap must be fined down. This is done in the first instance by scraping with a sharp knife, having drawn a line on the paper where the overlapping will come, and then finishing off with ink eraser. The piece that is to be uppermost must be scraped on the under-side only, and the undermost one on the upper side, so as to make, in fact, a scarph. This will lessen the appearance of a joint, and the inconvenience of ruling lines hereafter over it. **Joining Sheets.**

Drawing-paper is made of the following sizes :— **Sizes of drawing-paper.**

Demy	20 in.	by	15¼	
Medium	22¾ „	„	17½	
Royal	24 „	„	19¼	
Super Royal	27¼ „	„	19¼	
Imperial	30 „	„	22	
Elephant	28 „	„	23	
Columbier	35 „	„	23½	
Atlas	34 „	„	26	
Double Elephant	40 „	„	27	
Antiquarian	53 „	„	31	
Emperor	68 „	„	48	

Atlas, Double Elephant, and Antiquarian are most used in chart-making, and are the sizes supplied by the Hydrographic Office.

For all rough work, as sounding sheets, ordinary field work, &c., drawing-cartridge is used. It is quite good enough, and does not entail such expense as the use of an indefinite quantity of hot-pressed drawing-paper.

This cartridge-paper is mounted on the drawing boards by being wetted, and rubbed on to the well-pasted board in the manner described above for mounting on holland. It will dry quite flat. When this paper is done with, it must be floated off the board, which will cause it to distort and con- **Field Boards.**

tract considerably by the time it again dries; but as all the work on it will have been beforehand transferred to the tracing, this does not so much matter. The paper must be kept, however, as a record, for which it is just as valuable.

Where a field-board is wanted for delicate coast-line or other intricate work, a sheet of Atlas can be mounted, or white Bristol board used. The latter has many advantages, and can be *tacked* on to a board to keep it flat. If chart pins are used with thick Bristol board, they will not hold for long, and will give much bother by constantly falling out.

The Atlas paper, being good, may be mounted on to the board by merely pasting the edges, as described above. It can then be cut off without much distortion. If cartridge is treated this way, it is very apt to tear, being of a loose texture.

BOOKS.

Blank books of various forms and dimensions for boat work, field work, &c., are supplied from the Admiralty. These for the most part require no forms ruled in them; but there are a few purposes for which it is very convenient to have ruled forms bound up.

All such are now supplied by the Hydrographic Office, and it is not necessary to specify them. They are all enumerated in the Instructions to Surveyors. It is very necessary to record observations in these books in such form that they can be hereafter consulted as records.

CHRONOMETERS.

About the care of the chronometers little need be said. Full instructions are issued by the Admiralty on the subject, to which reference can be made, and Capt. Shadwell's "Notes on Management of Chronometers" * contains all that

---

* "Notes on Management of Chronometers and Measurement of Meridian Distances," by Captain C. Shadwell, C.B. Potter, London, 1861.

can be said on the subject. The box or "room" for the chronometers is now made after a fixed plan, the principle of which may be said to be that the solid block on which the chronometers rest, and which is, when practicable, bolted to the beams beneath, not the deck, can receive no blow or shock other than those communicated through the ship herself, which is done by surrounding it with a bulkhead, with a clear space between. Vibrations are lessened as much as possible by the interposition of sheets of india-rubber in building up the block, and by padding the partitions in which each chronometer rests with soft cushions.

The lid of each box is removed, and a general lid covers the whole.

A sheet of fearnought is laid over the chronometers, and has flaps cut over each one, so that they can be uncovered in turn, for purposes of comparison or winding. This is to assist to keep the temperature uniform, and also deaden the ticking of other watches when comparing. *Uniformity of Temperature.*

Winding is performed at the same hour daily, and comparing also. There is no necessity that these hours should be identical; but it is generally the practice that they should be. If they are both done at an early hour, there is more chance of the same officer being on board to do it, which is of importance. *Winding.*

Always wind until the mechanism is felt to butt, to ensure the watch being fully wound up.

To prevent butting too sharply, the turns can be counted, which will warn the winder; but if winding is done delicately, this is scarcely necessary. When, however, any other but the officer in charge of the chronometers winds them, he should do this; to enable him to know the number of turns each watch requires, a piece of paper with the information can be pasted on each box.

The watch has to be reversed to wind, and it must be eased gently back when the operation is complete, not allowed to swing back. This daily reversing of the watch is said to be a good thing, as it distributes the oil in the bearings.

Accurate comparing only comes, like most other things of the kind, by practice. The comparing of watches is gone into at page 225.

Records. A comparison book and chronometer journal are kept; the former being used to enter the comparisons at the time, with their checks, &c., the latter as a fair book for a permanent record, and contains rates, and all data for noting the performance of each watch.

A maximum and a minimum thermometer are placed in the chronometer-room, and the reading of their indices is taken, and recorded in the comparison book, at the time of comparing. This is of importance when it is intended to take account of change of rate from changes of temperature, and, in any case, will enable us to estimate how far our endeavours to maintain a uniform temperature in the room are succeeding.

MARKS.

It is of course necessary in making a survey of any description to have fixed objects, which are first plotted on to the sheet, and are technically known as "points." These vary, according to the description and scale of the survey, from mountain peaks, whose actual summits may be of considerable area, to thin staves.

Natural marks. It is a great saving of time to the nautical surveyor to find plenty of natural marks, as peaks, conspicuous trees, houses, church spires, &c., anything, in fact, which can be defined and recognised from the different directions it may be necessary to see them; but it is rare to find a sufficient number of these, properly placed, to be able to altogether do away with putting up his own marks for the details of the survey, and it is of these we now speak.

Whitewash. Whitewash is the great friend of the surveyor. It has the advantage of being portable, showing generally very well, being cheap, and obtainable all over the world; it cannot disappear by being blown over, or by being stolen or knocked down by jealous inhabitants, who very naturally do

not understand what the meaning of different objects dotting
their shores may be. Whitewash, therefore, is used wherever
practicable, as on rocky cliffs and points, or tree-stems,
angles of houses, &c. ; and is also used to whiten other objects
put up by the surveyor, as cairns, canvas, &c., where either
there is no solid substance to whitewash, or it is necessary
to see the mark from every direction, which it is evident cannot be the case when a cliff face is whitewashed, for example.

The nature of these marks must vary according to locality,
and the distance it is necessary to see them.

Where there are stones, nothing is better than a cairn. **Cairns.**
It is rather a question as to whether a cairn on a hill-top is
better whitewashed or not. If the sun strikes on it, or there
are higher dark hills behind, it shines like a star ; but on a
dull day, against the sky, a white cairn will be so much the
colour of it, that a dark object will show better.

A cairn on the beach should certainly be whitened.

Where there are no stones, tripods of rough poles or **Tripods.**
stakes, about eight feet long, round which a bit of old
canvas about six feet long, whitewashed, can be laced with
spun-yarn, will be found good. The poles are easy to carry
in the boat, and can be taken up hills without difficulty;
they are easily taken down, and can be used over and over
again. From their tripod form, they stand well in high wind,
though it is as well to give them spun-yarn stays. The
conical shape of the mark affords a capital object, and
rough poles of the kind required can be got anywhere.

Where bamboos can be got they are very useful, from
their lightness, to carry up hills, either to form tripods as
just described, or as flag-poles.

Pieces of wide coarse white calico are useful for temporary
marks, as where it is desired to see a station from another
station, when there is not sun to use the heliostat, and it is
not requisite to have a very large mark left for future use.

On a very flat low shore, where boat sounding has to be **Flags.**
carried out a long distance, flagstaffs with large flags must be
set up. Care should be taken in some parts of the world

that these flags are not national ones, or anything that can be mistaken for such, as difficulties have frequently occurred through such being hoisted. As the large old flags obtained from dockyards are always of this type, they should be cut up, and re-sewn with such an arrangement of colours as shall denote nothing.

The colours of red and white intermingled will be generally seen furthest.

Canvas.   On coasts lined with bush, like mangrove swamps for instance, square pieces of canvas, whitewashed and laced to the boughs, will be found to show very well.

In lacing these on, care must be taken to place them so that they will show as far round on both sides as possible, and always to have the lower part more to the front than the upper, as they will thereby catch more sun.

Whenever canvas is used, it is well to cut holes in it, sewing them round sufficiently to prevent tearing in the wind. This will make the canvas valueless for fishermen or natives of any kind, to whom, in all parts of the world, a good piece of stuff is a prize.

In preparing for a surveying cruise, therefore, provision of material for marks must not be forgotten.

BOATS' FITTINGS.

It is impossible to lay down any dogmatic rules for fitting boats for surveying work, as so much depends on individual tastes, and requirements of the locality; but a few points may be noted which have been found generally useful, and a list of articles which are always being wanted can be added, as some sort of guide.

Steam Cutters.   Steam cutters are of course the best boats for general sounding, as the engine never tires. The additional work that can be done with steam cutters at command is enormous, as they not only do their own work, but tow the pulling-boats to and from their stations, and thereby save many hours that would otherwise be spent in beating up, or pulling backwards and forwards to the ship morning and evening.

A little table may conveniently be fitted in the stern-sheets of a cutter, as there is plenty of room. The stern-sheet canopy, which will generally have to be in place when sounding, to prevent the spray injuring instruments, books, board, &c., should not be too high, so that the officer standing in the stern-sheets may be able to take his angles over it.

Fittings for a small wire sounding machine are necessary.

Steam cutters for surveying work should be fully rigged, as accidents will happen, especially as the boiler gets old, and it is awkward to find oneself broken down with the ship miles off, and probably out of sight, and nothing but a foresail, which is the present service-fitting for these boats. For this reason, unless working near the ship or in close harbours, the masts and sails should always be in the boat. Lumber irons should be fitted to carry these high up, so as not to interfere with the wash-streak of canvas, nor take up necessary room in the boat.

The usual fitting of a steam cutter, a canvas turtle-backed canopy forward, is inconvenient, as the leadsman cannot then stand in the bows, which is the best place for him. The canvas wash-streak must therefore be carried to the stem, and the stanchions on the bows must be higher than those amidships, to allow for plenty of pitching in a head sea.

It will be found absolutely necessary to use the bow and stern lockers for stowing gear, men's clothes, &c.; but care must be taken that the lids are screwed down, whenever the boat is at work in open water.

Steam cutters must have little skiffs of some kind, to tow **Light Skiffs.** as tenders for landing, as the boat is too heavy and draws too much water to be beached, and should always be kept off the ground, for fear of strains with the heavy boiler in her. When only sounding, the tender is of course not needed.

For pulling-boats, whalers will be found most generally **Whalers.**

E

useful; they employ fewer men, and have quite enough room in them.

The simpler the sail the better, as it may be often up and down, but a mizen is very useful.

Fixed wash-streaks forward and aft will keep out much water.

Crutches with a long shank, which will raise the crutch two inches, may be found useful in some types of whalers, as, with gear along the middle of the boat, and the low gunwale of an ordinary whaler, the loom of the oar cannot be depressed enough with a service crutch, when in broken water, for the blade to clear the tips of the waves.

Lockers, built in bow and stern, are useful for keeping gear and instruments in. These should be canvased over, for unless built of two thicknesses of wood, which is heavy, the tops will soon leak after a few months' hot sun. The top of the bow locker raises the leadsman, so that he can throw his lead well ahead, and he should have his foremost awning stanchion shipped, as a support in rough water.

The awning should be cut at the after-thwart, so as to enable the afterpart to be tipped, when it is necessary to stand up to take angles.

Cutters.   Cutters should also be fitted with bow and stern lockers, and a table can be arranged in stern-sheets if thought necessary. No other special fitting is required beyond those given below for all boats.

FITTINGS AND GEAR FOR ALL BOATS USEFUL FOR GENERAL SUVEYING WORK.

Keel Bands.   All boats should have stout iron keel bands. With the constant grounding and running over rocks, inevitable in surveying work, these save the boats enormously. With coppered steam cutters these must be of brass, fastened outside the copper sheets.

A galvanized-iron reel, under one of the foremost thwarts, to hold a 100-fathom line.

A small galvanized-iron davit, with snatch block to place **Sounding Davit.** sounding line in when in deep water, so that several men can assist in hauling up the lead. In steam cutters it will be found handy for the leadsman always to use this.

A *Massey's* patent log, with the stray line between fan **Patent Log.** and clock lengthened, so that the latter may be fastened outside the gunwale for convenient reference, while the fan tows in the water behind.

A small galvanized-iron nun buoy, with a light chain and **Buoy.** weight to moor it by, is useful when sounding out a shoal patch.

The boat-hook should be marked in feet, with the marks **Boat-hook.** slightly cut and painted. This is useful for sounding in shallow water, and many other purposes.

A box containing some spare tins of preserved meat, in **Spare Provisions.** case of accident detaining the boat beyond her time of return to the ship.

The following list of stores may be handy :—   **General Stores.**

Lead lines, 100 fathoms, 1.

	25	,,	2.
Leads,	11	lbs.	2.
,,	7	,,	1.

Anchor and cable. Latter should have a short ganger of chain in case of sharp rocky bottom, and be *always* made fast to crown, and stopped to the ring, in case of fouling.

Masts and sails.

Spare oar.

Awnings and stanchions.

Water barricoe.

Small portable ditto for carrying on shore, 2.

Axes, handy billy, 2.

Bag of lime.

Whitewash brush.

Box for arming for lead.

Tin pannikins.

Bag for biscuit.

E 2

Old canvas for mark.
Canvas cases for rifles.
(It is convenient always to have a rifle in the boat.)
Ensign, answering pendant, and signal book.
Tramping barricoes, 2.

Ammunition.

Ammunition case, containing—

3 rockets with rope tails.	20 blank cartridges.
2 long lights.	50 ball ,,
1 handle and primers.	20 pistol ,,
1 portfire.	2-feet slow match.

If a boat meets with an accident, this ammunition will come in handy to attract attention after dark.

Carpenters' stores.

Carpenter's bag, containing—

Hammer.	Fearnought.
Nails of sorts.	Lead.
Chisel.	Tallow.
Bradawl.	Strips of copper.
Gimlet.	

Boatswains' stores.

Boatswain's bag, containing—

Marlinspike.	Palm.
2 sail-needles.	Bits of canvas.
Twine.	Spun-yarn.

LEAD-LINES.

The first thing in a newly commissioned ship is to get the lead-lines well stretched.

Until this is done it is only loss of time to mark the multitude of lines wanted for surveying.

As soon as the ship leaves port, tow seven or eight hundred fathoms of the new line astern, with a heavy lead on it, for some days. The line will then have got to its normal length, but lead-lines will always want re-marking from time to time.

Hand lead-lines, for ships or boats, should be marked in feet up to 5 fathoms, and at every fathom between 5 and 25.

Ship lines for deeper sounding, and boat's 100-fathom lines, should be marked the same as hand lines, to 25 fathoms.

After that, at every 5 fathoms, up to 100 fathoms.

Over 100 fathoms, at every 10 fathoms, to 200 fathoms.

Over 200, a mark at every 25 fathoms will be sufficient.

For deep-sea lines, a mark at every 25 fathoms, throughout.

There is no recognised system for these marks, but a list is appended of that used by the author.

MARKS IN LEAD-LINES FOR SURVEYING.

Faths.	Feet.		Faths.	Feet.	
	2	line, 2 knots.	9	0	Blue.
	3	Blue.	10	0	leather, a hole.
	4	line, 1 knot.	11	0	„ 1 tail.
	5	„ 2 knots.	12	0	„ 2 tails.
1	0	leather, one tail.	13	0	Blue.
	1	line, 1 knot.	14	0	Red.
	2	„ 2 knots.	15	0	White.
	3	Blue.	16	0	Blue.
	4	line, 1 knot.	17	0	Red.
	5	„ 2 knots.	18		Canvas.
2	0	leather, 2 tails.	19		Blue.
	1	line, 1 knot.	20		line with 2 knots.
	2	„ 2 knots.	21		leather, 1 tail.
	3	Blue.	22		„ 2 tails.
	4	line, 1 knot.	23		„ 3 „
	5	„ 2 knots.	24		Red.
3	0	leather, 3 tails.	25		Red and White.
	1	line, 1 knot.	30		3 knots on line.
	2	„ 2 knots.	35		White.
	3	Blue.	40		4 knots.
	4	line, 1 knot.	45		White.
	5	„ 2 knots.	50		Blue and White.
4	0	Red, with line and 4 knots.	55		White.
	1	line, 1 knot.	60		6 knots.
	2	„ 2 knots.	65		White.
	3	Blue.	70		7 knots.
	4	line, 1 knot.	75		Red.
	5	„ 2 knots.	80		8 knots.
5	0	White.	85		White.
6	0	Blue.	90		9 knots.
7	0	Red.	95		White.
8	0	Canvas.	100		Blue and line, 1 knot.

MARKS IN LEAD-LINES FOR SURVEYING—*continued.*

Faths.	Feet.		Faths.	Feet.
110	1 knot.		180	8 knots.
120	2 knots.		190	9 „
125	Red.		200	Blue and line, 2 knots.
130	3 knots.		225	Red.
140	4 knots.		250	Blue and White.
150	Blue and White.		275	Red.
160	6 knots.		300	Blue and line, 3 knots.
170	7 knots.			
175	Red.			And so on.

BEACONS.

Floating beacons are frequently of great service.

These are now generally made on board. A useful and convenient form is depicted in Fig. 8, which pretty well explains itself.

The heads of the two 27-gallon casks should be filled up flush, and the planks above and below are screwed to the heads, the pole passing through the centre of each plank by a hole cut for the purpose. The planks can be hollowed out to fit the heads of the casks for further security.

Three casks can also be used if only small ones are available, by fitting the planks in triangle, with another plank across, through which the pole passes.

The strop for weighing should be of wire, which keeps well open from its own stiffness, and facilitates hooking on for hoisting in.

A slip by which the cable is attached to the mooring span assists in weighing the beacon.

Use a small kedge and light chain for anchoring, except in water, say, over 60 fathoms deep, when hemp should be employed, with some chain next to the anchor to take chafe. Beacons have been anchored in 3000 fathoms by means of sounding wire, and weight of 100 lbs.

In water of from 20 to 100 fathoms, about 1½ times the depth is necessary for the length of mooring rope. In deeper water, less.

This beacon will float nearly upright, and will carry in

Fig. 8.
Float'ng Beacon.

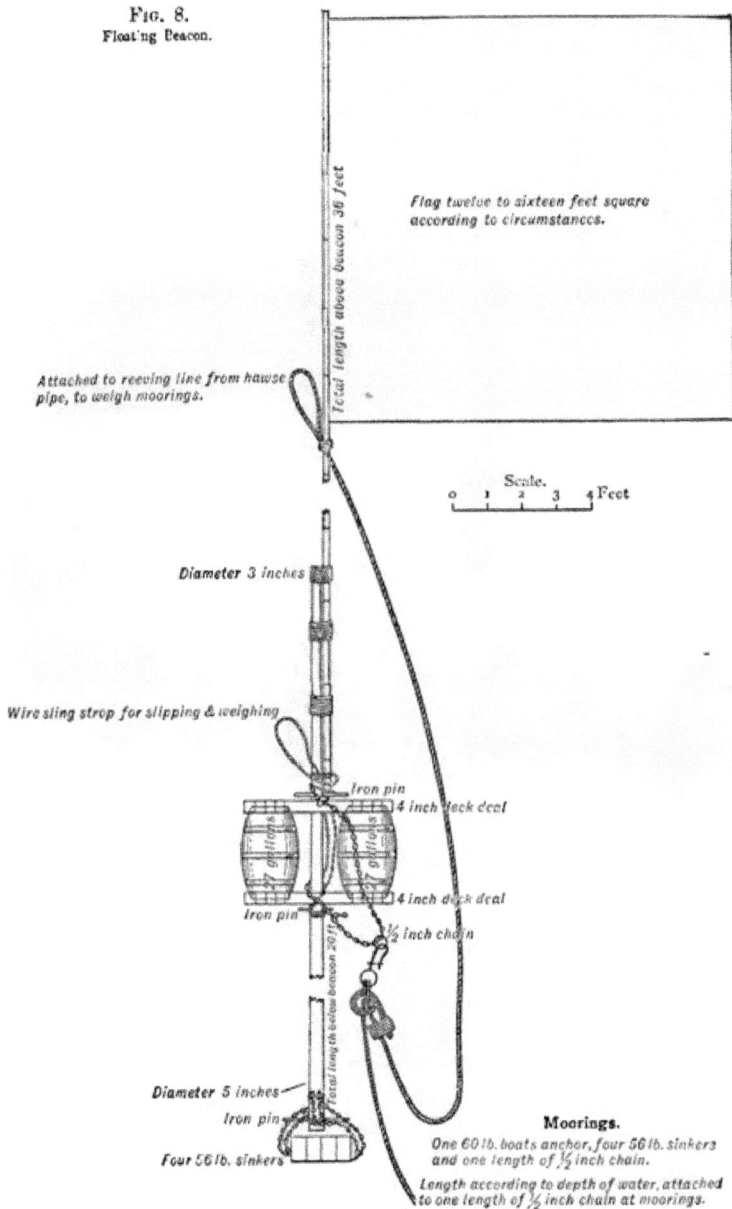

Total length above beacon 36 feet

Flag twelve to sixteen feet square
according to circumstances.

Attached to reeving line from hawse
pipe, to weigh moorings.

Scale.
0   1   2   3   4 Feet

Diameter 3 inches

Wire sling strop for slipping & weighing

Iron pin
4 inch deck deal

gallons

gallons

4 inch deck deal
Iron pin
½ inch chain

Total length below beacon 20 ft.

Diameter 5 inches
Iron pin
Four 56 lb. sinkers

Moorings.

One 60 lb. boats anchor, four 56 lb. sinkers
and one length of ½ inch chain.

Length according to depth of water, attached
to one length of ½ inch chain at moorings.

moderate weather a flag 12 ft. square, of calico, which is lighter than bunting, and will be visible from the ship 10 miles, with a 30-ft. bamboo.   Black, with other colours to distinguish one beacon from another, is recommended.

A piece of signal line should be fitted along the luff inclosed in several folds of the calico, and the flag is stopped to the bamboo round this.

Another form of cask beacon is made by woulding three casks to the central spar with rope, which tautens when wet, but the beacon above described is more quickly fitted when the parts are ready beforehand.

Slipping a beacon is best accomplished from the mainyard, but the foreyard can be used. The anchor being over on the weather side, and beacon lowered into the water, slip by means of a large well-greased toggle.

Weighing is accomplished best from the foreyard. Having hooked to the weighing strop, run the beacon up, and having made fast a line from the hawse pipe to the wire strop at the upper end of mooring chain, knock off the slip, when the beacon can be landed inboard, and the mooring run up to the hawse pipe.

Fixed Beacon. A fixed beacon can be erected in shallow water of from 2 to 3 fathoms by constructing a tripod of spars of about 45 ft. long. The heads of two of them are lashed together, and the heels kept open at a fixed distance by a plank about 27 ft. long nailed on about 5 ft. above the heels of the span. These are taken out by three boats, and the third tripod leg lashed in position on the boats, the heel in the opposite direction to the two others. The legs are weighted, and a gantline block lashed to the fork. The two first legs are let go first together, and the tripod hauled into position by guys. Weights can be added by slipping them down the legs, and the guys secured to anchors.

A vertical pole with bamboos can now be added, its weighted heel on the ground. It is placed by a jigger from the fork, to which it is afterwards lashed, and guys taken from the lower part to the tripod legs. A block and halyards from the bamboo permits a calico flag 14 ft. square to be hoisted.

CHAPTER II.

A MARINE SURVEY IN GENERAL.

THERE is a great variety in the methods that can be employed in making a marine survey, so much so that the task of describing any general scheme of operations is by no means easy.

In the first place, Marine Surveys may be divided into three heads. **Different kinds of Surveys.**

1st. Preliminary or Sketch Surveys.

2nd. Surveys for the ordinary purposes of navigation.

3rd. Detailed Surveys.

The boundaries between these are by no means strongly marked, although each differs considerably from the other, and a finished sheet as sent home is not unfrequently a combination of all three, comprising pieces of work done after very different fashions, according to needs and circumstances.

A preliminary survey does not pretend to accuracy. The time expended on it, and the means used, cannot ensure it, **Sketch Surveys.** and it only represents what our second name for it indicates, a sketch. A sketch survey will be founded on a base of some kind, but this will generally be rough, and in some instances, as in many running surveys, will depend solely upon the speed of the ship as far as it can be ascertained by patent log ; so that the whole affair from beginning to end is only a rough approximation.

The necessity for sketch surveys may be said to be getting less and less every year. Most parts of the world have their coasts mapped at any rate as far as this; but there still remain portions of our globe of which the coast-lines are not marked at all, or are extremely hazily delineated, and to these any sketch survey will be an improvement.

Ordinary Surveys. The second head comprises the majority of charts now published, and many of those in course of construction in the present day, *i.e.* they are constructed on such a scale, and with such limitations of time, &c., as to make it impossible either to show small details of land or sea, or to be perfectly certain that small inequalities of the bottom, or detached rocks, may not exist, unmarked. Everything, however, shown in such a survey should be correct, and it is only in its omissions that it should be imperfect.

Detailed Surveys. A detailed survey is accurately constructed from the commencement, on a scale large enough to admit of close sounding, and time is given up to working out all minutiæ.

Detailed surveys are mostly confined to the more civilised shores of the world, where there is much trade, and to such ports, harbours, and channels as are largely used in navigation.

The necessity for these surveys increases to an enormous extent every year, with the prodigious strides trade, more especially trade by means of steam-vessels, is taking.

A steamer works against time; her paying capabilities largely depend on her getting quickly from port to port, and captains will take every practicable short cut that offers, and shave round capes and corners in a manner to be deprecated, but which will continue as long as celerity is an object. A channel which a sailing vessel will work through in perfect safety, from the obvious necessity of keeping a certain distance off shore, for fear of failing wind, missing stays, &c., will be the scene of the wreck of many a steamer, from the inveterate love of shortening distances, and going too near to dangerous coasts only imperfectly surveyed. Better charts will not cure navigators of this propensity, but will save many disasters by revealing unknown dangers near the land.

Time, and the comparative scarcity of marine surveys, do not permit of keeping up to the rapid advance required in this style of survey; and unless the countries of the world interested in ocean traffic largely increase their expenditure on these matters, it seems as if charts will get further and further behind requirements as years roll on.

Having settled of what description a chart is to be, there is still much diversity in the method of undertaking the details of it. The extent of the work, whether simply a plan of limited extent, or a large piece of open coast; the scale on which it is to be done; the nature of the coast and sea; the time and means at disposal; the number of assistants; will all be considered in determining exactly how to set about the work. Detail always variable.

All this makes it very difficult to lay down rules for marine surveying. Experience alone can dictate what should be done in each particular instance. Though a plan may be produced, the time employed, and the result of the labour expended, will greatly vary according as to whether the work has been undertaken in the right way or not, apart from any personal qualities of the assistants, and nothing but the possession of the true surveying "knack," combined with experience, will point out this right way.

All surveys are, however, alike in this respect. They are, as it were, built up on a framework of triangles of some kind, the corners of which are the main "points" of the chart, and to obtain this framework is always the first thing to do, and how to set about it the first thing to consider. Triangulation.

The construction of this "triangulation," as it is termed, is of various kinds; ranging, from the rough triangles obtained in a running survey, where the side is obtained by the distance it is supposed the ship has moved, and the angles are sextant angles, taken on board from a by no means stationary position; to the almost exactly formed triangles of a detailed survey, when carefully levelled theodolites observe the foundation of a regular trigonometrical network, which covers the whole portion to be mapped.

The term triangulation would seem to infer that this system of triangles would be always apparent; but in surveys irregularly plotted, and when working on a sheet previously

graduated, it will seem that there is no triangulation, and in
the strict sense of the word there is none, but the framework
of the chart is still built up on the system of triangles, and
it is difficult to find any other name for the process.

For the present we will speak only of the second and third
kinds of surveys, leaving Sketch Surveys to be described
separately.

The system employed in Ordinary Surveys and Detailed
Surveys is the same, and they really only differ in the scale
of the chart, and the amount of time that is spent on them,
especially with regard to closeness of soundings.

In a detailed survey, time must be subservient to the
necessity for exactness, and for exploring every foot of the
ground.

In an ordinary survey, judgment has to be exercised
as to how far we must be satisfied with what we can get
for triangulation, and how much time we can spend on
details.

It is by no means necessary in an ordinary survey to
observe the angles at each corner of the triangles. The happy
fact that the sum of the three angles is 180°, enables us to
manage whenever we have two of them, though it is of course
more satisfactory to actually observe all three for the more
important triangles.

**Accuracy
necessary.**   The accuracy necessary in many details of a chart depends
very much upon its scale.   Over-accuracy is loss of time.
Any time spent in obtaining what cannot be plotted on the
chart is, as a rule, loss of time ; but it cannot be too strongly
impressed upon the young nautical surveyor, that his work
should be as correct as his scale will allow.   Nothing should
be put down of which he is not sure, and it is no loss of time
to repeat angles to prevent mistakes.   It is better to be over-
accurate than to err in the opposite direction, and experience
will soon show him when he must be very exact, and
when a little latitude is permissible without interfering with
the result.

The accuracy of the main triangles of a chart is most important, everything depends on them, and if they are incorrect, nothing will be satisfactory afterwards.

The general plan of a survey may be said to be this :— **General plan of Survey.**

1st. A base is obtained, either temporary, as in the case of an extended survey; or absolute, as in a plan. This is the known side of the first triangle.

2nd. The main triangulation, that is the establishment by means of angles of a series of positions, at a considerable distance apart, from which, and to which, angles are afterwards taken, to fix other stations. These are the corners of our framework, and are known as the "main stations," the two ends of the base being the first two, on which everything is built.

3rd. The fixing by means of angles from these main stations of a sufficient number of secondary stations, and marks, to enable the detail of the chart to be filled in between them. In most cases angles will be required to be taken from the marks themselves as well.

4th. All these points, or those embracing a sufficient area to work on, being plotted on the chart, they are transferred to the field boards, either by pricking through the plotting sheet with a fine needle, or, what is a better way when carefully done, by making a tracing of them on tracing-cloth, and pricking through that on to the boards.

5th. Each assistant then has a certain portion told off to him to do. It must depend upon circumstances, but as a rule it is more satisfactory to have the coast-line put in first, and the soundings taken when this has been done. The topography, or detail of the land, can be done at any time.

6th. Each piece of work is inked by the assistant on his board, with all detail, and when complete, is carefully traced on the above tracing of the "points." All bits are thus collected together, and the total is retransferred to the plotting sheet by means of transfer paper, and inked in as the finished chart.

These details must not be taken as unalterable. Some prefer plotting everything on to the same original sheet, and when a surveyor is by himself, or with one assistant, he would probably do this, but the method described is calculated for a number of assistants, and has been found to work well.

It is not absolutely necessary to get a base before starting a plan. Circumstances may make it imperative to wait a day or more for this, and in the meantime, a distance between two stations, to be finally measured, can be assumed and plotted, and the whole system of triangulation built upon this. But it must be remembered that no heights can be measured by means of angles, until a scale is obtained.

If an extended survey, and plenty of hands, some will carry on the triangulation and marking on ahead, while others are putting in the detail, and sounding the part already marked.

The deeper soundings will be taken from the ship, to a sufficient distance or depth off shore.

When to obtain the Astronomical Positions.
It will depend upon circumstances when the astronomical observations for latitude and longitude are taken. If only an isolated plan is being done, the observations to fix some definite point on it can be taken at any time.

When an extended survey is in progress, that has been commenced on a measured base, they can also be taken when convenient. In this case the final scale of the chart will always depend upon the observations taken at either extremity of the chart, and they must consequently be done very carefully. Circumstances of weather, time of year, &c., will therefore influence the choice of the best time for these.

Sometimes an extended survey will be originally plotted on a base obtained by the astronomical positions, and in this case of course they will be the first thing to undertake. At any rate it will nearly always be convenient to obtain a true bearing at once, in order to have the meridian of the chart placed squarely on the paper from the commencement.

These separate steps in a survey will now be described in detail, following the order in which they will generally come, as far as can be done.

CHAPTER III.

BASES.

By Chain—By difference of Latitude—By Angle subtended by known
length—By Measured Rope—By Sound.

BASES for marine charts or plans upon which to build **Different** **kinds of** the triangulation are obtained in several ways, according as **bases.** circumstances permit and accuracy requires.

1. By means of the 100-ft. chain or steel tape supplied for the purpose.

2. By difference of latitude, or difference of longitude.

3. By measuring with a micrometer or sextant the angle subtended by a known length, as two poles a measured distance apart, the ends of a long pole, or the masts of a ship.

4. By a measured rope, as a lead-line; or by the wire from a sounding machine.

5. By sound.

CHAINED BASES.

The ground for a base, to be measured either by chain or rope, must be as level as can be found. Its length will be partly determined by the extent of the work to depend on it, varying from say 9000 feet to 1000 feet, or even less for a small harbour.

While it is certainly convenient to measure the base in one straight length, if convenient ground can be found, it is by no means necessary.

If several short lines making angles with one another are measured with the angles carefully observed, the terminal

points being visible from one another, the resulting distance
calculated between the terminal points should be just as
correct as if it had been measured direct.

**Extension
of base.**
It is seldom that a chained base can be found, even for a
small plan, long enough to plot from directly, *i.e.*, the
measured length when protracted on the paper would be
generally so short, that by placing that on the sheet first,
and making it the starting line, errors would be sure to creep
in, in increasing the size of the triangles, any little error
being multiplied. It therefore is usually necessary to extend
the base, as it is termed.

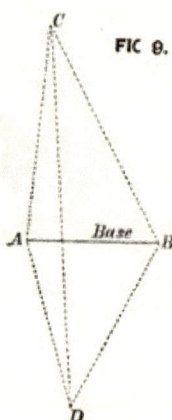

FIG 9.

This consists simply in calculating a **sufficient number of**
triangles, conveniently arranged, to obtain a side long enough
to form a good start, so as to plot *inwards* as much as
possible, when **any** little errors will be diminished, instead of
increased.

As a commencement of this process, the base to be
measured should, if possible, be placed so that there are two
stations, one on each side of it, which can be used for the
first triangles and consequent extension of the base.

Here, Fig. 9, A B is the measured base, C and D the two
first stations. Angles are observed at A, B, C, D. The other

two sides in the triangles A B C, A B D being found, C D can be found in both the triangles A C D, B C D, which will check the result, and C D will be the extension of the base for further triangulation.

Of course this desired convenience will not always be found, but it is a thing to look out for.

It is by no means necessary to measure a long base, provided that convenient triangles can be found for extending the base by calculation. If the angles of these are of the necessary number of degrees, and they are carefully observed with theodolites, a short base, measured on flat smooth ground, will give a truer result than a longer one measured over inequalities. With a sextant survey it will be well to have as long a base as possible.

The ground having been walked over to ascertain it will *Planting* do, and that the base stations (the ends of the base) are so *staves for measuring* placed that they see as much as possible on all sides: set *by.* up the theodolite at one end, and at the other a flagstaff or another theodolite, and let a man plant staves (boarding pikes make good ones) exactly in line between the two stations, giving him the position for the first two or three, by looking through the theodolite directed to the other station. After these are in place, he can plant the others in line by guiding himself by them.

Having the staves placed and in line, begin to measure from *Method of* one end. If two persons are to measure, begin from opposite *measuring.* ends. A man is required for each end of the chain. The man at the foremost end of the chain carries ten pins, and the surveyor attends with his book to see the chain fairly placed in line between the staves, and to note down each length of chain measured. Do not let the men stretch the chain too tight, but it must lie straight on the ground between the two ends.

The chain being down for the first length, measuring from under the centre of the theodolite, put a pin in the ground, at the foremost end, *inside* the handle, and touching the flat side. Make a mark in the note book, and walk on together,

F

the man at each end lifting the chain as much as he can, until the hindermost comes to the pin. He must then place the *outside* of his handle so as to touch the pin. Another pin is put in at the foremost end inside the handle, the second note made in the book, the first pin taken up by the hindermost man, and on you go.

The lengths are best noted by strokes, crossing every fifth over the four, as in ordinary tallying.

Check at every ten lengths by the number of pins. When the tenth stroke is made, the foremost man should have no pins left in his hand, and the other man should have nine, the tenth having been just put in.

The odd feet and inches in the last length are measured by counting the links, which are each a foot long.

In walking forward, take care that the hinder man does not overwalk the former, or the chain will have a bight dragging on the ground, links will catch in something and get bent, and the error of the chain will be very different when retested, to what it was before landing.

Repetition necessary. The number of times a base must be measured depends on circumstances. If for a harbour plan, only twice, if they agree to a foot or two, will be sufficient. For a survey of greater extent, three or four times will be more satisfactory, unless the two first measurements agree very well.

Inequalities. Perfectly level ground can seldom be found, and the surveyor must make an allowance for inequalities by his judgment, which will be of course always subtracted from the measured length.

The chain must be tested for length, before and after measuring the base, to ascertain the error.

BASE BY DIFFERENCE OF LATITUDE.

When two stations are available from twenty to thirty or forty miles apart, visible from one another, and bearing not more than two points from the meridian, having also a few intermediate points visible from both, a very good base can be got by latitudes, and careful true bearings.

The base will then be diff. lat. × sec. Mercatorial bearing.

Similarly, if the stations bear nearly east and west from each other, the diff. long. may be obtained by chronometer or rockets. The true diff. long. by observation is converted into spherical diff. long. from which the departure is found. The length of base will be Dep. × Cosec. Mercatorial bearing.

By means of the intermediate points, triangles can be calculated down to a workable length of side for fixing marks.

Where no smooth ground for measuring a base can be found, and we want our scale to be near the truth from the first, this method is valuable.

The only drawback to it is the effect of local attraction on the pendulum, or, in other words, on the mercury in the artificial horizon. With high land behind a station and a deep sea in front it may result that there will be considerable error; and the difference in the distance between the terminal points of the survey, if it covers much ground, as deduced from such a base, and as determined from the observations at either end, may be much more than if starting from a measured base.

BASE BY MAST-HEAD ANGLE.

This consists in measuring, with a micrometer or sextant, the angle between the mast-head of the ship and the hammock netting, or some other fixed line on the ship's side; not the water-line, as that varies.

The vertical circle of a theodolite being only marked to minutes, unless it be a much larger one than is generally available, is not sufficiently accurate for this.

It is well to use two sextants to check errors, and read them both on and off the arc.

The height of mast-head above the line must be accurately known to give a good result.

Working out a right-angled triangle gives the distance required.

A table should be formed of the distances corresponding to different angles of the mast-head of the ship, as this will be frequently used in sounding banks.

BASE BY ANGLE OF SHORT MEASURED LENGTH.

Where the ship is not available, a base for a small plan can be obtained by measuring the angle between two well-defined marks placed in the ground at a carefully-measured distance apart, or that subtended by the ends of a long pole.

This must also be done with the sextant or micrometer.

If staves in the ground are used, care must be taken that they are at right angles to the required base. Similarly, if a pole is used, care must be taken to hold it at right angles to the observer, which can be ensured, either by having a pointer nailed on to the centre of the pole projecting at right angles, and which must be directed towards the observer by the man holding the pole in both hands horizontal, or by simply waving the pole, held in this position, backwards and forwards gently, when the observer will register the largest angle he observes as the correct one.

The angle observed should not be smaller than 1°, which with a distance of 20 feet, will give a base of over 1100 feet. It would be better, however, if practicable, to get a base by means of a longer distance, and larger angle than this, when a very trustworthy result will be obtained ; or to be content with a shorter base, and extend it by angles, as already described, to a longer working base.

Measurements must be made on and off the arc, and it would be well to use more than one sextant.

Small lengths of this kind may also be measured by a micrometer, but a sextant will give just as good results, and is in a ship always handy.

No appreciable error will be introduced by taking distance = length of pole × cot. angle.

MEASURED BASE BY ROPE.

Measuring by a rope is of course not accurate. It is difficult to avoid stretching it more at one time than at another, and if it gets wet, it alters its length considerably. If measuring over ground where it is sure to get wet, it will be better to wet it well beforehand. Test it in that condition, and keep it well wet all the time of measuring.

BASE BY SOUND.

This consists in counting the interval of time which elapses between the sight of the flash of a gun, and the arrival of the sound of the explosion, the gun being at one end of the required base, and the observer at the other.

Recourse is had to this method of obtaining a base when no flat ground can be found on which to measure. Its accuracy is not great by any means, but, if the final scale of the chart is to depend on astronomical positions, it is quite near enough for working out details such as heights, small parts measured with ten-foot pole, &c. *Final scale should not depend on base by sound.*

Its value is much increased by observing from both ends, which should always be done if possible, and a surveying vessel should have two small brass Cohorn mortars supplied for this purpose, which can be sent away in a boat, and tumbled overboard without damage. *Useful hints.*

The ship is often used at one end of a base by sound, especially in work amongst small islands, and it is also necessary sometimes to have a boat at the other, but if at any rate one shore station can be obtained it will be better. If choice of direction can be had, measure with the wind across the base, as, though the error from increase and decrease of velocity is eliminated by measurement from both ends, the sound may be difficult to hear against the breeze, if at all strong.

For either end choose positions for the hearers as much out of the wind as possible, as it is the whistling of it in the ears which disturbs the receiver more than anything.

A base of three miles is a very good length, but the surveyor will generally not have much choice in this matter. Needless to say, on a calm day the sound will be heard farthest and easiest, but the choice of days is seldom possible in practice. If we waited for the best opportunity for every detail of a survey, it would never get on, and the utmost that can be done is, when there is alternative work for which the day or opportunity is more suited, to take that in hand.

Signal to be made. The guns from the two ends should be fired alternately, at regular intervals, and at some preconcerted signal, as dipping from the ship a flag visible from both stations, which should be hoisted a minute or half a minute before as warning, or rehoisting a dipped flag steadily, the gun being fired as the flag reaches the masthead. It is distracting to the receiver to be waiting an indefinite period for the flash.

Watch to be used. A chronometer watch is the best, beating five ticks to the two seconds. An ordinary watch which beats nine ticks in the same period, goes at such a pace as to be rather confusing, especially when not in practice, though, if the observer is used to the process, he will measure as accurately with an ordinary watch, and possibly more so.

Preparation for counting. When awaiting the flash, hold the watch to the ear and count to yourself—nought, nought, nought, &c., continually, keeping time with the ticks; you will then be ready to commence—one, two, three, &c., as soon as you see the flash or smoke of the gun.

If going to use a telescope to watch for the warning signal, tie the watch over the ear with a handkerchief, which will leave both hands free.

Count only up to ten or twenty, and mark off each ten or twenty by putting down a finger of the unoccupied hand, or by some such means.

If time allows, three or four measurements should be made Repetieach way, or more if they do not agree with one another. A tions. signal must be arranged to ask for more than the number previously settled, if it be wanted.

In meaning the result, the arithmetical mean is not strictly Calculatcorrect, as the acceleration caused by travelling with the ing the mean. wind is not so great as the retardation caused in the opposite direction, as in the latter case the disturbing cause has clearly acted for a longer period. The formula used is

$$T = \frac{2 \, t \, t^1 *}{t + t^1}$$

when T is the mean interval required,

    t the interval observed one way,

    t^1 the interval the other way.

The mean interval thus found, multiplied into the velocity of sound for the temperature at the time, will give the required distance.

The velocity of sound varies considerably, and an accurate Velocity of law for all its causes of variation has not yet been discovered. sound. The main cause is, however, temperature, and for this it can, to a certain extent, be corrected.

The most trustworthy experiments made, show that sound travels about 1090 feet in a second of time, at the temperature of 32° Fahrenheit, and increases at the rate of 1·15 foot for each degree of temperature above the freezing-point, decreasing in the same proportion for temperatures lower than 32°.

This is the only correction that can be made, and a base measured in the manner described, with these data, will give an approximation sufficiently near for all practical purposes.

As an example, let us suppose A and B the two ends of the Example base to be measured. of Base by Sound.

---

* See Appendix F.

At A have been observed

> 44 beats with watch beating 5 beats to 2 seconds
> 45
> 44
> —

Mean 44·33 beats = 18·532 seconds.

> 81 beats with watch beating 9 beats to 2 seconds
> 82
> 83
> —

Mean 82 beats = 18·222 seconds
>   Mean at A = 18·376 seconds.

At B have been observed

> 85 beats with watch beating 9 beats to 2 seconds
> 87
> 88
> —

Mean 86·66 beats = 19·258 seconds.

> 47 beats with watch beating 5 beats to 2 seconds
> 47
> 48
> —

Mean 47·33 beats = 18·932 seconds.

>   Mean at B = 19·095 seconds.

Then working $T = \dfrac{2\,t\,t^1}{t + t^1}$

we get T = 18·728 seconds.

Temperature is 80°, at which velocity of sound is 1145·2 feet per second.

This multiplied into the interval, gives 21448 feet for the length of our base.

The temperature must be taken in the open with the thermometer shaded from the direct rays of the sun, but not in too cool a spot, or it will not give the true temperature of the free air.

CHAPTER IV.

THE MAIN TRIANGULATION.

General—Making a Main Station—False Station—Sketch—Convergency
—Calculation.

THE main triangulation has been already defined as "the establishment by means of angles of a series of positions, from which and to which angles are afterwards taken, to fix the secondary points of the survey." Definition

All positions from which angles are taken, with the intention of fixing other objects, are called "stations," the symbol for which is △, but the ones with which we are immediately concerned, that is, the first and important positions, are distinguished as "main stations," and these collectively form the "main triangulation." Main Stations.

The first object of main stations is to see other main stations, and with this in view their positions are chosen accordingly; but angles to everything useful, secondary stations, marks, &c., are, of course, taken as well.

Secondary stations are those from which angles are taken solely to fix the smaller marks and details, &c., of the survey. They will be nearer together than the main stations, and may often be perforce so placed as to be useless for any other object. Secondary Stations.

All objects fixed and plotted on the skeleton chart are known as "points." A "point" may be a main station, a secondary station, or simply a mark; but when fixed and plotted on the sheet, with the intention of using them in the survey, they are one and all spoken of by the generic term of "points." Points.

Main triangulations may be divided into two kinds:

Varieties of Triangulations. "calculated," in which the triangles are all worked out, so that the length of any side, or the distance between any two main stations, can be found; and "plotted," in which the main stations are simply the first points laid down on the paper.

Calculated Triangulations. A calculated triangulation is used in any detailed survey, in plans, or whenever from circumstances it is convenient to have different parts of the same survey on separate sheets, which can therefore afterwards be put together in the engraving, without any fear of their not fitting into one another.

Bases for plans, on a larger scale than the rest of the chart, can often be taken out of a calculated main triangulation without measuring separate small lengths.

Plotted Triangulations. Plotted triangulations may further be subdivided into "regular" and "irregular."

A plotted regular triangulation will be when triangles have been obtained which could, if requisite, be calculated trigonometrically. As, however, a calculated triangulation is of great service as a record, and for future resurveys, it is expected to be furnished with every chart.

It is more satisfactory that triangulation should be regular if possible, but it very much depends upon the nature of the coast to be surveyed, in what manner it can be carried out.

Irregular Plotted Triangulations. In many extended surveys, where, for instance, the land is low and densely wooded, or perhaps bordered by reefs to a great distance from the shore, a regular triangulation is hardly possible, or would entail so much loss of time as would not justify its being undertaken.

The main points must be plotted in these cases by all sorts of means. The ship enters largely into the scheme, and frequently boats also. Stations may have to be fixed solely by angles observed at them. True bearings are freely used in the construction of the chart, and any regular system of triangles disappears.

A large proportion of existing charts have been, and man

more are now being, constructed, by means of irregular plotting.

A survey can often be commenced with a regular triangulation, when it will be found necessary, after having advanced a certain distance, to have recourse to irregular means to fix main stations.

Here it is, when ordinary rules and systems fail, that the skill of the chief of the survey is shown in overcoming these difficulties in the readiest and best method, and these are the circumstances on which we can give the fewest hints. Such as we do mention will be found in the next chapter on Plotting. In the present one we shall confine ourselves to regular triangulations.

The angles of the first few triangles in a triangulation, commenced on a measured base, will require to be extra-carefully observed, and the theodolite must be carefully placed exactly on the spot of the mark which will distinguish the station. For, as we shall be increasing our distances in each triangle, until sides long enough to carry on the triangulation without further enlargement are arrived at, any little error in an angle will give a larger error in the resulting side. These first triangles will nearly always require to be calculated, as already remarked under the head of Bases, in order to get a side long enough to plot from, whatever it may be the intention to do afterwards. *Great accuracy in first Triangles.*

Although we are about to speak of triangulation from shore stations, as carried on by means of the theodolite, as this instrument is always available in a surveying ship, it must be understood that, *with care*, an excellent triangulation may be obtained with that invaluable instrument, the sextant. *Triangulation by Sextant.*

The point on which care is principally needed, is that the angles measured should · be horizontal angles. A practised surveyor will usually be able to note some small natural mark, directly above or below the object whose angle is required, and *at his own level*, to which to measure his angle, and in most cases of using the sextant this will give a *Horizontal angles with Sextant.*

sufficiently near result, but if forced to use the sextant for triangulation, another means may be used.

From the end of a longish pole (boat-hook staff will do), planted at a slight angle from perpendicular, let a plumb line fall, and getting the object transit one point in the line, the angle can be taken to any other part of it. The plumb line must not be too close to the observer, or it will be difficult to keep the transit on, and parallax will creep in.

It is a question of circumstances as to whether the main triangulation is to be carried on by itself first of all, or in combination with the secondary stations and marks. This in noways affects the principle of the work, but only the detail of what is done when the angles at the main stations are observed.

The main triangles should be as large as possible. The fewer triangles there are, the fewer are the chances of errors of observation.

MAKING A MAIN STATION.

Choice of Zero.   Observing angles at a station is technically called "making" it. Let us suppose a surveyor making a first station, probably one of the base stations.

He has been previously furnished with a list of the main stations visible from him, and has been told how many times his angles to them are to be repeated. He has also received instructions about the secondary stations and minor marks, if any have been selected and marked.

Having levelled the theodolite, the first thing is the choice of an object from which to measure all the angles, which is called the zero.

A zero should be, if possible, another main station. It must be at some distance, but not so far as to be easily obscured on a hazy day; well defined; so placed that the rays of the sun, when it moves from the position in which it happens to be when the station is commenced, will not obliterate it. It should be a fixed object, *i.e.* not likely to be removed, or to tumble down, and not so high as to be covered with clouds, as a mountain peak.

A great deal of trouble is given when a zero has to be changed, or when on a subsequent visit to a station the same zero cannot be used. Attention to the above mentioned points is, therefore, of importance.

The bearing of the zero by the theodolite compass should always be entered in the book.

The zero fixed upon, and the theodolite directed upon it, observe the main angles, or those to the main stations, first, repeating them the required number of times, by either of the two methods described under "Theodolite." Observe main Angles first.

These completed, observe the secondary stations a sufficient number of times, as well as all marks and conspicuous objects.

It is important to remember that the position of the sun has a great effect on the visibility of objects, and therefore that those stations and objects on which the sun is shining should be secured first, because later, when they fall into shadow, they may be wholly invisible.

In most instances a sketch will be also necessary, on which the angles to conspicuous objects, tangents, &c., will then be recorded, instead of in the book.

All angles should be read twice, in order to prevent mistakes; but to ensure accuracy when required, the angles must be repeated on different parts of the circular arc, for the following reasons :— Repetition of Angles.

A theodolite, however well turned out, is seldom exactly centred, hence arises error; as no matter how uniform the graduation of the circular arc may be, a slight deviation of the axis from true centring will give a difference of reading for an angle on different parts of the arc.

The sum of the readings of the two verniers is supposed to correct errors of centring, but for remarks on this, see "Theodolite."

The reading itself of an angle can never be considered as perfectly correct. Slight parallax frequently exists, especially when an instrument has been some time at work, and is getting worn. In small theodolites the marking of vernier and arc at any given angle will often not coincide exactly, and judgment may assign the wrong reading.

By multiplying readings, then, a mean will be obtained which will to a great extent eliminate these errors, and this must always be done in observing main angles.

Excepting for main angles, forms ruled in the angle book are unnecessary, and in this case the form is simple, consisting of columns to keep the figures separate, as under.

♭ July 4th, 1881, at Pagoda △, Theod. 77.

△ˢ·	∠ Observed.	Difference.
	° ′ ″	° ′ ″
⊕ Patero △	360 00 00	
Mango △ (flash)	25 14 30	25 14 30
	50 29 30	25 15 00
	75 44 00	14 30
	100 59 00	15 00
	Mean	25 14 45

♭ July 4th, 1881, at Pagoda △ ⊕ Patero △, Theod. 77.

⊕	Prince △	Flag △	Snow △	
°	° ′ ″	° ′ ″	° ′ ″	
360	24 18 00	29 10 30	48 26 00	Z. O. K.
100	124 19 00	129 11 00	148 27 00	Z. O. K.
200	20 00	10 30	27 00	Z. O. K.
300	19 00	11 00	27 00	Z. O. K.
Means	24 19 00	29 10 45	48 26 45	

The first of these forms is adapted for the observation of main angles by repeating round and round singly; which is done when a solitary angle is required to be observed

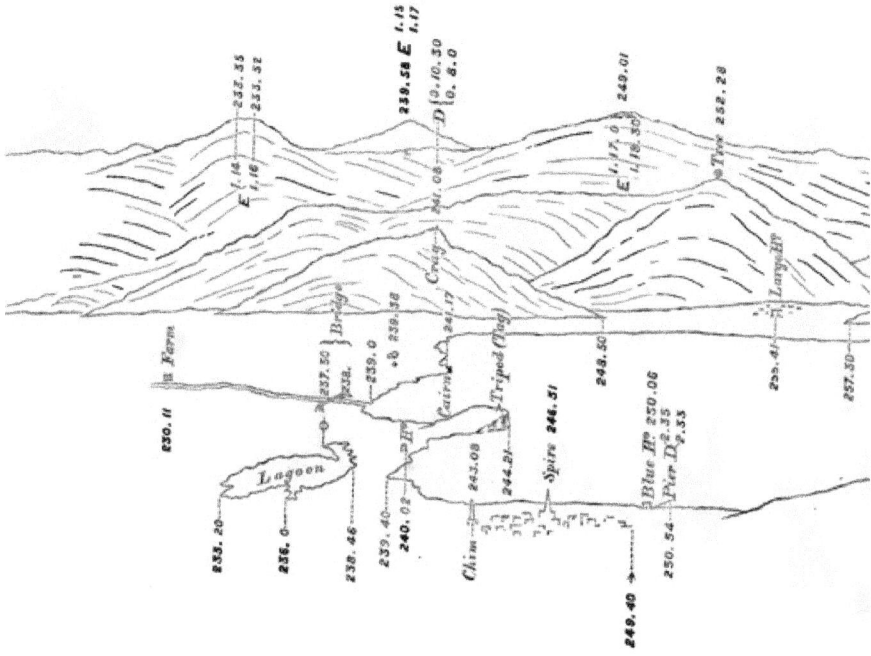

237.35
233.32

238.58 *E* 1.15 1.17

D {0.10.30 {0. 8.0

249.01

Two 252.28

E 1.14 1.16

E 1.17.0 1.18.30

LargeH?

255.41

257.50

Craig 241.08

230.11 *Farm*

237.50 } *Bridge*

+b 239.32

+b 239.38

Cairn 241.17

239.0

248.50

Triped (Tug)

Lagoon

233.20

236.0

238.46

239.40

240.02

243.08

244.21

Chim

Spire 246.51

Blue H? 250.06

Pier D? 2.35 2.33

250.54

249.40 →

277.14.30 E 1.20 1.24.30

E 0.07 0.09

274.03

Hav. 275°

268° Vall.

272.50 Vall.

273.05

277.12

278.0 Vall.

← 278.50

High 281.10

→ 283.32

281.01 D 2.20.0 1.0

283.17 Highest " 48 Outer

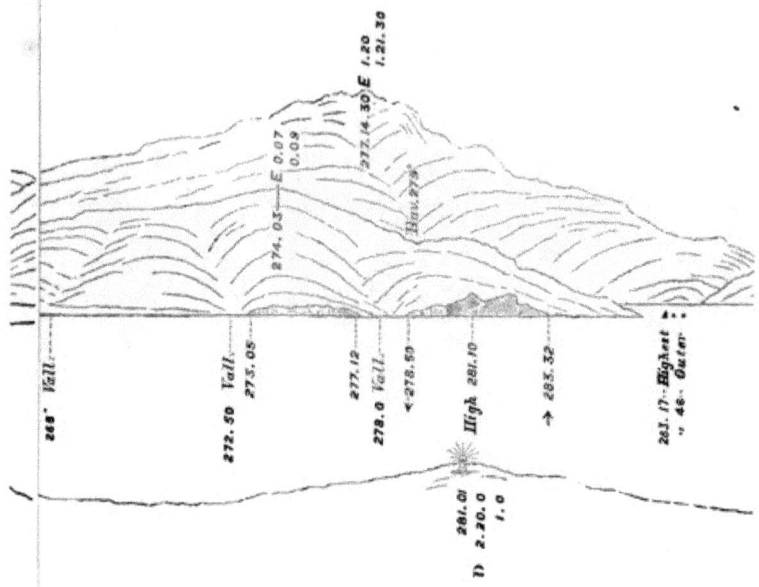

fro~~n~~ a △ showing method of marking a~

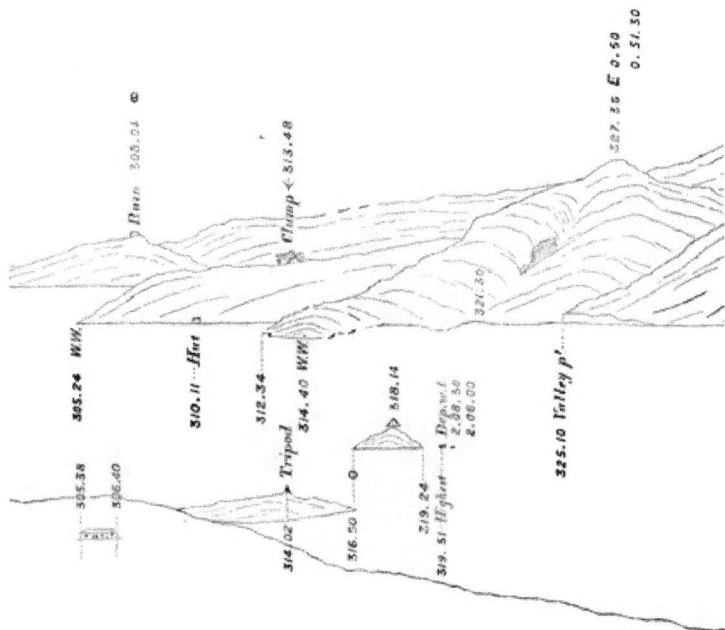

depressions, &c., and also mode of tu

333.46 *E* 1,10.30 +1. 30

337.50 *E* 0.46 0.48

354.08 — WW.

335.30

R*k* 346.45

344.09 II° tall Chim ←363.30 →345.53

Mole

354.24 — *Cairn*

Fill Pt. *Lt HE & Dep base* 1.04.30 1.02.30

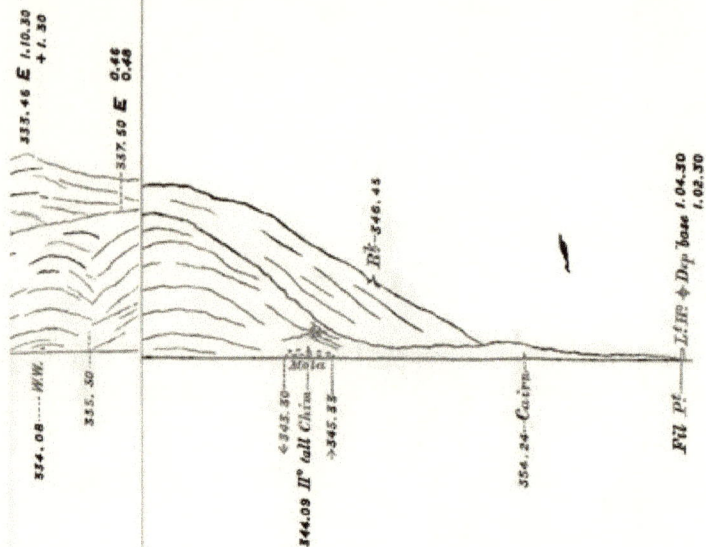

ing dow paper   for   continuous   panorama.

accurately, but to obtain great accuracy the reading should be repeated right round the arc.

The second is for ordinary main angles. This method saves much time when there are a number of angles required, and is as correct as is generally necessary.

The weak point of the first method is that the zero cannot be referred to, but, as only one angle is taken each time, a theodolite must be very much out of order to introduce error. *Comparison of Methods.*

If the angle to be observed is small, this method will not answer the purpose, as the theodolite will only be rotated through a small part of the circle, unless an inconveniently large number of repetitions be made.

The weak point of the second method is that any slight error in setting or reading the zero, affects every station observed; whereas in the other, the vernier being once *set* at the commencement, is afterwards *read* only.

By either method, the observer will see if his different observations of each angle are agreeing together, and can take more if requisite.

In all observations of angles with the theodolite (except the case referred to above), the zero must be looked at from time to time, and invariably at the conclusion of the set of angles, to make certain that the direction of the instrument has not changed by any unnoticed touch or shock. On every occasion of doing this it must be noted in the book, so as to know, in the event of the zero presently being discovered to be wrong, how far back the angles must be recommenced. A common form of notation of this is, Z.O.K., or Z.K., for zero correct. *Verifying the Zero.*

If the zero is found continually getting displaced without any apparent cause, something is loose, and this must be looked to at once, or nothing will be satisfactory. The parts most liable to go wrong have been mentioned under the head of "Theodolite." *Defects of Instrument.*

If using a heliostat, it must be placed in front of the theodolite, in the direction of the station to which you mean to flash. When the stations are distant one from the other, it is desirable to arrange who shall flash first; the receiver of *Arrangement for using Heliostat.*

the flash, say at A from B, then takes his angles to it, and
does not direct his flash to B until he has got the requisite
number of repetitions. When he does flash to B, the latter
will know A has done with him, and can direct his flash to
some other station, while he observes A. When B in turn
has finished with A, he must give the latter another flash to
acquaint him with the fact. A's turning off his flash will
show B he understands.

Heliostat invaluable.   As already remarked, the amount of time saved when the
sun is visible, by the use of a heliostat, is incalculable. It
is useful for long distances, and short also, and on all sorts of
occasions, and is, in fact, one of the surveyor's greatest friends.

FALSE STATION.

It will often happen that a beacon having been erected, the
theodolite cannot be placed exactly on the spot, at any rate
without a great deal of trouble; or if a building or tree has

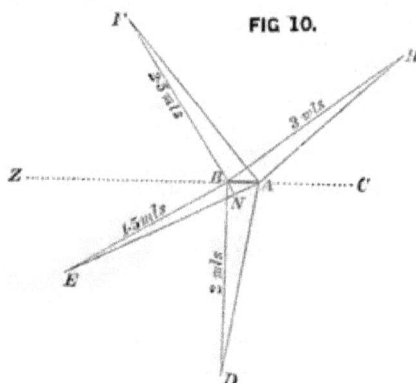

FIG 10.

been selected as a station, that the observer finds on going
there, that he has to make his station on one side of it in order
to see what he wants, or has to make a supplementary station
to see a few objects obscured by the building, &c. This is
called "False Station," and if the object is already plotted, or

it is desirable to plot it instead of the actual theodolite spot, the angles taken there must be corrected for the distance the theodolite was from such object.

The correction will vary according to the direction of the objects with regard to the true station, as the figures annexed will show.

In Fig. 10 let A be the true station at which angles are required; B the false station; D, E, F, H objects so placed as to illustrate all positions of false and true stations with regard to them. We have observed at B the angles to D, E, F, H, and A, and measured the distance B A.

Firstly. Required the angle EAD.

Produce B A towards C and Z.

$$\text{Now } EAC = EBC + BEA$$
$$\text{and } DAC = DBC + BDA.$$

Subtracting we have

$$EAC - DAC = EBC - DBC + BEA - BDA$$
or $\qquad EAD = EBD + (BEA - BDA).$

Secondly. Required the angle E A F.

Here $\qquad ZAE = ZBE - BEA$
and $\qquad ZAF = ZBF - BFA.$

Adding, we get $ZAE + ZAF = ZBE \times ZBF - (BEA + BFA)$

or $\qquad EAF = EBF - (BEA + BFA).$

Thirdly. Required the angle D A H.

Here $\qquad CAH = CBH + BHA$
and $\qquad CAD = CBD + BDA.$

Adding, we get $DAH = DBH + (BHA + BDA).$

These small angles, B D A, B E A, &c., the angles subtended by the distance between true and false station at each object observed, must be either calculated in each triangle, having two sides and the included angle (for the rough distances B D, B E, &c., will answer the purpose), or else, which is simpler, have a table* made of the angles which are sub-

---

* Appendix, Table O.

tended by different lengths at different distances, and take
the required angles out, thus:

Calculating correction by Table. Let us suppose the theodolite angle in our book corresponding to A is 60°, D 160°, and E 220°. A B is 12 feet. E B
is 1½ miles, and B D 2 miles, measured roughly on the sheet.
Required B E A by the table.

It will be evident that the angle B E A is that subtended
by a chord drawn across to E A from B. This chord we get
near enough by considering B N as at right angles to both E B
and E A, and looking out in the traverse table with B A, or
12 feet, as a distance, and B A N or 20° (180°—160°) as a
course, and taking the departure for the length of B N, which
in this case is 4·1 feet.

We then turn to our table, and see that four feet at
1½ miles subtends 1' 31", which is the angle B E A. In a
similar manner we can deduce any of the required angles,
quite near enough for ordinary purposes.

Arrangements for working by the Table. Now this process becomes far simpler, and much time is
saved if, in making a false station, a zero for the theodolite
is chosen in a direction exactly opposite to the true station,
as for example, in our figure at Z; for then each angle taken
can easily be corrected separately for the error of the false
station, and the true angle entered in the book. Difficulties
as to whether the ultimate correction is + or — will be
avoided, as in correcting the angles the error is subtracted
from all theodolite angles up to 180°, and added to all angles
between 180° and 360°.

Thus, in the case as in the figure above, the angles will
stand in the book—

Object.		Observed Angle.	Correction.	Angle at true △
		°	′ ″	° ′ ″
Zero, Z	..	360	0 00	360 00 00
F	.. ..	50	2 08	49 57 52
H	.. ..	130	1 42	129 58 48
D	.. ..	280	3 23	280 03 23
E	.. ..	340	1 31	340 01 31

In using the traverse table, take for the course—
Up to 90° the observed theodolite angle itself
Between 90° and 180° ... 180°—the observed angle
   ,,   180° ,, 270° ... observed angle—180°
   ,,   270° ,, 360° ... 360°—observed angle
—and the departure is looked out in each instance.

The table of angles subtended by different lengths is useful **Table generally useful.** for other purposes. As when an angle is taken to an object, and it is afterwards decided to plot a station made near that object instead of the object itself, the angles to the station can be corrected by it, in precisely the same manner as described above, the distances and direction of the station from the object being known.

Distances or lengths, greater than those included in the **Extension of Table.** table, can be got by multiplication or division.

Thus, if the angle of 18 feet at 5 miles is required, it is double that of 9 feet.

Again, if the angle of 12 feet at 10 miles is wanted, it is half that at 5 miles.

SKETCH.

A sketch taken from a station is made with the object of more easily identifying details to which it is necessary to take angles. By having a view of hills, islands, houses, trees, &c., from two or three stations they can, if fairly placed in their proper positions, be easily recognised in the different sketches when plotting. No description in the angle book will do this so well unless, of course, there is something very remarkable in the object, but even then the position of it as shown in the sketch will assist materially to prevent mistakes, and a curt description is also written against it on the sketch.

Sketching to this extent is within anybody's reach. A fairly correct outline is all that is absolutely necessary, and a very little practice will enable the least likely draughtsman to make a sufficient sketch for practical purposes.

Checking Scale by angles.
It is well for a beginner to commence by taking some rough angles to check his scale, or, until he is used to it, he will probably have one part of his view two or three times as big as the other, which is confusing afterwards, although the proper angles will be written against the prominent objects when the sketch is finished.

Preservation of Scale.
Always put the most distant outline on the paper first, as it is far easier to keep the scale uniform if this is done.

Begin on the extreme left of your view, or if it is an all round view, choose a point, in the direction least required, to be the left, and always work to the right.

Useful hints for sketching.
If the sketch is too long for one double page of the sketch book, when the right-hand end of it is reached, turn over, and turn one or two inches of the last page down, so as to show on the fresh page; this will give a commencement for the part to follow, and the sketch will be continuous.

Commence by settling whereabouts on the paper some two well-defined points of the distance are to be, and use these after as a scale from which to measure by eye the proper position of everything else.

If taking angles to assist correct drawing, as suggested above, a scale for the sketch must be decided on, say about one-third of an inch to a degree, but this will vary according to the complication of the sketch. If no divided scale is at hand, mark the edge of a strip of paper by eye, which will answer the purpose perfectly.

Take an angle from some definite point of the distance on the extreme left to some other, say about 20° to its right. Make a dot for the first object, lay the scale or strip of paper on the sketch, and dot again at the proper number of degrees, and at the proper height, with regard to difference of altitude, for the second object. Other angles can be taken to other objects between these, and the view sketched in between these dots, commencing as already said with the outline most distant, and therefore highest in the sketch.

In sketching for this purpose, it is well to rather exaggerate the height of objects, as, where there are hills, range upon range, or many objects, as houses, trees, &c., at different altitudes, they will get so crowded up as to make the sketch difficult to decipher, unless this course is adopted.

The great thing in a sketch is to place objects fairly correctly with regard to each other horizontally considered; *e.g.*, if there is a hill with a point nearly underneath it, take care that the latter is drawn on the correct side of the hill, right or left. Nothing is more calculated to confuse anybody plotting angles from a sketch, than to find that an object drawn apparently to one side of another object has an angle which shows it should have been on the other side. Doubt is at once thrown on the angle, when it is probably the drawing of the sketch which is incorrect. *(Important point to observe.)*

When the sketch is finished resume the theodolite, using the same zero, and mark the angles on the sketch itself, noting what the object is, when it may be doubtful, as for instance—Chimney of red house, Right of two fir trees, Big white boulder, &c. See example of sketch attached. *(Description of objects in Sketch.)*

PREPARATION FOR CALCULATING THE TRIANGULATION.

It is well that a true bearing be obtained between two distant stations, before plotting; but the method of doing this will be described under observations, and as far as absolute necessity goes, a good compass bearing from a shore station is quite sufficient to begin on. The bearing is only wanted to plant the meridian fairly square on the paper, and the compass bearing will give us this, near enough to be able to lay off any bearings which may be taken in the course of mapping the detail. The compass will never be used in any of the important part of the chart, unless our survey partakes of the nature of a sketch or running survey. *(A bearing required for Orientation of Chart.)*

If, however, regular triangulation is likely to fail, true bearings in the course of the work may be necessary to carry it on, and in this case we *must* begin with a careful true bearing.

Prepara-
tion of
Triangles.
In preparing the triangles for working, they will of course never be found exactly correct, *i.e.*, the three observed angles will be either more or less than 180°.

Spherical
excess.
In dealing with this theoretically, the sum of the three theodolite angles taken at the corners of any triangle will be greater than 180°, in consequence of each angle observed being in a different plane. This is known as the spherical excess, and in extended triangulations for topographical purposes, as the survey of India, &c., must be taken into account. For practical nautical work we need not regard it, as our instruments are not large enough to measure angles so exactly, nor is our work of sufficient extent.

Correcting
the
Triangles.
In dealing with the amount the triangle is in error, for the three angles of the triangle must be corrected to make the precise 180°, before using them for calculation, circumstances must guide its distribution among the angles.

An angle observed with a large theodolite should have more value given to it than others. One station may have been more exposed to the wind than others, which would depreciate the value of the angles observed there.

Without any indications of this kind to guide, it is as well to divide the error equally among the angles; but it must be remembered, that an alteration in the small angle will make more difference in the resulting position than in either of the other two, so that if this angle at all approaches the limit which should be used for a receiving angle (30°) it is perhaps well to put the smallest amount of change into it, but it is of course impossible to *guess* where the error is. If the angles have been repeated often enough, the resulting error any way will be very small.

Error ad-
missible in
Triangles.
No rule can be laid down with regard to the amount of deviation from the 180° that can be admitted, it so much depends on the degree of accuracy required, but in an ordinary theodolite survey the error should not be more than three minutes, and ought to be under two, working with five-inch theodolites, and repeating the angles three times if satisfactory, or more if they vary much.

In the first few triangles, the error should not be more than one minute.

Having corrected the triangles we come to the calculation. Calcula-tion simple. The working out of the triangulation is the very simple affair of plane triangles which every naval officer understands. The rule of sines, and the rule to find the third side,* when two sides and the included angle are given, are all that are required.

Logarithms of all angles must be taken out to seconds, Loga-rithms exact. so that the possession of tables giving these for every second of arc, will save much time and chance of mistake.

Into the final calculation of an extended calculated trian- Conver-gency. gulation some other considerations enter.

The actual working of the triangles will be the same ; but here we want the bearing of every side, as well as the distance, and the "convergency of the meridians" must be considered. This convergency will be explained before proceeding further.

CONVERGENCY OF THE MERIDIANS.

The true bearing of any two points on the earth, taken one from the other, in both directions, will be found to differ by a quantity which is called the convergency, and varies with the latitude, distance apart, and position of the points in bearing, or in other words, with latitude and departure.

Thus, if R and L are two stations lying roughly N.E. Illustra-tion of Conver-gency. and S.W. of one another, R being nearest the pole, in this case the North Pole, the true bearing of L from R will be found to be a greater number of degrees and minutes as measured from the meridian than the reverse bearing of R from L.

This results from the form of the earth. The true bearing Explana-tion. of one position from another, is the angle which the arc of a great circle drawn between the two positions makes with the

---

* The rule where sines only are involved must be used.

meridian of the observing position. As meridians are not
parallel, but converge at the poles, the great circle will cut
each meridian it passes at a different angle, the amount of
difference, for equal meridians, depending on the latitude.

To further the comprehension of this, let us consider the
method of projection of the sphere used when graduating a
map, made from the original data of angles and measurements.

Projections of the Sphere.
It will be evident to any one who considers the subject
that as our globe is a sphere (speaking roughly), a portion
of its surface cannot be shown on a flat piece of paper
without distortion, more or less, according to the extent so
shown.

There are a variety of methods used to delineate a portion
of the earth's surface on a map, which are called "projections." Into this variety it is not proposed here to enter, as
but one can be used when actually making a survey, which
is the " Gnomonic Projection."

Gnomonic Projection.
This projection is the only one on which great circles
of the earth are shown as straight lines. As it is on the
chord of a great circle that we see one object from another, it
is evident, that in graduating a map on which we have laid
down, or are going to lay down, one position from another by
drawing straight lines, we *must* use this projection.

A chart on the Gnomonic Projection is supposed to be
drawn on a flat surface laid against the earth, touching it at
the central point of the flat surface, and there only. From
the centre of the earth lines are supposed to be drawn, passing
through the different points to be shown on the map, until
they pierce the flat surface.

The positions so indicated on the upper side of the flat
surface, are those corresponding to the points required.

Here, in Fig. 11, P Q S is the globe, and A B C D a flat
surface laid against it, touching at the point J, the centre of
the flat surface, the *under* side of which is shown. P is the
pole. M F are points taken on the same meridian as J.
Imaginary lines drawn from the centre of the earth through
these points will touch the flat surface in N and G, and the

line joining them, the central meridian of the chart, will be a straight one.   K, another point on the globe east of the central meridian, will be projected at L, by the same method of drawing a line from the centre through K.   X is the point

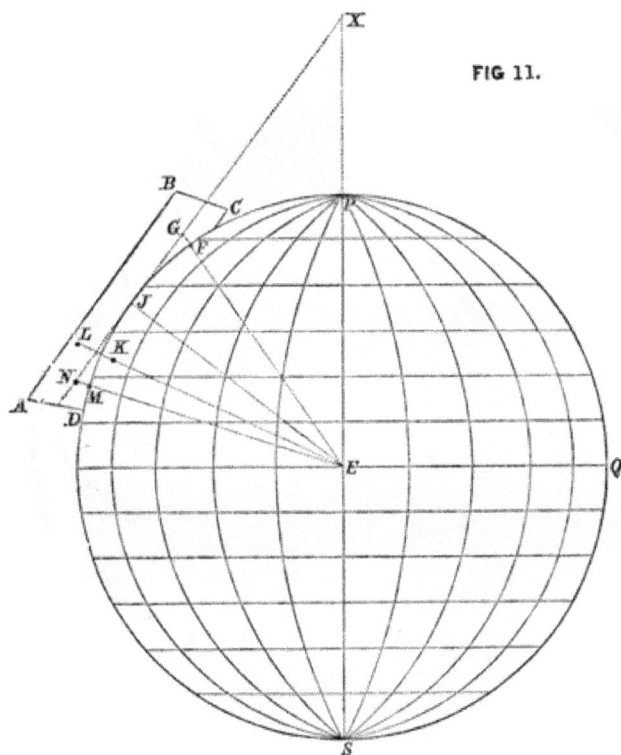

FIG 11.

in which the axis of the earth, produced, meets the central meridian of the chart also produced.

Let us again look at our flat surface, which we may now call the chart, from a different point of view, *i.e.* from a point in the extension of the line joining the centre of the earth and the central point of the chart.

In Fig. 12, A B C D is the chart as before, touching the spherical earth at the central point J. G and N are the positions on the chart of the points on the earth's surface, F and M in the other figure. G J N is then the central meridian of the chart. X is, as before, the point where the extension of this meridian meets the extended axis of the earth. L is the position on our chart of K (see other

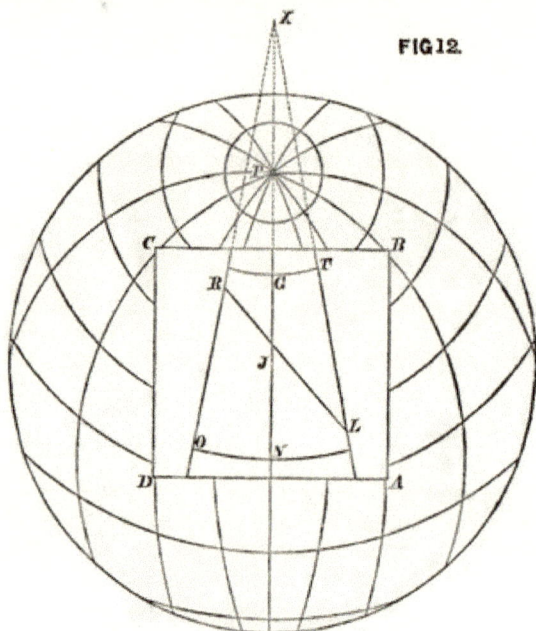

FIG.12.

figure). It is the position of a similar point, invisible in first figure, being on the other side of the earth. Meridians passing through L and R are projected on the chart by the same method as before, *i.e.* by drawing imaginary lines from the centre of the earth through different points in the required meridian; they will be found to lie as T L, R O, and their extension will also pass through X, making an angle R X L, which is the Convergency of the meridians; and this will be seen at once to be equal to the difference of the reverse bearings of R and L, for,

$$\angle O R L = O X L + R L X$$
or
$$O X L = O R L - R L X$$

i.e. Convergency = Bearing of L from R − Bearing of R from L.

A little consideration of this last figure will show, that the further towards the pole the central point J is, the greater will be the convergency of two meridians a fixed number of degrees apart; that when the pole P and J coincide, the meridians will radiate over the chart from that centre, and the convergency will equal the distance between the meridians; and that when J. is on the equator, the meridians will be parallel, and convergency will be nothing. Convergency nil at Equator and equal to diff. long. at Poles.

Parallels of latitude will appear on the chart as curves, concave towards the poles, and cutting each meridian at right angles. Parallels.

The equator being a great circle will be a straight line, and, generally, the further from the equator, *i.e.* the higher the latitude, the greater will be the degree of curvature in the parallels.

More consideration will show, that, the farther a part of the flat surface is from the surface of the earth, the greater will be the distortion of the positions resulting from this method of delineating the globe; or in other words, that the distortion increases from the centre of a gnomonic chart, and will become very considerable towards the edges, if a large area of the earth is attempted to be shown on a flat surface. But in practice, a marine survey does not extend over a sufficient area to make this distortion in any way apparent. Our diagrams are of course much exaggerated in this respect. Distortion.

It will be understood that the convergency is an actual fact, and does not result merely from the method employed in this projection. We have only considered it in connection with the projection, as it is thought that by so doing the nature of the convergency becomes more plainly apparent.

Mercatorial Bearing.

The mean of the two reverse bearings, or either one of them, plus or minus half the convergency, will give the Mercatorial Bearing, so called from being the bearing which each station will be from the other in a Mercator's chart, where, the meridians being all parallel, the line joining the stations will cut them at the same angle, this angle being also the one at which the line on our gnomonic chart will cut a meridian midway between the stations.

The actual observed bearing of a distant object, if protracted on a Mercator's Chart, will not pass through its position, in consequence of the existence of convergency. Mercator's charts are generally on such a small scale that, for navigating purposes, the error of taking the bearing swallows up the error introduced by not allowing for convergency.

The formula for Convergency is—

Convergency Formulæ.

Tangent Convergency = Tan departure × Tan Mid. lat. (1)

Or in anything but high latitude, or when the departure is great, it is correct enough to say—

Convergency (in mins.) = dep. (in mins.) × Tan Mid. lat. (2)
which can be converted into
Convergency = d. long × Sin Mid. lat. . . . . . (3)
   „ = dist. × Sin Merc. Bearing × Tan Mid. lat. (4)

any of which can be used as convenient.

The proof of the formula is given in the Appendix A.

Convergency by Spherical Triangle.

The convergency can also be found when the latitudes of and difference of longitude between the two stations is known, by working out the spherical triangle, with the pole, and the two stations, as the three points. Here we have the two colatitudes as the sides, containing the difference of longitude as the polar angle, to find the other two angles, which will be the bearings of each station from the other. The difference of these will be, as before, the convergency.*

---

* See following article for application of Convergency.

CALCULATING THE TRIANGULATION.

We now resume our remarks on working out a calculated main triangulation. All sides being calculated by the ordinary method of plane triangles, we now want the bearing, the mercatorial bearing, of each side, or, at any rate, a considerable number of them, in order that we can take any triangles or sides to work up details on, on a separate sheet, and that such sheet may be complete in itself as to bearing, distance, and position, with regard to other portions of the main triangulation.

We will take as an example the following :—

Lat. A, 49° 30′ 24″ N.

True bearing observed B from A.   N. 69° 05′ 00″ W.

Angles Observed and as Corrected.

		Observed.			Corrected.		
		°	′	″	°	′	″
A	..	86	06	35	86	06	19
B	..	38	52	02	38	51	47
H	..	55	02	04	55	01	54
		180	00	41	180	00	00
B	..	59	33	10	59	33	27
H	..	80	27	51	80	28	09
C	..	39	58	14	39	58	24
		179	59	15	180	00	00
C	..	56	58	08	56	58	44
H	..	46	26	22	46	26	58
D	..	76	33	43	76	34	18
		179	58	13	180	00	00
C	..	96	50	21	96	50	27
D	..	11	17	06	11	17	13
F	..	71	52	13	71	52	20
		179	59	40	180	00	00

		Observed.				Corrected.		
		°	′	″		°	′	″
B	..	72	40	17		72	40	31
C	..	62	46	39		62	46	53
E	..	44	32	22		44	32	36
		179	59	18		180	00	00

FIG 13.

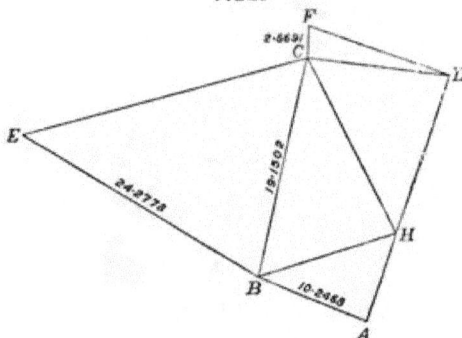

Example of Calculated Triangulation.   We have in the annexed figure (13) a portion of a triangulation, where all the angles have been observed at each station. The latitude of A is known, A B is the original long side obtained by extending the base, and the true bearing of B and A have been taken from one another, from which we have deduced a mean bearing of B from A with which we intend to work. The length of each side has been calculated by ordinary trigonometry. We now want to calculate the bearings of the different sides, so as to be able to break up the triangulation into different sheets. We shall want also the latitude, and difference of longitude from A, of F, which is a station in a plan on a large scale we have made.

For the purposes of this plan we have obtained the side F C in the triangulation, which will serve as our base instead of measuring another.

We shall commence by calculating the convergency for ten miles of departure at the average latitude of the chart, as we shall want it directly.

In this case we find that—

Convergency for 10′ of departure = 11′·92.
Or for each mile of departure = 1′·2.

We then find approximate latitude of B by the formula—

Diff. lat. = A B × Cos rough mercatorial bearing.

We obtain the bearing, near enough for this purpose, by finding the rough convergency and applying half of it to the observed bearing of B from A, thus—

Take departure from the traverse table, in this instance 9·5. Multiply it by the convergency for a mile, just found to be 1·2, which gives us 11′·4 as the rough convergency. Adding half of this to the bearing of B from A, we get rough mercatorial bearing N. 69° 11′ W., and working out the difference of latitude, we find it to be 3′ 38″, which gives for the latitude of B, 49° 34′ 02″, and for middle latitude 49° 32′ 13″.

Then convergency = dist. × Sin merc. bearing × tan mid. lat.
Using the rough bearing just found, we get—

Convergency for A B = 11′ 13″·8.

This convergency, and half of it, added respectively to the bearing of B from A, will give the reverse bearing of A from B, and the mercatorial bearing, thus,

B from A = N. 69° 05′ 00″ W.
A „ B = S. 69 16 14 E.

Mercatorial bearing $\frac{\text{N.}}{\text{S.}}$ 69 10 37 $\frac{\text{W.}}{\text{E.}}$

If this differs much from the rough mercatorial bearing, we must recalculate the latitude of B before proceeding further, but this should not be necessary.

Then to calculate bearing of B E, we have the bearing of A from B, just found, to start from. Adding the three angles, A B H, H B C, C B E, to it, we shall get the bearing of E from B. The convergency for B E is calculated in the same manner as above, and we shall then have mercatorial bearing of B E.　Thus:—

A from B	S.	69° 16' 14" E.
A B H		38 51 47
H B C		59 33 27
C B E		72 40 31

E from B	S. 240	21 59 E.
Or	N. 60	21 59 W.
½ convergency . . .		12 26

Mercatorial bearing of	$\frac{N.}{S.}$ 60° 34' 26	$\frac{W.}{E.}$
B E		

Application of Convergency.
In like manner we must calculate the mercatorial bearing of all the sides we require, remembering that of the reverse bearings, the bearing of the station nearest the pole from the one farthest from the pole, is the smallest. In this case then, being in the northern hemisphere, where a bearing is measured from the north point, the convergency is added to obtain the reverse bearing.

Having obtained the bearing of each side, we can calculate the relative position of any two stations by working out the traverse between them.

Thus to get position of F we have,—

A B	N. 69° 10' 37" W.	10·2468 miles.
B C	N. 12 20 56 E.	19·1502 ,,
C F	N. 1 24 17 W.	2·5691 ,,

From which we calculate difference of latitude and departure in the ordinary manner.

We thus get the mercatorial bearing of A F, N. 12° 32' 45" W., and distance 25·5269 miles.

Calculation of Triangulation not generally necessary early in survey.
It will be understood that it is by no means necessary to work out all the triangulation as just described when commencing the plotting. All that is then required is as long a side as we can get on which to begin. The main triangulation can be calculated afterwards, and in many instances must be,

as the whole of the angles will not be obtained till later on. In some nautical surveys it will not be necessary to calculate any triangulation at all.

In the example of triangulation we have given we have supposed ourselves to be working from a measured base. If the survey is extensive, the ultimate scale of the chart will depend upon the astronomical positions. It is very unlikely that when these are obtained, the distance between the extreme points depending upon them will agree exactly with that deduced from the short side, and therefore all the sides will want correction in probably both bearing and distance. *Correcting Triangulation for error of temporary base and bearing.*

The readiest way of doing this is to get a proportion between the two total distances, as found by the triangulation and by the astronomical positions respectively, in the shape of a logarithm, and multiply each side found by it, which will give the true value as dependent on observations. The bearing of every side will have to be corrected by the difference of the bearings of the extreme points.

Thus referring again to our example, (which, for the sake of brevity, we have confined to only a few sides,) let us suppose we find by observations that A F is N. 12° 36' W. 26·248 miles.

Dividing this distance by the former one, we get a proportion whose logarithm is 0·012097. Adding this to the log of each side required to be corrected will give us the true value.

The difference of bearing is 3' 15" more to the westward. The bearing of each side will then have to be corrected by this amount. Thus the bearing of A B will stand N. 69° 13' 52" W.

This difference is somewhat exaggerated. It should seldom, when true bearings have been well observed, amount to so much, but in some climates it may be unavoidable.

In a case of this kind the result of both triangulation and astronomical observations would be transmitted home, as their concurrence or otherwise will form a good test of the value of the work generally.

In stating as we have that the ultimate scale of the chart

of an extended piece of coast will depend upon the astronomical positions at either end, it is not intended to lay down
a too hard-and-fast rule. The conditions of each element,
triangulation and positions by astronomical observations, must
be considered. Both, under the ordinary circumstances of a
marine survey, are liable to error. In a rigorous trigonometrical survey, the triangulation is more likely to be correct
owing to the unknown error in the astronomical positions
due to local attraction of the pendulum, or in other words
of the mercury in the artificial horizon; but this very local
attraction makes it necessary in a marine survey to regulate
the distances by observations, as other surveys have to start
from the same position, and subject to the same error, and
moreover, in an ordinary marine survey, the triangulation is
carried on under conditions which prevent the possibility
of ensuring freedom from errors in it. Nevertheless, should
the discrepancy in bearings be large, when we know our true
bearings have been well observed and our triangulation to
have carried it on within the limits of the discrepancy, it is
desirable to adjust the astronomical positions so as to reduce
the discrepancy in bearings.

Triangles containing small angles not always ill-conditioned. It will be observed that we have a triangle C D F with a
very small angle. This not being a receiving angle does not
matter in the least. We are obtaining the position of F from
C and D, which are already fixed, and the angle of intersection at F being nearly a right angle, the change of position in
F, resulting from a small error in the angle at either C or D,
will be as small as is possible, and much less than if the
angle at C being the same, that at D was 60°, which would
result in the intersection at F being more acute, and any
error would consequently change the position of F to a
greater degree.

If we were obtaining D from C and F, such a small angle
would not be admissible for a moment, as it is evident that
any small error at C or F would result in a great change of
position in D.

It would be awkward and inconvenient to have many such

triangles in the main framework of the triangulation, as the small side is of no use in carrying on the chain, and we should be forced to multiply triangles in consequence; but we are, notwithstanding, sometimes obliged to include some such in our work, from the lie of the land and other causes, and as long as we use them as in the example they will not affect the result, as far as chance of accuracy goes, and should not be under these circumstances considered as "ill-conditioned."

In working out the diff. lat. and diff. long. of two positions from the triangulation geodetically, we have been treating the earth as a sphere. This is not strictly the case, as the form of our globe is that of an oblate spheroid; but the error introduced by assuming it to be a sphere is small, and can often be disregarded in hydrographical work, as being swallowed up in the larger errors incident on imperfect triangulation. *Correction for the Spheroid.*

When, however, a triangulation has been carefully done, and we wish to get the difference of longitude as near as we can, either for the scale of the chart, or for purposes of comparison with that deduced from the astronomical positions, or in latitudes far from the equator, the necessary correction for the spheroid should be applied.

This correction is 2 Cos² Mid. lat. × compression.

The compression of the earth is the proportion that the difference of the equatorial and polar diameters bears to the diameter, and can be taken as $\dfrac{1}{300}$.

The formula for correction for a given difference of longitude will then stand,—

$$\text{Correction} = \text{diff. long.}\ \frac{\text{Cos}^2 \text{ Mid. lat.}}{150}$$

This is subtractive from the calculated difference of longitude by the triangulation.

In the latitude of 20°, this correction for a difference of longitude of 100′, amounts to 35″, as will be seen by the following example:—

Example.   In latitude 20° the departure deduced from a triangulation was found to be 94′, required true difference of longitude.

$$
\begin{array}{llll}
\text{Dep.} & \dots & \dots & 1\cdot 973128 \\
\text{Sec. lat.} & \dots & \dots & 0\cdot 027014 \\
\hline
\text{Spherical d. long} & 2\cdot 000142 & \dots & \dots & 100'\cdot 0327 \\
\text{Cos}^2\text{ lat.} & \dots & \dots & 9\cdot 945972 \\
\hline
& & 11\cdot 946114 \\
150 & \dots & 2\cdot 176091 \\
\hline
\text{Reduction } \overline{1}\cdot 770023 & \dots & \dots & -0\cdot 5889 \\
\hline
& \textit{True} \text{ diff. long.} & 99\cdot 4438' \\
& \text{or} & 1° \ 39' \ 26''\cdot 6
\end{array}
$$

The true difference of longitude can also be calculated from the tables of lengths of a minute of latitude and longitude in the Appendix M as follows :—

True diff. long. $=$ dep. $\dfrac{\text{No. of feet in minute of Lat.}}{\text{No. of feet in minute of Long.}}$

Working out the above example this way, we have—

$$
\begin{array}{llll}
\text{Dep.} & \dots & \dots & 1\cdot 973128 \\
6053 & \dots & \dots & 3\cdot 781971 \\
\hline
& & & 5\cdot 755099 \\
5722 & \dots & \dots & 3\cdot 757548 \\
\hline
\textit{True}\text{ d. long} & \dots & 1\cdot 997551 & 99'\cdot 44 \\
& \text{or} & 1° \ 39' \ 26''\cdot 4
\end{array}
$$

which gives the same result as the other method.

CHAPTER V.

PLOTTING.

THIS chapter will comprise, besides a description of the Subjects comprised in chapter. method of placing the points on the paper, which is more generally understood by the term "plotting," an account of the different manners in which those points may be obtained, other than by a regular chain of triangles. This is, perhaps, more correctly, a part of triangulation, and for some reasons should be described under that article, but it is thought that it will tend to clearness of comprehension, if it is taken in connection with the mode of laying down the points as obtained, as it is not easy to separate the two steps in many instances.

In discussing the general question of Plotting, therefore, we will first take the placing of the points of an ordinary triangulated survey on paper, and then consider some other systems to be adopted when regular triangulation fails us.

Plotting the points is a most important operation, and one requiring great care. Great care requisite in Plotting.

No matter on what scale, or on what system, a survey is being made, equal pains must be bestowed on plotting the points. Indeed, it may almost be said that in proportion as the elements of a survey approach to the least accurate form, viz., a sketch survey, so does the necessity for careful plotting increase, as the numerous checks, which in a detailed triangulation will instantly make any error in plotting apparent, will be more or less absent in proportion to the departure from such regular triangulation; and not only will the minor details of such a chart be inaccurate, which we expect, but the main and prominent points may be unnecessarily out of place unless care is bestowed on the plotting.

Plotting by chords.   Before describing in detail the different methods in plotting, it is necessary to understand the system of laying down angles by chords, and why this is done.

It will easily be seen that, where lines are to be drawn of considerable length, a protractor whose radius will be much shorter than the desired line, can hardly give the angle exact enough to ensure the extremity of the line being precisely placed; for the straight-edge, perhaps six feet in length, by which the required line is to be drawn, will only be directed by two pricks in the paper, which, with the largest protractor, will not be more than eighteen inches apart. However exactly the protractor has been placed, and the pricks made, the mere laying of the straight-edge so that the line drawn will pass *precisely* through the centre of the two pricks near together, is almost an impossibility, and an error, quite imperceptible at the pricks, will be very appreciable at the end of the straight-edge.

For this reason, we want our directing prick as far along the straight-edge as we can get it.

We accomplish this by using chords.

If two radii of a circle of given length of radius, containing between them a given angle θ, be drawn to cut the circumference of the circle, the chord to the arc of the circumference thus cut off is 2 radius $\sin \dfrac{\theta}{2}$.*

Thus, by reversing this and describing from the centre A, Fig. 14, an arc of a circle of any radius, drawing the line A C, and measuring the chord C B (which will be done in practice by describing a short arc of a circle with the required chord as radius, from the centre C), the point B, where the chord cuts the circumference (or the two arcs intersect) joined to A, will give the required angle θ.

---

* *Vide* proof of this rule in Appendix **C.**

A table of chords for a radius of 10 inches is given in **Table of Chords.**
Appendix,* which saves much time and chance of errors, as
the chord to the angle required can be taken from the table,
and multiplied by the radius with which it is meant to lay
off the angle, divided by ten ; but in case this is not at hand,
we must calculate our own chords.

Tables of natural sines are not included in Inman's, the **Calculating Chords.**
tables generally in use at sea, and logarithms of sines are in
that work only given for every fifteen seconds, and we may
want to take the angles out exactly. Moreover, by using the
logsine, three logarithms will have to be taken out, and the
process is somewhat longer. It is simpler, therefore, to use

FIC 14.

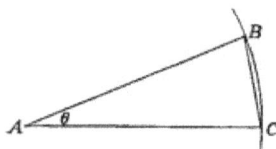

the table of natural versines, which are given in Inman to
seconds.

As $\sin. \dfrac{\theta}{2} = $ versine $(90 + \dfrac{\theta}{2}) - 1$, our required chord
will be 2 radius (vers. $(90 + \dfrac{\theta}{2}) - 1$).

Versines are given for a radius of 1,000,000, so we have
to divide the versine taken out by that number. This
reduces the rule in practice to this.

Look out the natural versine of $90° +$ half the required
angle, leaving out the left hand figure 1, and putting a
decimal point before the remaining six figures. Multiply
this number by twice the radius, and the result will be the
chord required.

Let us take now an example in practice. **Example.**

---

* Appendix J.

At A, Fig. 15, the angle between B and C is 35° 14′ 30″. The line A L from A passing through B is already drawn. We want to lay off this angle, and requiring accuracy, we take a long radius, *i.e.*, 45 in.

Forty-five inches must be carefully measured, by the brass diagonal scale, on to a pair of beam compasses, with the two steel points shipped. Flattening the paper down by placing the straight-edge close to the line A B, and putting weights on it, with the centre A describe a short arc of circle D E, scratching lightly the surface of the paper. Then moving the straight-edge into the direction of C (which can be ascertained roughly by a protractor), and again weighting it, make another small scratch F G. With the assistance of a reading-glass, and by means of a needle mounted in a handle, and

FIG 15.

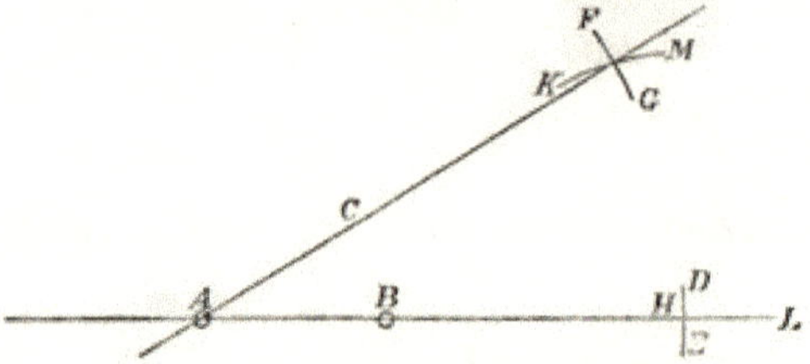

spoken of as the "Pricker," make a fine prick at the intersection of the lines A B, D E, *i.e.*, at H.

Look out the versine of 107° 37′ 15″ (90 + half the required angle), which is 1,302,717. This becomes ·302,717, which multiplied by 90, gives 27·244 inches as the chord.

Measure this distance on the beam compass, and flattening the paper as before, draw, with H as a centre, a short arc K M crossing F G. The point of intersection is to be pricked carefully as before, and the straight-edge can now be laid on A and it, and the line ruled will be at exactly the angle required. This seems a tedious operation, but it is the only way in which points can be got to go down satisfactorily, and in the end much time will be saved.

It may be noted here, that it is preferable to make a mark

with a steel point instead of a pencil, from the practical Steel points to be used. difficulty of measuring accurately the required distance on the beam compass when the pencil point is used, as, when the pencil point is cut sharp enough to make a fine line, it is almost impossible to prevent breakage in applying it to the brass scale divisions. It is also cleaner. In marking, the point must be held sloping, so as only to impress, and not actually to scratch the surface of the paper, which it will do if held perfectly upright.

Of course, if the paper is stretched on a board instead of being loose on the table, the time and trouble of seeing the paper flat is saved, but this is seldom used in our work.

If the table of chords is available, look out the chord for 37° 14′ 30″ and multiply it by 4·5, as the table is made out for a radius of 10 in. This will give the same quantity of 27·244 inches as found above.

C may be of course anywhere on the line A B, and supposing ourselves to be plotting from an original base A B, will probably be much nearer to A than to F G, but by taking such a long radius we get a straight line in the true direction of the angle laid off, and when we want to measure another angle on to another object, perhaps three times the distance of C from A, we have a long line we are certain of, to do it from.

Here let it be impressed upon the surveyor that all lines Always draw long lines. drawn for plotting the main points, and indeed all points (except very minor ones, on which the position of nothing else will depend), must be drawn as long as possible, and with more or less long chords, if we desire correctness. If we have a line drawn between two stations which lie, say six inches apart on the paper, and it only projects a few inches beyond each, and we hereafter require to lay off an angle from one, having the other as zero, to a station which will be, say two feet or more distant, we cannot do it correctly, as this longer line will have to be directed by a prick which cannot be farther off than the length of the zero line; but by drawing long lines with long chords, we are ready for

anything, and it will not matter whether the station we take
for zero be near or far, as we use, not it, but the long line
ruled through it.

Length of radius. In no case should a line to a **station** be laid off with
a protractor or chord whose radius is less than the distance
of the station, excepting in a rough plan which we want to
do rapidly, or in most parts of a running **survey**, where pretensions to accuracy being thrown to the winds, we get
points near enough for our purpose down with a protractor.

Lengthening a line. It is difficult to extend correctly a short line once drawn,
by simply ruling on with the straight-edge. If a longer line
is wanted, it is better to lay off the angle to it again from
some other long line, with a sufficient radius.

Ruling a straight line. To rule a true straight line which will pass exactly over
the centre of the pricks is by no means an easy thing. The
ruling **pencil**, which should be of the hardest lead manu-
factured, should be cut to an edge, not a point, and the
straight-edge being placed in position, and weighted to keep
it in contact with the paper throughout its length, the flat side
of the pencil is placed against it, and tried at both points, to
see whether the line will pass truly over them. Care must
then be taken to hold the pencil in the same position while
drawing the whole line.

Angles over 60°. In laying off by chords an angle over 60°, or a little under
60°, it will be found best to mark off 60° first, and measure
the remainder of the angle from the 60° prick. This is done
by drawing short arcs with the radius used, from the station
from which it is desired to lay off the angle, and from the
radius prick (H in last figure), the intersection of these must
be pricked off as 60°, and another short arc being drawn with
the originating station as centre, the chord of the difference of
the angle from 60° is measured from the 60° prick to the last
short arc, as in Fig. 16 (p. 107).

This is done not from any incorrectness of the principle if
the angle were laid off at once, but because it is inconvenient
to be measuring long distances as chords, as there is a greater
chance of some little inequality of the paper causing error,

and also, the longer the chord measured, the more acute will be the angle between the two intersecting arcs, and consequently the greater the difficulty of pricking in accurately at the intersection.

Understanding then how to lay off angles by chords, and having obtained by calculation as long a side as we can for a plotting base line, so as to plot as much as possible *inwards,* or with decreasing distances, and not *outwards* to stations farther distant than the original two, and having settled whereabouts on the sheet this base line shall be placed, draw a meridian line, parallel with the side of the paper, and passing

FIG 16.

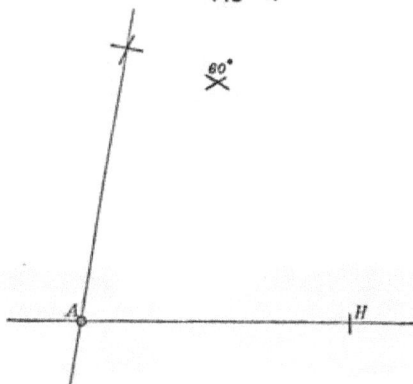

at one end of where the base is to be. Make a prick on this line for one end of the base, using, as always for pricking, a reading-glass, to ensure getting the prick exactly on the line. Let us call this A.

From A, lay off, with as long a chord as can be commanded, the true bearing of the base, and having ruled this line of bearing as long as possible, make another prick on it, at the required distance from A, for the other end of the base. From the two base stations lay off angles to two other main positions, and choose the one of these where the intersection of the lines makes the nearest angle to 90° as the third station to prick in, doing so with great care on the intersection of the two lines. Then from this third station lay off an angle to the fourth, and if this, when ruled, passes exactly over the intersection of the two lines from the base stations, it can be pricked in. All four stations are correct, and the groundwork of the chart is laid ; but if there is any little triangle visible with the reading-glass, all must be plotted over again, for unless these first four stations are exactly right, nothing will ever go right afterwards.

These four stations settled, proceed in like manner with other main stations ; but now we shall of course have three intersecting lines for each station, and care must be taken that these lines do truly intersect, and no station must be pricked in, that has not got three such converging lines through it.

The main stations down, smaller chords may be used for secondary theodolite stations, and the protractors will come in in plotting the marks and other minor points, the necessary angles for which we may suppose some of the party are getting, whilst the first main points are being carefully plotted.

As the chart fills, there will be many lines from which the angle to a new point can be measured, and it is well to remember that as a standing rule the smallest angles both give less trouble, and prevent least chance of error.

Marking "Points." The ordinary way of marking the points is to ring a small circle of carmine round them. Larger circles can conveniently be used to distinguish the main stations.

Stretching of the Paper. It will be found in the course of plotting that the paper will vary so much, expanding at one time and contracting at another, that the arcs of radius once measured and scratched on the paper, cannot be considered as so done once for all. If some hours have elapsed since marking any radius, it must be remeasured, to ascertain if it has altered.

Calculating third Angle. In getting angles for plotting stations of all kinds, it must be remembered that two angles of a triangle will always give the third, and that as far as mere plotting goes, it is not necessary to waste time in observing the third angle. If the two observed angles have been got fairly accurately, the double error which will be thrown into the third angle deduced from them should not be enough to show in plotting, and if it does, it will soon make itself apparent by not intersecting. An angle from a fourth station will show which of the other three angles is wrong.

Thus if we have observed at a station C, which we want to plot, the angle between A and B, and also the angle at A between B and C, the angle at B which is wanted to draw a line to C can be calculated without the trouble of visiting B. It is indeed a blessed circumstance for the marine surveyor that the three angles of any triangle equal 180°.

Some surveyors have preferred to plot main stations by distances. In this case the triangulation must necessarily be calculated beforehand. We do not consider that much is gained by this method. Three distances must be measured to obtain an intersection, as three angles must be laid off for the same result. A distance is sometimes useful as a check. **Plotting by distances.**

IRREGULAR METHODS OF PLOTTING.

We have up to the present been considering the plotting of stations for a regularly triangulated survey. Let us now look at some other methods.

In plotting the points of a chart which is being constructed on the principle of do-with-what-you-can-get, which is very often what has to be done in marine surveys, it is frequently found necessary to plot a position by its own angles, as, for instance, where the ship, anchored or moored off a low coast, has to be a main station, and only angles from aloft can be obtained to objects inland, such as hills, conspicuous trees, &c., already fixed. **A Position by its own Angles.**

A station pointer, generally, has some small errors of centring, &c., that prevent it being used where exactness is required, and, moreover, only two angles can be laid off at a time by this instrument. In this case then it is better to **Use of Tracing-paper.**

plot all the angles obtainable on to tracing-paper, using
chords for the purpose, and being very careful to make a very
minute hole at the centre from which they radiate. If the
objects are fairly well placed, a very exact position will be
obtained, by laying this tracing on the sheet, and pricking
through for the position. This will be much assisted if but
one line can be got from a fixed station, as the angles can
then be plotted on this line, supposing that in this case back
angles cannot be calculated.

Only two Angles available. Again, it may sometimes be found necessary to carry on
the main stations with a point plotted by only two angles ;
but if this happens, efforts must be made to check this, by
getting an angle back from stations plotted on by means
of this doubtful position, to some old well-fixed station, as a
distant mountain; or if this is not to be had, a regular
beginning must be made again by plotting two stations with
two angles, pricking one, and then laying the angle from
that to the fourth, as practised at the commencement of
the chart, which will give a certain amount of check.

Mountains invaluable. A well-defined mountain, though miles inland and never
visited by the surveyors, will often prove the very keystone of
a chart that cannot be regularly and theoretically triangulated. When once well fixed, it will remain to get angles to,
long after all the other first points of the survey have sunk
below the horizon as the work progresses.

Use of True Bearings. The bearings of this will often be useful, and these can be
laid off from the mountain by applying the convergency.

Let us take an example, which will perhaps explain what
is required easier by means of a diagram.

We hope that we have made it plain, by what has gone
before, that if a distant object bears, say, N. 47° 20' W., we
do not bear from such object S. 47° 20' E., but so much less
or more by the convergency; and that in all cases of fixing
ourselves by means of true bearings observed from our own
position, the amount of convergency, due to the bearing and
distance of the object, must be calculated and applied to our
bearing, before we can use it as a bearing from the object.

Here, Fig. 17, let B A be the original meridian drawn
at the commencement of plotting through any station A.
M is the distant mountain. At X our main points are falling
short from some reason or another, and we are obliged to
have recourse to a true bearing of M, which we accordingly
obtain. Required to draw this true bearing from the fixed
point M. If we have the sheet graduated, it will not much
simplify matters, as it is a great chance if a meridian passes
close enough to M to use it without further correction ; but let
us suppose that we have no other meridian on the chart but
A B. We must lay off the true bearing from M., with A as

FIC 17.

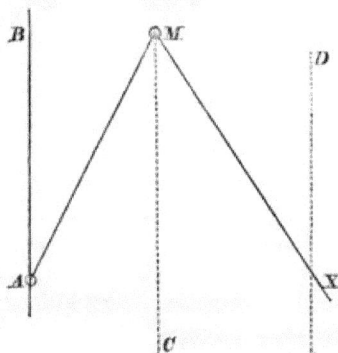

the zero, so we require the angle A M X. If M has been
observed from A, whence we had a true bearing by which
the meridian A B is directed, we have the bearing or angle
B A M. If not, we must measure it from the sheet by
reversing the chord method ; drawing a line from A to M, and
measuring the chord to the line A B at a given radius with
beam compasses, and calculating the angle which corresponds
to it, or B A M.

Now consider the figure again, M C, X D, being imaginary
meridians to assist conception.

The bearing of A from M = bearing of M from A + the
convergency, as M is nearer the pole than A, or

C M A = B A M + convergency for difference of departure of A and M.

In like manner :

C M X = M X D (the observed bearing from X) + convergency for difference of departure of M X.

Adding, we have C M A + C M X = B A M + M X D + convergency for A X.

Or A M X = bearing M from A + bearing M from X + convergency for A X.

To get convergency in this case, we must assume a position for X, which we can roughly plot for the purpose, and measure the distance A X and bearing B A X.

We can then from this calculate the convergency required, knowing roughly the latitude of A, for

Convergency = distance × Sin merc. bearing × Tan Mid. lat.

Drawing true Meridian. If M is likely to be used much in this way, it will be worth while to lay a meridian off through M by plotting the bearing A M C or B A M + the convergency for A M; from which meridian subsequent bearings can then be laid off, duly corrected for convergency, for the distance between M and the station from which the bearing is observed.

Neglect of Convergency. Of course it will depend on the latitude how much error will be introduced by neglecting the convergency; but when it is considered that in latitude 45° the convergency is equal to the departure, it will be seen that a large error will result by not applying it; for in this latitude, supposing A and X are 30 miles apart, an error of half a degree would be made by drawing a meridian parallel to A B, and laying off the bearing observed at X from M.

As a rule, therefore, it can never be safely neglected except very near the Equator.

If it is intended to lay off the true bearing of an object from a station plotted on the chart, the convergency must likewise be borne in mind, and the meridian to be ruled through X (in this case considered as fixed) from which to measure the bearing, must be, in transferring it from A B, corrected for the convergency due to the distance A X, by, after ruling a line through X parallel to A B, laying off at X, from the parallel just ruled, towards the pole, and on the side of A, an angle equal to the convergency required, which will give the direction of the true meridian.

The system of true bearings may be used in many ways whilst carrying on an irregular triangulation. It is impossible to give instances of all the difficulties which may be surmounted by this means, but an example, taken from actual practice, will show the style of use to which true bearings may be put.

Further use of True Bearings.

Let us suppose ourselves employed in the survey of a piece of coast which offers no facilities for obtaining a base by

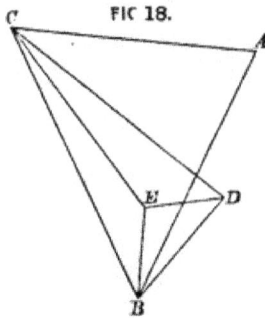

FIC 18.

measurement; but it is the season for observations, and we have points so placed that we can work directly from the astronomical base, instead of obtaining a base by sound or other doubtful methods, which we should otherwise have to do.

In Fig. 18, A and B are two positions invisible from one another near the confines of our chart; C is a distant inaccessible mountain visible from both A and B; D is an elevation visible from B, but not from A, and from which C can also be seen.

To utilise this arrangement we take observations for latitude at A and B, and run the meridian distance; we also get the true bearing of C from A, B, and D.

I

Calculate the bearing and distance A B astronomically, and place this line on the paper. The lines B C, A C can be now drawn by the difference of the bearings observed and calculated from A and B, which will give us C with two cuts. B D is drawn from B, and the back bearing of D from C (calculated from the observed bearing of C from D, with convergency applied) drawn, by which we shall get D, also with two cuts only.

If we can find a point E which can be seen both from B and D, and from which C can be seen, we can lay it down with three cuts, as the angle from C can be calculated in either triangle C E B or C E D, and the intersection of these three will prove the exactness of our work. B E will then be our base for working, as we are supposing B A to be about 60 miles, which, as we have drawn it, will make B E about 15 miles, which is a workable base.

In the case which we have put it is very unlikely that, after all these different bearings, the intersection of the three lines at the point E will be a perfect one. If it is not good, the best way to obtain the base B E may be to calculate it in as many triangles as we can command, and, taking the mean of these results, to commence the actual plotting from this mean base. This would depend however upon circumstances. It is impossible to lay down any hard-and-fast rule with respect to this kind of work, and the case is simply given as an instance of the uses to which true bearings may be put.

Gradua-
tion of
Sheet
before
Plotting.

In some extensive surveys on a small scale it may be necessary to graduate the sheet first, when positions can be placed on it by their latitudes and longitudes, and the intervening parts plotted or triangulated by means of bases measured at each of these astronomical positions. This will be done when coasts are low and marks scarce. We can scarcely hope that when these different bits meet, they will agree exactly; but with a small scale, say half an inch to the mile, the discrepancy ought not to be sufficient to introduce much error, if we square in five or six miles of the points worked up from either end, when they meet and disagree.

This undoubtedly partakes of the nature of " cooking ; " but when we undertake to map a coast on such a small scale we cannot pretend to much accuracy in detail, and shall only do this when it has been considered advisable to lay down a large extent of coast in the time available, with the intention of presenting its more salient features as correctly as we can.

Work amongst islands (as portions of the Pacific) would be done in this manner.

FIXING MARKS.

It is not possible to lay down any dogmatic plan for fixing the marks which have to be erected. In many cases it is well to put them all up first, and then get angles to them afterwards ; but if non-surveyors are deputed to make the marks, they will seldom be placed in the right spots. A whitewash, for instance, will be so placed that it cannot be seen in certain directions. A tripod or pole will not be in the most convenient position for the officer who afterwards puts in the coast-line, and numerous small errors of this description will be made by one who is not capable of taking in all the little requirements.

It is therefore more satisfactory to send a surveyor to do this, and while he is there he may just as well take angles, so that the writer has found it saves time in the end, in general, to have a surveyor at some main or secondary station, whence he can see most of the marks, and let the officer who erects the mark take angles at it to the above station, which we may call the "shooting up" station, and to a sufficient number of other stations which can be seen from the "shooting up" station also, to fix himself. The angles from these other stations can then be calculated. In this way two or three officers can be at work putting up marks, and fixing them at

Systematic fixing of Marks.

I 2

the same time. The officer who erects a mark gives it a name, and notes the time by his watch when he is there. The officer at the shooting up station also takes the time, and notes the position and kind of mark put up, to which he takes his angles, writing the name against it in his book when he returns 'to the ship and meets the other officers.

Officer marking responsible for sufficiency of Angles. The officer marking must think for himself whether he has enough angles to fix the point; and in case any mark cannot be seen from the shooting up station, he must get an angle from some other of his marks, which will be then used to calculate the other angles in the same manner.

Use of Heliostat. A heliostat is invaluable here. In hazy weather, and when the shooting up station is distant especially, a flash will be seen when neither mark, nor boat, nor anything to direct where to look for the mark, will be visible. The officer shooting up should also return the flash, to show he sees the station, as well as give a well-defined object to get the angles to.

Of course circumstances may not render this system advisable, but it is here suggested as having worked very well in many places, a long extent of coast being " marked," and all marks fixed in a short time.

Triangulating, and fixing Marks by Ship. Frequently the minor marks must be fixed by angles from the ship, or a boat at anchor, as on a straight coast where nothing behind can be seen from the marks. When this is necessary it will often be also necessary to carry on the main triangulation as well by means of ships and boats, so that a description of one serves for the other.

The ship, anchored short, or moored if necessary, should be shot up from one or more shore stations. If the angles taken from the ship are indispensable to fix her own position, try calculating the back angles from other objects first, and lay them off as cuts to the position, as if they agree it will be the most satisfactory manner; but often back angles, calculated from sextant angles, will not be correct enough to give a good intersection, especially if the points are distant. In this case, let all the angles taken at the ship or boat be plotted on tracing-paper as before described, and the position pricked through on the guiding line from the shore station. A signal should be made when the angle to the ship is to be observed, and the angles from the ship taken at the same time.

The ship angles should be observed from the fore part of the ship, and frequently the foretop will be found the best place. Whatever spot is used it must, of course, be arranged beforehand, so that the observer's exact position on board may be taken from the shore station.

From the ship the main angles, that is the angles to the **Taking Angles from Ship.** positions already plotted, which are to be taken for the purpose of fixing the ship, must be observed first, using some well-defined station as zero, and measuring all the main angles from this with the sextant. Some other station must be chosen as the zero with which to measure the angles to the marks, and the angle to this second zero observed from the main station zero.

This second zero is wanted to be in such a position with regard to the marks, that any slight movement in the ship will make the least possible difference in the angles to be observed between it and the marks. It must be, therefore, at about the average distance of the marks. It will not do to choose some object miles away behind the marks, as the least swing of the ship will at once alter the whole of the angles. Generally speaking, the central mark to be fixed will answer the purpose best, but in many cases it will be found necessary to change this zero for some marks, measuring from some other object at an equal distance from the ship.

When the minor angles have been taken, repeat the main **Repeating Angles.** angles to see if the ship has moved, giving another signal to the shore station for another angle from it. All mark angles should then be observed again to check errors.

It need scarcely be said that the more rapidly these angles are taken, the less the chance of any error arising from variation of ship's position, by change of direction of current, wind, &c.   An experienced hand should therefore be chosen for this work.

Telescope of Sextant.   A sextant with a telescope of high magnifying power is most useful.   On this head see page 10.

CALCULATING A POSITION FROM TWO ANGLES TO THREE KNOWN OBJECTS.

It may be sometimes required, in the course of a survey not regularly triangulated, to calculate the distance of the observer from an object, from the two angles he has observed between three known "points," one of them being the object whose distance is required.   Or he may require the angle, at the object observed, to him, from the same data.

This is, perhaps, best accomplished by using the one-circle method, so called in contradistinction to the method of protraction by three circles already explained under "Station Pointer."

The three figures 19-21 give the three possible positions of the objects, viz.: When the observer is inside the triangle formed by the objects; when he is outside, and the centre object is nearer than one of the others; and when, under similar circumstances, it is the farthest.

If the angles between the three objects are known, which is most probable, the calculation of the second formula will be unnecessary.

Let $A\,B\,C$ be the objects observed.   X the position of observer to be determined. $A\,B = c$, $B\,C = a$, $A\,C = b$, are the sides known, $A\,X\,B = m$ and $B\,X\,C = n$, the angles observed. Required $X\,A$ and the angle $B\,A\,X$.

At A, in A C, draw, on the side remote from X, A D, making $C\,A\,D = n$.   At C, in A C, draw in like manner C D, making $A\,C\,D = m$.

When X is inside the triangle (Fig. 19) $C\,A\,D$, and $A\,C\,D$ must be drawn to equal $180° - n$ and $180° - m$ respectively.

Describe a circle to pass through the points A, D, C.

Join D B, and produce it until it cuts the circumference of the circle in X.

Then X is the position required.

FIG. 19.

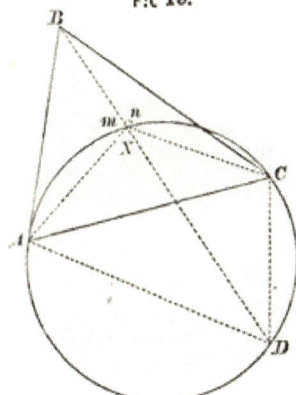

FIG. 20.

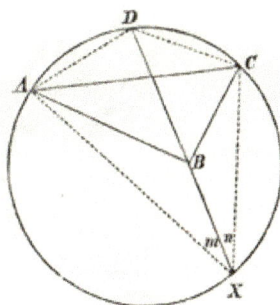

FIG. 21.

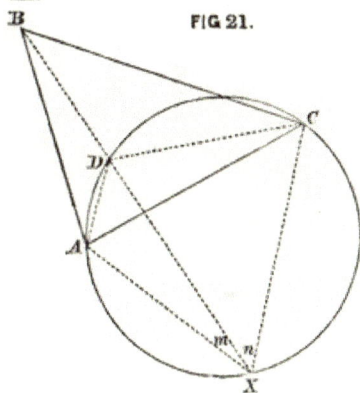

For A C D, A X D, being angles in the same segment, are equal, and A C D is drawn = m.

$$\therefore A X D = m$$
$$\text{or} \quad A X B = m$$

Similarly B X C = n.

Then $A D = b \, \text{Sin} \, m. \, \text{Cosec} \, (m + n)$ . . . . (1)

$\text{Cos} \, \dfrac{B A C}{2} = \sqrt{\dfrac{S (S - a)}{b \, c}}$ . . . . (2)

$B A D = B A C \pm C A D$ . . . . (3)

$\text{Tan} \, \tfrac{1}{2} (A B D - A D B) = \dfrac{c - A D}{c + A D} \text{Tan} \, \tfrac{1}{2} (A B D + A D B)$ (4)

$X A = c. \, \text{Sin} \, A B D. \, \text{Cosec} \, m.$ . . . . (5)

$\text{Sin} \, A B X = \dfrac{X A. \, \text{Sin} \, m}{A B}$ . . . . (6)

$B A X = 180 - (m + A B X)$. . . . . (7)

X can now be plotted by the angles from A, B, C, if required.

DRAWING RECTANGULAR LINES.

The methods of drawing a line perpendicular to another line are well known, but are here repeated.

FIG 22.

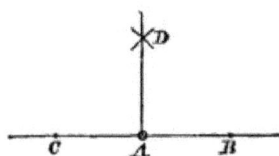

Measure from the point A with the beam compass any equal distances right and left of A, as A B, A C.

Erecting a Perpendicular to a line from any point not near its extremity.

FIG 23.

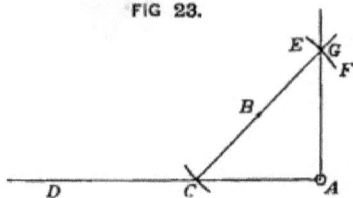

From B and C draw, with a radius about half as much again as A B, short arcs intersecting one another. A line drawn

through this intersection D, from A, will be at right angles
to A B.

Take any point B, Fig. 23, in a direction about 45° from
A, and from it as centre, with the radius B A describe a
short arc intersecting A D in C, and likewise a short arc E F
in the opposite direction.   Join C B and produce it to intersect E F in G.   A line joining A and G will be at right
angles to A B.

In all careful work, these operations should be checked
by repetition, with different radii.

Erecting a
Perpendi-
cular from
the end of
a line.

CHAPTER VI.

RUNNING SURVEY.

A RUNNING survey, the least accurate form of "sketch" survey, is one where the best part of the work is done from the ship running along the coast, fixing points, sketching in the coast-line and prominent parts of the land, and sounding, at the same time.

It is capable of many modifications, more especially with regard to the fixing of the main points.

Roughest form of Survey.   The rudest form of running survey is where, beginning upon nothing, everything is eventually put on paper by observations, angles, and soundings taken from the ship without anchoring.

Modified Running Survey.   At the other extreme comes a running survey made upon some main points already fixed by triangulation of some kind, and which has for its object only the sketching of coast-line and detail of an inaccessible coast, which is assisted by occasional anchoring, and where sounding would be carried on in the boats as well as the ship, after enough natural objects have been fixed by the angles from ship stations.

Graduating beforehand.   In making an extensive running survey of the simplest kind, *i.e.* where we commence on nothing, and only run past the coast once, it is well to have the paper graduated (see p. 270), as astronomical observations from time to time will fix the scale of the chart, and it is easier to plot these positions when the sheet is graduated.

The course and the distance run by the ship between each

position where series of angles are taken, as given by patent logs, will form a series of bases, which will have to be, however, modified afterwards to agree with the positions astronomically fixed, which must be taken as the fundamental points of the chart.

A running survey must be roughly plotted, and everything sketched in, as we go on, putting down position after position by course and distance, and cutting in the objects we choose for marks, giving them names by which to recognise them, and to record in the sounding book. Assistants should be told off for separate duties. One to look after the sounding; another to sketch in the coast-line and hills between each object chosen, on another sheet or sheets of paper; the chief and some assistants getting the angles; one writing down; another plotting the stations and drawing the lines to the points, so as to see what angles are wanted at the next station to objects already chosen, and how far on the next station should be. Bearings should be taken of all prominent points in transit.

At each position, as laid down by course and distance, commence plotting by laying down the bearing of the object we have selected for zero for the round of angles. From this, the other angles can then be laid down.

It follows that a bearing must be obtained, as a necessity, from each position. This should be taken to the zero selected.

Distant hills are a great help in a running survey, as, when replotting from the astronomical positions, if these hills can be fixed by bearings (true or compass) from them, the angles taken to the hills, at a position now and then, may possibly be used as fixes, which may be plotted by station pointer, and so get intermediate positions independent of the patent log positions, which are so liable to error by the action of currents.

A running survey will nearly always have to be replotted, as the astronomical positions and those by patent log will never agree.

FIG. 24.

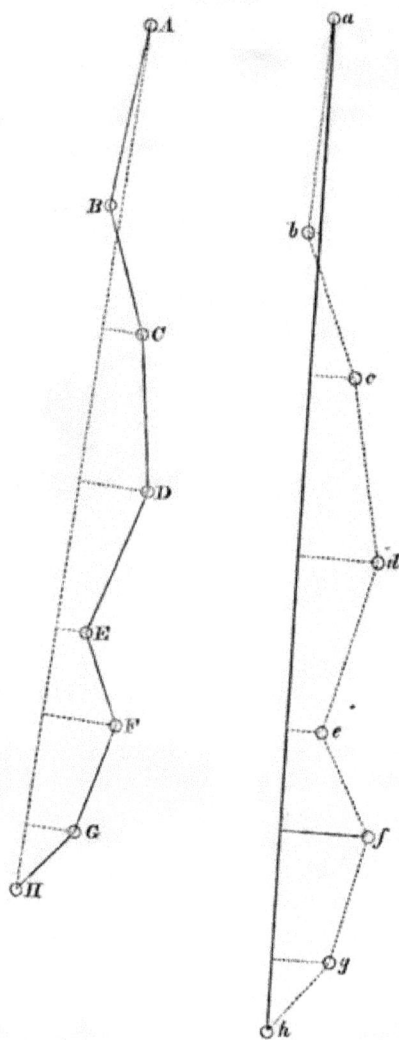

Having plotted the positions where astronomical observa- Replotting and Squaring in.
tions have been taken, if the intermediate stations are to be
put in by bearing and distance, they must be squared in so
as to agree in total distance and bearing with the astronomical positions. Thus, in Fig. 24, let A be the position
from which we start; B, C, &c., to H, are positions of the
ship as plotted by course and distance on the rough chart;
a, *h*, are the same positions as A, H, but as given by the
astronomical observations.

To bring the intermediate positions to agree with *a*, *h*, as
plotted on the graduated sheet, we join A H and *a h*. Drop
perpendiculars from B, C, &c., to the normal line A H. With
the proportional compasses set to correspond to the different
lengths *a h*, A H, measure the corresponding distances along
a h for the points where the perpendiculars will cut, and lay
off perpendiculars along which the corresponding distances
can be measured, and so we obtain *b*, *c*, *d*, &c.

If any mountains have been observed both from A and H,
their positions should next be put down by these two bearings.
The angles taken from the first positions are now laid off, and
as objects are fixed, they can be used as checks to the next
positions. If we can rely upon the bearings taken to the
mountains we shall use them to fix the intermediate positions
in preference to course and distance, so that *b*, *c*, &c., may be
again shifted, especially if the ship has not been accurately
steered on her courses, or we have reason to think currents
have varied at different parts of our run.

Nothing will agree exactly in a running survey of this No exactness expected.
kind, but a very fair approximation to the relative positions
of conspicuous objects may be got.

The amount of detail possible will not be very great, but Amount of Detail.
will vary with the quickness and accuracy of eye and hand
of the officer sketching it in. There is nothing that requires
the knack which distinguishes a good surveyor so much as
this sketching in fairly accurately of a coast-line in a running
survey, and good judgment as to depth of bays, and other
points that must be mainly put in by eye, is most valuable.

It is well to have one officer aloft, who will be able to get a better view of river mouths, &c., and make little sketches of bits not seen from deck. He can also take angles to objects that have sunk, or not yet risen above the horizon of the deck.

Compass-bearings are of great use, as direction of valleys, &c., may be noted without making a position.

The whole course of a running survey will have to be one of compromise between discordant results, and only long practice will enable the surveyor to decide what to throw out, and what to accept.

Modified Running Survey. It may often occur in a survey, that a portion of the coast is inaccessible for landing by reason of heavy surf; or the shore is so cliffy or densely thick with jungle, that stations cannot be made without loss of more time than they are worth. A running survey of this piece may be as much as is requisite, but the probability is that we shall be able to fix on some main points from the triangulation of the other and more important part of the survey, and these will greatly help us to make the best chart of the portion we can under the circumstances.

In such a case, the best course to pursue is to pass along the coast at some distance, stopping at convenient positions, where the ship can get station-pointer fixes by the main points, anchoring, if possible, for this purpose, and cutting in from these positions other secondary points nearer together, and nearer the coast than the first. Then pass along again closer to the land, and fix points on the shore itself, using the secondary points to fix the ship with. Boats may then be sent to sound, if required, or to sketch in more details of little bays, &c., if they can get near enough. Compromise will be required here too, probably, in plotting the points, as, unless the ship is absolutely motionless, it is unlikely the angles will intersect exactly, but it is astonishing what good results can be obtained with a number of officers taking angles at the same time, with the ship's way stopped, each being told off to take two or three angles as quickly as possible; the most important angles, and those that change most rapidly, being taken first.

Advantageous use may be made of beacons in a running survey.

As an example we will give some details of a method employed with success on the south-east coast of Africa * on an open coast, beacons dropped in from 20 to 60 fathoms.

At a distance of from 2 to 4 miles from the shore drop a beacon abreast of some conspicuous object, called the First Breastmark (1 B in Fig. 25), which should be if possible some 3 or 4 miles back from the coast. Note the time, and shape a course parallel to the coast. Put over logs, and steam about 10 miles down it, sounding and fixing with objects selected, until the beacon gets indistinct from aloft, and you are abreast the second Breastmark. Stop, haul in logs, note time, and drop Beacon II. During the run down the coast, three primary objects and other secondary ones have been selected and named, and on arriving at II. a Provisional Breastmark, 10 miles or so ahead must be selected, and also the middle Primary Point of the next fleet.

At II. simultaneous angles are taken between I., the First Breastmark, A, B, C, Primary Marks, Second Breastmark, Provisional Breastmark, and Middle Primary of fleet 3. The Secondary Objects are next taken, using any of the above-mentioned points which are most conveniently situated as regards distance, so that any small change of position of the ship shall make the smallest possible alteration in the angle. Each officer is told off to a primary object and some secondary ones, and is responsible that his secondary objects are taken to a suitable zero.

On taking angles, the rough bearing of Beacon II. (which will be as close to as is safe), and also its distance by elevation of its staff, is noted. Take also compass bearing of Breastmark I. as a check.

Now steam straight out for, say, a mile or so. Turn ship's

---

* Communicated by Captain A. M. Field.

Fig. 25.

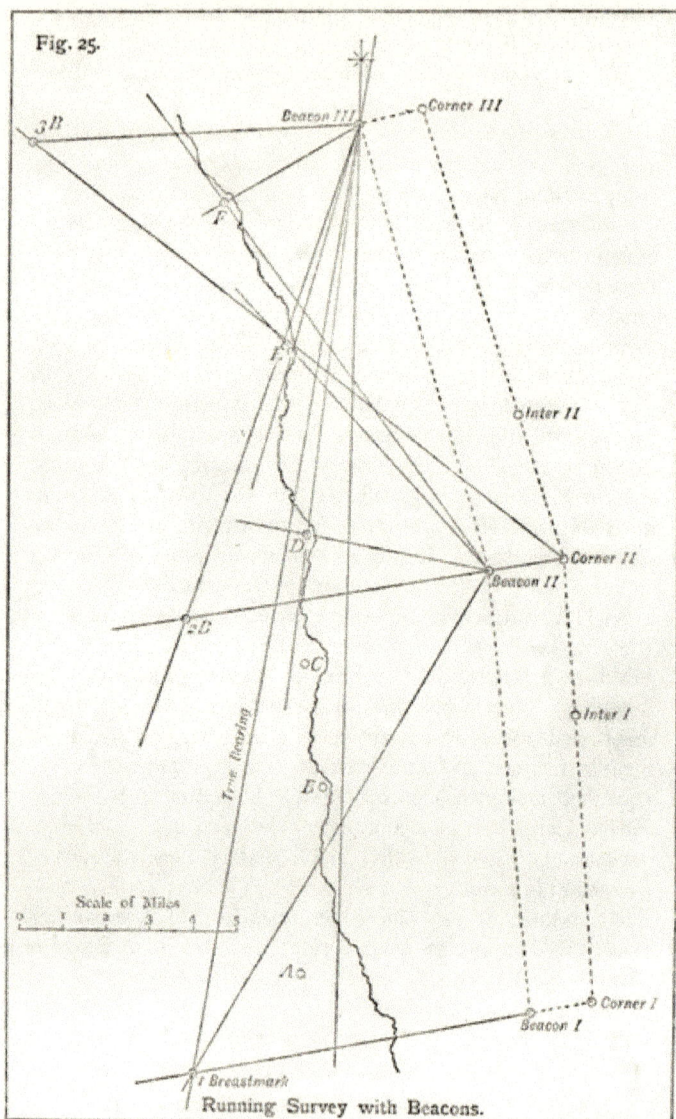

Running Survey with Beacons.

Walker & Boutall sc.

head ready for the run back, and stop. Take simultaneous angles as before, at Corner II., Beacon II. being observed instead of Beacon I. which there is no occasion to look for.

Note time and bearing of II., put over logs, and shape a course parallel to 1st Run.

Run back about one-third of the distance. Stop, and make Intermediate Fix I.

When I. is on the same bearing as was II. when logs were put over, note time, read logs, and stop for the Corner Fix I.

Then run into I., take angles, &c., as at II., and pick up Beacon I.

The primaries and breastmark can be plotted roughly from what we have whilst beacon is being picked up, and will furnish enough to sound upon, and ensure filling up properly and not crossing the old line, for the distance between Beacons I. and II. is obtained from the runs up and down.

Whilst sounding up to II., the coast and topography are shot in and a rough sketch of the coast and hills are put into the deck-book, which at the end of the fleet is sent down to the officer in the chartroom for plotting.

On passing close to Beacon II., put over logs and shape a straight course roughly parallel again to the shore, until abreast of Breastmark III., when stop, put over Beacon III., and take angles as described at II.

Turn outwards again and get Corner Fix III., and run back parallel to old course to Intermediate Fix II., which, and all subsequent intermediate fixes, should be at two-thirds of the distance back. At this position it is important to be able to get a good fix on the points of Fleet 2, in other words, that the angle between the Breastmark 2 and Beacon II. shall be over 30°. It matters not whether Beacon II. is outside or inside of Breastmark 1.

On dropping a beacon, the essential angles are the primary and breastmark of the fleet, and the other beacon, the provisional breastmark ahead and the middle primary ahead.

At the corner fixes, obtain the same angles as in dropping

the beacon, only using the beacon just dropped instead of the other beacon. Take elevations for heights.

At the intermediate fixes, get the primary and breastmark of the fleet, the beacon towards which you are running, the breastmark of that beacon, and the next primary behind.

On picking up a beacon, get the primary and breastmark of the fleet ahead, the other beacon, the breastmark abreast of you, and if visible, the breastmarks behind, as well as primaries.

It will not always of course be possible to distinguish the breastmarks of fleets so far ahead and astern, but whenever possible they should be taken, as points so far distant give excellent zeros for plotting, and the bearing is preserved. Any conspicuous object, whether used as breastmark or not, will answer this purpose.

A continuous series of true bearings is necessary, and it is more convenient if they are taken at the beacons.

Simultaneous angles must be observed at the same time to correct the bearing to points as far as possible up and down the coast.

With a Thomson's compass in a favourable position, i.e., a position where it is not liable to change, a compass bearing will be admissible now and then.

The first true bearing should be taken from the last beacon from which the first breastmark is visible. No other bearing is necessary till the points fixed from this position are passed, when another is wanted to carry on and preserve the bearing.

When the coast trends nearly north and south, latitudes by twilight stars north and south are advisable every 30 or 40 miles to check the scale; and similarly when coast is nearer east and west, longitude observations east and west.

Obtain shore-observations when possible.

To plot, begin when the first true bearing is attained. Draw the meridian through the beacon from which it is observed, and lay off the true bearing of the first breast-mark. Let us suppose we begin at Beacon III., and we are going to plot just the Fleet III.-II., and then II.-I.

Lay off the angles to II. and breastmark 2 B, drawing long lines. Calculate the patent log base III.-II. from the two runs, and prick off Beacon II.

Lay off all angles from III., first making sure that the whole angle between first breastmark and the breastmark ahead is the same both at dropping and picking up the beacon.

Then lay off the angles from II., and prick in the two breastmarks and primary points.

Lay off on tracing-paper the angles from intermediate Fix II., and all points should intersect, and be pricked in. Angles to points in Fleet II.-I. can be laid off from this position.

Plot intermediate Fix I. by the points already fixed, and lay off angles.

Lay off angles from II., using III. as zero.

Prick in Beacon I. by distance, and test it by a station pointer fix, using II. and Breastmark 2, the centre being on the line drawn from II. to I. If this agrees, there should be an intersection at Breastmark I. of five lines, viz., from III., II., the intermediate fixes, and from I.

Plot the corner fixes and lay off all angles to secondary points.

All points being down, the soundings can be fixed and coast-line and topographical features sketched in, getting additional angles when necessary.

It is desirable to use two deck books, entering all angles for each alternate fleet in one or the other. The plotting can then go on in the chart-room, while the angles for the next fleet are being obtained, and recorded.

It is necessary to remember that on obtaining the true bearing of an object a long way ahead, such object is eventually plotted on that line.

If care be taken over the details described, the objects should plot very closely, when the tide is not too strong, and precautions are taken not to run the bases when the streams are changing.

At the extremities of such a survey shore observations

should be obtained if possible.   From these positions and
from the shore true bearings obtained, corrections to scale
and bearing can be made as already explained on p. 97.

In calculating the length of the different patent log bases
the following formula should be used :—

$$X = \frac{t\,(b-a)}{t + t'}$$

Where X is the current in time t with the stream,

      a is distance shown by the log with the stream,

      b is distance shown by the log against the stream,

      t' is time occupied in run against the stream.

From X the true distance can be deduced.

CHAPTER VII.

COAST-LINING.

WHEN and how the putting-in of the coast-line is done, must depend much upon circumstances.

If making a chart with pretensions to accuracy in the details, it is better to do it before the soundings are taken, as, for the inshore soundings, the little points and bays, not distinguished by marks, will be very valuable. In this case, too, every yard of the coast that can be walked over should be. If the surveyor pull along the coast in his boat, from one spot to another, he will be liable to miss little details, such as stream entrances, which may be blocked by the sand beach in summer; lagoons behind the shore, etc. The boat should therefore only be used to pass rocky points and cliffs that cannot be walked along, or to make stations in, at anchor off the coast, if it is necessary to do so, to shoot up the details. *In a detailed Survey.*

The method of putting the coast-line on to the sheet also varies. The angles can be taken, and the details between subsidiary fixes on the beach sketched into the angle book, using always a larger scale than that of the chart, and then these fixes and angles plotted on to the chart after return on board; or the surveyor can take a field board, with the points on it, with him, and plot the coast as he goes along it on to his board. *Plotting the Coast-line.*

Of these two the latter method is by far the best, and should always be employed as a rule. There is no chance of having necessary angles omitted if the fixes are plotted at *Plotting on the ground.*

the time, and any little error is easier detected on the spot than when plotting afterwards on board. Of course rainy weather or other circumstances will sometimes prevent the work being plotted at the time, but unless some good reason exists, it should be done.

Instruments required. If conveniently situated marks are plentiful, the coast-liner will only want his theodolite or sextant, or both, to take his angles, and a station pointer and tracing-paper for plotting, with protractor, etc. But if the coast has no objects off it to seaward, and landward marks are also short, or invisible from the shore, he will require, very probably, a pole of measured length, whereby to ascertain, by observing the angle subtended by its extremities, the distance of points, etc., from one another.

A convenient form of this pole is described under "Ten-foot Pole," page 39.

Each assistant should have a copy of the Ten-foot Pole Table,* on a piece of cardboard, always in his angle book, ready for reference in the field.

General method of Coast-lining. Let us suppose an officer landed with his board of points to do coast-line.

He will start at some point already plotted on the chart, and will take angles from it to all the objects he can distinguish between him and the next fixed point, and beyond, if necessary.

He will then walk on to another spot, where he will make a supplementary station, fixing himself by angles to known points, either by theodolite or sextant, according to circumstances.

He will then plot this, his No. 2 △, on his board, by station pointer or tracing-paper, taking care to check his position by his line from the 1st △, or by a third or " check " angle from his present position. His No. 2 plotted, he will sketch in on the board, the coast-line between that and the first, having noted any peculiarities as he walked along.

---

* Appendix R.

The scale of the chart will largely influence the distance between the subsidiary stations to be made by the coast-liner, as will also the character of the shore line, and the intended nature of the chart as to exactitude of detail.

If the work is to be plotted on return on board, the system is precisely the same, only the detail of coast between the stations must be sketched in the angle-book, instead of directly on to the board.

When the coast-liner sees that at the next station he will **Using Ten-foot Pole.** not be able to fix himself by angles, he must use his ten-foot pole, sending a man on with it with instructions where to stand, or going on himself to some point solicited, to which he will first take the angle ; leaving the pole behind with a man at his present station, with directions, when signalled, to hold the pole horizontal, and at right angles to the observer.

To ensure the latter, either a rough pointer of some kind can be attached to the centre of the pole, so as to project at right angles, in which case the holder will be directed to point this to the observer, or, he will be told to sway it gently backwards and forwards, and the observer will read the largest angle he can measure.

The angle observed, and the corresponding distance looked out of the table, the latter is measured on the scale of the chart, and applied by a pair of compasses, as a distance from the last station along the line laid off from that last station in the direction of the required station.

If necessary, the whole coast can be carried on in this way ; but if the marks are a long way apart, great care must be taken in observing the angles on to the positions to be measured, as there is no check on the work, and each error will be accumulative. In this case the man must be sent on, and must mark the exact place he stood when the angle was observed to him, and the coast-liner must make his next station precisely on that spot.

The azimuth compass may sometimes be employed in this work with advantage. Any little error, when a properly fixed station is reached, can be squared in.

It will be understood that this ten-foot pole method is only used for the smaller detail, where sufficient angles to fix cannot be obtained. It is especially useful in delineating the shores of islands, or of small bays which have no fixed point in them.

For instance, in Fig. 26, let us suppose the two points, marked Ash and Lime, are fixed, but in between them is the small bay shown.

At Ash we obtain the angle between Lime and A, the next point visible, and also the distance by our ten-foot pole. If we can make out that B is a point, and can see any prominent spot on it, we shall get an angle to that also.

We then go to A, sketching in between on the way. At A

FIG. 26.

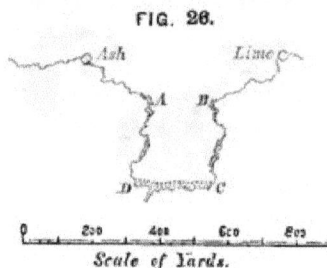

Scale of Yards.

we become aware of the little bay, and we send the pole over to C, pointing out to the man with it where to stand, and telling him to put a stick or stone there, when he is signalled to go on to B.

At A we get all we can, angles from Lime as zero, to Ash, B, C, tangent of bay on towards D, and anything prominent, and the distance to C by the pole.

Leaving a little mark at our station at A, we go to Lime, and take angles from Ash to A, B, and distance to B by the pole now there.

We then go back to B, and send the pole over to D, and again get all angles we can, and distance to D.

We now sit down and plot our data. We have two angles to A from Ash and Lime, and a distance to A from

Ash. These ought to agree, and we prick in A. We have
the line to B from Lime, and perhaps from Ash as well, but
we will suppose not, and will plot B by the distance from
Lime. Then placing our protractor on B, lay off the angle
observed there to Ash, which ought to go through, and make
a check for B.

We plot C and D by their distances on their respective
lines from A and B. We then walk round the bay, sketching
it in, and can get an angle at C, from A to D, as another
check, and any other angles to assist in sketching in details.

The coast-liner will generally be responsible for all the
details of topography close to the coast such as follow, the
scale of the chart being taken into consideration as to with
what degree of accuracy detail can be laid down.

Coast-liner to do topography near to the shore.

Heights of cliffs must either be measured with a lead-line,
or by getting an elevation to some definite point, which must
afterwards be fixed, from one of the stations, or may merely
be estimated and entered in the angle book.

The height of a cliff can be readily calculated on the spot
from an angle, by the formula :—

$$\text{Height in feet} = \frac{\text{Angle in seconds} \times \text{distance in miles.}}{34}$$

Cliffs have generally to be exaggerated on the chart, to
show distinctly. The height in feet should be written against
them.

The directions of lower parts of streams, or rivers, must
either be walked up, and fixed, a certain distance back, or
can merely have their entrances fixed, and an angle taken
up for their general direction.

Lower spurs of abrupt hills must be sketched in, assisted
by angles to them from different points.

Houses standing back from the shore must be put in.
These can usually be fixed by angles to them without
visiting them, unless it is necessary to get their dimensions,
names, etc., or perhaps to ascertain if a good well or spring
of water may be near, that would do for watering on an
emergency.

Swamps near the coast should be sketched in as far as necessary, and a look out kept for evidences of any extension of their area in winter. Information on these points can be picked up from passing inhabitants.

Angles should be got also to any conspicuous objects farther inland, as they will be very useful when the topography is sketched, and the surveyor should always look ahead, and seize any opportunity of the kind for helping on other parts of the work than those he may be immediately engaged in.

Roads near the coast should be walked back to, and fixed here and there, sketching in between.

Rocks above water, or breaking, should be fixed. Though these come into the province of the sounding, it is often useful to have them down first; and in the case of a break only, it may be very much so indeed, as it may be an isolated head, which a boat sounding near high water may miss.

Low-water Shore-line. Though it is the high-water line that the coast-liner is more immediately concerned with, he should mark at low water the position of the dry line, especially where this runs off a long way at points, etc.

In a detailed survey on a large scale, it may be necessary to send some one round the water-line at low tide to get it accurately, but this is more usually obtained by the soundings, for by reducing these to the low-water level of springs, a series of points will be obtained, where each line of soundings crosses the low-water line, which can then be drawn in as a line passing through these points.

Elevations of hills. Angles of elevation for heights of the hills should be taken when getting the angles for fixing the points of the chart, from main and secondary stations, or any well-fixed points; but if the coast-liner gets some more elevations from marks on the water-line, they will never come amiss, as long as the position is well fixed.

General Information for Directions. The officer coast-lining will make note of anything worth recording in the sailing directions, as little nooks for landing, convenient places for watering, etc., letting his captain know on return on board, in order that they may be, if necessary, again looked at, or entered in the latter's notes.

It may be convenient to keep a book for the purpose, in which any useful information can be entered.

As an instance of the application of the ten-foot pole method, we may mention the following, which is adapted for use on shores with fringing coral reefs, or broad sand or mud flats, which dry sufficiently at low water to enable people to walk on them, and when either the steepness of the hills or the denseness of the vegetation prevent marks being fixed on the coast. *Further application of Ten-foot Pole Method.*

FIG. 27.

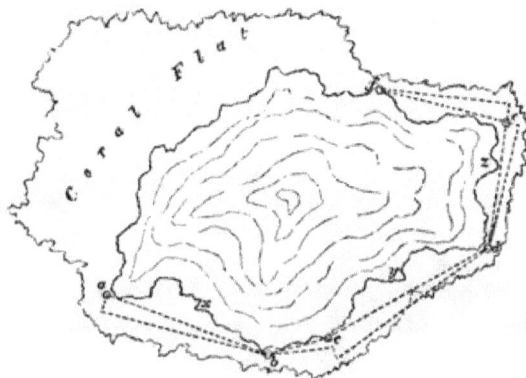

Let annexed diagram, Fig. 27, represent an island of this kind.

A long measured lead-line, say of 500 feet, is provided. This is taken by an officer we will call B, who has a prismatic compass. Another officer, A, is provided with theodolite, or sextant, or micrometer, and prismatic compass, according to circumstances, sextant and compass being quite sufficient.

Starting at *a*, B remains there while A walks to *b*. B stretches his line out at right angles to *a*, *b*, and plants a flag at the extremity. A observes angle subtended by flag and

$\triangle$ a, with his micrometer or sextant, and both A and B observe the bearing of $a b$.

A waves to B, who goes on to c, when the operation is repeated.

A then moves on to d, B pivoting his line round c, so as to be rectangular to $c d$; and so on, until f is reached. We will here suppose that, from a to f, we have been able to triangulate, the reef being broader. We have therefore the correct bearing and distance of $a f$.

To plot this, the mean compass bearings and distances a, b, c, etc., will be put on a separate sheet of paper on a larger scale than the chart, and the positions $a f$ being joined on both, the other stations will be squared in on to the chart.

Marks will be left at each station, if required for sounding, or delineating the outer edge of the reef. Subsidiary marks can be made at other points, as x, y, z, and fixed by angles from b, d, etc., with distances measured by the angle of the line.

The shore line can either be sketched by A, as he walks from station to station; or can be put in afterwards, if greater correctness is required, using the ordinary 10-foot pole to fill in between a, b, c, etc.

If a theodolite is used, which it is well to do in a case where we have not been able to get any measured base at all, and must consequently work back to a, it must be set up first at a, and the angle to b taken from some fixed object, whose true bearing we should obtain, as we in this case must not be dependent on the compass. B will be at b with his line, and when A has finished, will walk on to c, so that A, when he arrives at b, can take the angle from a as zero, to c. With a theodolite, then, A must visit every station, unless B has one also.

At every new position, the last $\triangle$ will be used as zero.

The readiest way for B to direct his line so as to be at right angles is to use the so-called "cord-triangle," which is simply a triangle formed of a piece of line whose sides are in the proportion of 3, 4, 5, the angles being marked by knots.

When stretched on the ground, with the corner between 3 and 4 at the △ , and the 4 side coincident with the direction of the other △, the direction of the 3 side is at the right angle required. Any similar contrivance will serve the purpose.

NOTE.—This method was largely used by Lieutenant W. U. Moore in the survey of the Fiji Islands, and is a good example of the dodges that have to be improvised to meet circumstances.

CHAPTER VIII.

SOUNDING.

Boat Sounding—Ship Sounding—Searching for Vigias.

Import-
ance of
Sounding.

IT is difficult to say that any one step in the construction of a chart is more important than another, as each is necessary for the completion of the whole, and an error anywhere may cause a disaster; but if any particular item *is* to be picked out, perhaps the sounding should rank in the highest place.

The operation of sounding is the least pleasant part of a marine surveyor's work, especially when the weather is against him, and the sounding uninteresting, that is, where the depths are regular, and there is no excitement in the way of discovering, and working out, shoals and reefs; but the notion that it is therefore always to be relegated to the juniors of a survey, is not only hard upon them, but may introduce errors into the very part of the chart which, as we have already said, is the most directly important.

As soon as the points are down, *i.e.*, plotted, the sounding can be commenced; but, as before remarked, on an intricate piece of coast it is better if the coast-line is put in first.

Ordinary
Method of
Sounding.

The ordinary main plan of sounding is thus. The boat proceeds in straight lines in a direction, of a length, and at distances previously decided on, with a man in the bow constantly sounding. Every so many soundings, as the case may be, the officer takes angles with a sextant to fix the position of the boat, always doing this at the beginning and ending of every line.

It is evident that this main plan may be largely varied in its details.

In the first place rises the question as to whether it is better to plot fixes, and enter soundings on the sheet, regularly, in the boat, or leave them until return on board, merely putting down an occasional fix to see where you are. The writer says, certainly, as a rule, plot them at once. It can be done in ordinary circumstances just as correctly, and gives more information to the officer sounding as to little bits which may want additional casts, and it also gives the men at the oars a little rest from time to time. In very rough water it of course cannot be well done, and must be left till return on board to the comparatively motionless ship; but when you can, plot at once. In harbour work on large scales, again, it will be better to plot afterwards, as great accuracy will be required.

The extent to which the soundings themselves can be entered at the time on the chart, depends of course upon the state of our knowledge of the tide. If the tidal range is small, or the motions of the tide are sufficiently known to form a table of reduction beforehand, the reduced sounding can be written on the board at once. If not, the soundings as taken can be written down, and reduced on inking on return on board, or, only the sounding taken at each fix can be written against the prick of the fix, and intermediate soundings left to be entered on board. The latter will generally be found most convenient.

The pace at which the boat may go, and the necessity, or not, for stopping at the casts, will depend on the depth of water and the capacity of the leadsman.

Whether it is necessary to stop to get the angles depends upon the convenience and visibility of the marks, and the quickness of the angle-taker. A beginner will of course do everything deliberately, until he feels capable of combining speed with correctness.

Whether each fix shall be plotted at once, or whether to wait until two or three have been got, and then lay on oars,

Circumstances guide many details.

or anchor for a few minutes, must also vary with circumstances.

If laying on oars, keep the lead on the bottom with a slack line, and let the coxswain keep the boat in position.

What the distance should be between each fix will depend largely upon the scale of the chart, and the nature of the bottom. On an evenly sloping bottom many soundings can be got without another fix; but where depths vary or increase rapidly, the fixes must be closer together.

The soundings which will be joined together on the finished chart by fathom lines, *e.g.*, the three, five, ten fathoms, etc., should always be fixed, and in doing this it must be remembered that it is the *outer* sounding of any of the same depth that will be on the fathom line, and also the tide reduction must be taken into consideration. This latter will of course be in many cases only approximately known, so that exactly the right sounding may not be fixed.

Direction of lines. The sounding lines should be in ordinary cases at right angles to the coast, and parallel to one another, as not only will a better line be got for tracing the fathom lines, but the boat will easier be kept in her right direction by observing two objects which have been seen to be in transit, in the right direction, at the commencement of the line.

Marks in transit for directing lines. In nice work on large scales it is generally necessary to place two marks in line for this purpose; but, for ordinary surveying, changing them from one line to the other will take far too much time for the purpose, and marks to answer all practical purposes may usually be found placed by Nature already.

Before starting in towards the coast on a line, it is frequently desirable to take off by the protractor the angle to the point on the shore where the next line out will commence, and placing this on the sextant, try to find some object on the shore which can be utilised as a mark to warn when to turn out. When close in shore it is often impossible to fix, and the lines may therefore be irregular, without some such assistance.

In sounding out a small harbour, circumstances must guide the direction of the lines.

The depth to which the boat soundings are to be carried will depend upon circumstances. When soundings of over 20 fathoms are taken from a boat, it gives a great deal of labour, unless a small sounding machine is carried. **Depth to which boat's soundings should be carried.**

When the boat gets to the end of her line, and turns to pull along to the end of the next one to return, soundings should still be carried on, as before.

The method of using the station pointer has been explained under the head of " Station Pointer."

It only remains to note that it must be recollected, in getting the fix, that the right or left angle (according to whether a right-handed or left-handed station pointer is in the boat) must be observed of a sufficient number of degrees to be measured on the instrument, if possible. If this cannot be got, recourse must be had to tracing-paper for plotting the position. **Construction of Station Pointer to be remembered.**

The sounding book need not be ruled. There are several ways of writing down the objects used for fixing and the angles between them, but the best, if space permits, as it does in the sounding book supplied by the Admiralty, is to put them down as you look at them, the right-hand object to the right, the middle one in the middle of the page, and the left one on the left-hand side. The hour, in Roman numerals, and minutes should be entered from time to time, to know the reduction for tide ; the sounding at the fix goes on the extreme right, and subsequent soundings up to the next fix, in a row underneath, thus— **Entering soundings in Book.**

X 14   Pagoda 28° 31'   Mat 62° 14'   Can   7½

                                                s

7½   8 × ×   8½   9 × ×   10 ×

        s   m   m

    „   23° 02'   „   60° 08'   „

                Pea 41° 17'   „   11

                                    m

The cross (×) signifies the same sounding as before ; and

L

All casts to be entered. it may here be mentioned that *all* soundings must be put down, even though there may not be room for half of them eventually ; as, the man heaving regularly, if all his casts are not registered, the change of fathom will not come in its true place when interpolating between the fixes.

Space for reducing. Space must be left under each line for the soundings, as reduced to low water, to be written in in red ink.

Check angles. A check angle should be taken, from time to time, to make certain that things are right, as is noted above at the last cast, in the example Can to Pea. This is especially necessary at the commencement of work with new points, as mistakes *will* occur in plotting points occasionally. A check will show at once if points are true, and if the angles have been taken correctly.

Nature of bottom. The nature of the bottom must be taken every few casts, and recorded, the officer having a look at it from time to time himself, to make certain that the leadsman is calling the stuff he brings up by its right name. For instance, many men will insist on calling "stones," rock, which is of course quite a different thing.

Same "Points" to be used. The same objects should be taken for the fix as long as possible. It tends to check errors in reading off, as the angles at each fix will bear a definite proportion to the last set. For instance, if we are pulling off shore with both Mat and Pagoda astern of us, the angle will be less each time, and a reading of say 33° instead of 23° would be at once detected as erroneous, before the disjointing of the line when the fix was plotted showed there was "something wrong somewhere."

The variation in the angles will also enable us to see if the "fix" is remaining good. This plan also saves time in setting the station pointer verniers.

Necessity for assistance. When assistants are not thoroughly used to the work of sounding, it will be necessary to have two in each boat, to ensure no mistake ; but when not only officers, but men get used to it, one officer will in most cases be able to carry on the work by himself, with the assistance of a man to write down for him. Now that seamen are all taught to write, there is seldom any difficulty in finding one of the boat's crew, the coxswain if possible, to write down fairly. The same man will steer generally, and so permit the officer to keep his eyes for other matters.

In deep water the boat must of course be stopped, and the leadsman will only heave when told. The interval can be timed by watch, or, in very open deep soundings, by the Massey's log towing astern, fitted as described on page 51.

The distance between the lines of sounding will depend upon the scale and the character of the survey, also upon whether the place is inhabited or not, for where there are natives, information can be picked up as to shoals, &c., from the fishermen. The value of this, however, largely depends upon the intelligence of the informant, and often cannot be trusted. *Distance between lines of soundings.*

If the coast or harbour be unknown, and the land of certain geological formations, it takes a great deal of sounding to be certain no stray rocks exist undiscovered; and, as was pointed out in our preliminary remarks, the majority of marine surveys are not on a sufficient scale, nor will time at disposal allow us, to sound so close as to be absolutely certain nothing is missed. The surveyor must make up for this by keeping his eye ever on the look-out for discoloured water, and by examining every suspicious spot.

It must always be remembered that in the ordinary scales used for surveying, figures may look close together, and yet be, in nature, quite far enough apart for a rock or bank to exist, without giving any indication in the lines of soundings passing on either side of it. On a scale of 3 inches to the mile, each figure will occupy a space of 50 yards nearly.

It will depend upon the orders received from the chief of the survey whether suspicious ground is searched at once, or merely pointed out on return on board for further examination. As a general rule, whenever the soundings, in pulling off shore say, *decrease*, it is suspicious, and the spot must be examined by intermediate lines, which in many cases should *Suspicious ground.*

L 2

be at right angles to those previously run, and looking out
sharp with the eye as well.

In calm weather, when there is a tide, a sharp eye may
detect a pinnacle rock by the ripple it may form.

It is in looking out for, and utilising such small indications
that the genius of the true surveyor displays itself, and many
are the rocks that have been missed for want of such sharp
intelligence.

Small buoy. A small nun buoy, with light chain and a weight to anchor
it by, is useful in the sounding boat, to drop over on a shoal
spot, so as to guide a boat working round and round while
trying for still shoaler water.

Doubling the Shoaler lines. In many cases it is convenient to run double the number

FIG 28.

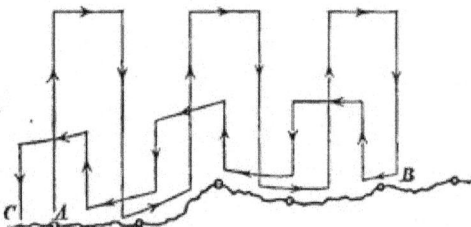

of lines in shoal water, (say out to 7 fathoms,) that are
required in greater depths. In this case, one set of lines will
be run first, and when the boat gets to the end of her allotted
space, she will return in the opposite direction, and run intermediate lines.

See Fig. 28, where we suppose the boat to start at A, work
along the long lines to B, and then return to C along the
intermediate lines, crossing the old work at every line, and
thereby getting a check on it.

Sounding sections. In sounding a harbour channel on a large scale, it is often
convenient to stretch a lead-line across from side to side, and
sound at regular distances apart by this line, shifting it for
each section required.

Sweeping. Sweeping for a reported pinnacle rock is resorted to when sounding fails to discover it. Two or more boats, pulling abreast, tow a lead-line between them, well weighted under the stern of each boat. If one weight in the centre is used, the rock may very likely be missed. The size of the boats will govern the length of line between them. An iron bar is still better. It is by no means an easy thing to do efficiently, so that all the ground shall be traversed without unnecessarily going over it again and again. If steam-cutters are used, care must be taken not to go too fast for the weights attached, or the bight of line will be towed nearer the surface than is intended.

Shoal banks, out of sight of land, or too far off to use marks, can be sounded by starring round the ship, at anchor on it, or off its edge. For these, compass bearings of the ship taken from the boat, with distance measured by the masthead-angle, will probably suffice in accuracy, the boats sounding in lines radiating from the ship in all directions. *Sounding shoals out of sight of land.*

A large canvas ball or cylinder, on a light framework of iron and painted black, will be found very useful at the masthead when taking the angle for this purpose, as it will clearly define the masthead, and also indicates, "Ship in position."

Boats or beacons can be moored in convenient positions, and fixed by angles to one another, and to and from the ship, also at anchor, and the base obtained by masthead-angle, if it is necessary to sound a bank a little more accurately. These will then be used as marks, and the soundings fixed by angles in the ordinary way.

When surveying a large bank, where accuracy is desired, the beacons should be placed on a regular plan, and nothing is better from every point of view than anchoring them in two lines, so as to form equilateral triangles and a series of parallelograms, the beacons being about 5 miles apart. This distance permits the corners of the parallelograms being seen from one another. Bases are obtained by patent log, and astronomical observations fix the extreme points.

If sounding out the triangles by boats, a mark boat, flying

a flag from a bamboo lashed to the mast, can be moored half-way between the lines to aid fixing.

Reef Sections. Sectional lines off coral reefs are sometimes now required to show the exact slopes for scientific purposes, or for cables.

It is not an easy operation, and cannot be hurried.

The soundings must be close to show the exact slope, when it is, as in many cases, steep.

The section must be run on a transit line, and there are many ways of fixing the distance.

A boat anchored on the edge of the reef for the outer transit mark, with a long bamboo or other light spar stepped, will, by means of vertical angles, afford a means of ascertaining the distance up to perhaps half a mile, but beyond it will be necessary to have another boat or mark on the reef at a fixed distance from the transit line, to which horizontal angles can be taken, making, in fact, an exaggerated ten-foot pole. Other methods will suggest themselves to the surveyor.

The diagram should be drawn on a true scale, *i.e.*, the vertical and horizontal scales equal, an inch to 30 fathoms, and *the slope to the left*, so as to facilitate comparisons with other diagrams.

Measuring "Lead-lines." In all sounding, the lead-lines should be measured on return on board, and a note made in the book, "Lead line correct," or so much out. When the line has not been used for some time, it should be measured before leaving in the morning also; but if it has been examined the evening before, this will not be necessary.

While on this subject, it may be noted that new lead-line should never be used for boats' soundings. At the beginning of the commission it may be necessary to do so, but afterwards make lead-lines out of old well-stretched stuff that has been used for deep lines for ships sounding, and measure and mark them when wet.

Necessity for fractions. The soundings must be put into the book to the exact depth obtained, but it will depend upon the scale, the general accuracy of the chart, and the thickness of the soundings, how far halves and quarters will be placed on the sheet. As a rule, fractions should be retained up to 6 fathoms, and over that depth only the even fathom, taking of course the fathom under the depth. Thus a sounding which, when reduced to low water, is 9¾, will appear as 9 fathoms.

The necessity for accuracy in reducing soundings to low **Reducing Soundings.** water will also very much depend on the scale of the chart and the depths. It is evident that with soundings of over 6 fathoms at low water, if we are using a small scale, where the size of the figure placed on the chart will, in reality, cover ground on which we have taken five or six soundings, any nicety of reduction is an absurdity, and labour thrown away; but in *shallow* water the reduction will be just as necessary in a small scale as a large, as a sounding of 5 fathoms will be a danger or not, according to what amount of reduction we apply.

It is usual in surveying vessels to depart from the time- **Calling Soundings.** honoured habit of calling soundings, and to call simply " six and three-quarters," " five and a half," and so on. This is simpler, and saves time. The men should also be trained to call out sharply, and on no account allowed to drawl.

There are, however, two exceptions to this. " Seven " and " Eleven " have a great similarity when called from the chains, and to prevent mistakes, " Deep eleven " should be called. Similarly " Nine " and " Five " sound much alike, and " Deep nine " should be given. " Five " and " Seven " are given simply.

On all occasions, whether in ship or boats, when the leads- **" Shoal water."** man suddenly gets a shoaler cast than expected from his previous soundings, he should call out " Shoal water," without waiting to complete his usually fruitless endeavours to gather in the slack line, and find out the depth. The author has been on shore from the neglect of this, the leadsman being foolish enough to wait until he had repeated his cast, so as to give the correct depth, and gave no warning to the officer on the bridge until too late.

Belcher proposes a plan for ascertaining the depth on a bar **Belcher's " Sounding a Bar.** which it is desired to cross, without risking a capsize, which

may be quoted, though we have no knowledge of its having
been practically tried. He suggests anchoring the boat as
close to the bar as is safe, with the tide at flood, and veering
away a barricoe with a grapnel hanging at a given length of
rope. The barricoe is permitted to drift freely over the bar,
when the anchor catching, will give a shock to the barricoe
that will be seen by the watcher in the boat, and will indicate
that a less depth than the length of the cable allowed to the
anchor is on that part of the bar.

The line attached to the barricoe, with presumably a tripping
connection with the grapnel, will bring the apparatus back to
the boat, when she can test another part of the bar in the
same manner.

SHIP SOUNDING.

The soundings over a certain depth, about 20 fathoms, can
generally be most advantageously done from the ship.

Usual plan. Where a steam winch is fitted, soundings can be got with
great rapidity; and by dropping the lead from forward and
heaving it up to a davit fitted on the taffrail, up-and-down
casts can be got in 40 fathoms at a speed of about four-and-
a-half knots without stopping, with a 100 lb. lead.

If a long spar be fitted as a derrick aft, soundings can be
obtained in water up to 20 fathoms, by merely swinging the
lead and letting it go without heaving forward.

Arrangement for expeditious sounding. For deeper water a variety of methods have been devised
for getting the lead forward and dropping it rapidly.

The following is now generally used.

The lower boom is got out and topped to an angle of
about 40°.

An endless rounding line of lead-line is carried through a
block at the end of the boom, by leading blocks to the steam
winch, and to the derrick or davit aft. A slip is attached to
this with a broad projecting palm of sheet iron to the catch.
See Fig. 29.

To this the lead is attached, and hove forward by the winch.
When up to the boom, the rounding line is let go, and on

striking the water the palm releases the catch, and the lead
falls free to the bottom.

The rounding line is at once rounded aft again, ready for
the slip to be again attached when the lead comes up.

There are varieties in the detail of the fittings, according
to different ideas.

By dropping the lead well away from the ship, the chances

Fig. 29.

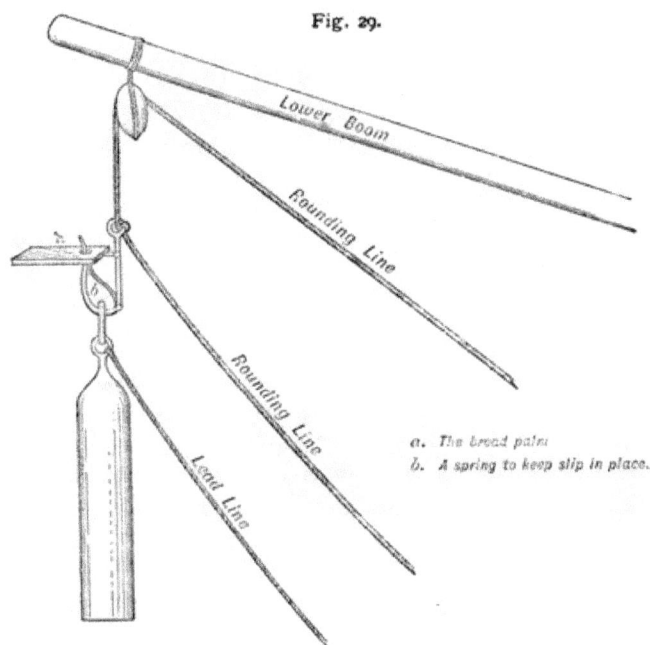

a. The broad palm
b. A spring to keep slip in place.

of the lead-line fouling the screw, if the helm is over, are
much lessened.

A variety of instruments have been invented for giving Instru-
the accurate depth when the line cannot be got up and down; recording
some depending on a fan which works a series of cogged depth.
wheels, as Massey's; others, on pressure at different depths.
These are all useful up to a certain point, and when their

errors have been obtained, may sometimes be attached to the lead with advantage.

Recorders, however, of great value to navigation, are of no use in surveying operations, and the majority of these navigational inventions are liable to small errors, which we must not have in depths which are to be placed on charts.

Burt's bag and nipper are useful when the ship drifts away from the vertical position over the lead, and one should always be handy when sounding, but great care is necessary that it bites the line.

Perfect machine yet to be made. It is evident that a *perfect* machine is more trustworthy than the record of an up-and-down cast with the ship in motion, as given by a fallible man; and when such perfect machine is invented, it will be gladly adopted by surveyors; but, up to the present time, the machines are more liable to error than a trained man, under most circumstances.

Long lines of soundings off shore. In localities where currents are prevalent and vary, when we are running long lines of soundings in the ship off shore, out of sight of land, it is very important to get, on the return line towards the shore, a fix as soon as possible. The soundings we are obtaining may be hereafter used, especially where fogs are frequent (as, *e.g.*, British Channel, Bay of Fundy), to give vessels a notion of their position, and we must therefore use every dodge to get our true position at the earliest opportunity, so as to depend upon dead reckoning as little as we can.

Two theodolite stations, from which a large flag at the masthead can be observed as soon as it appears above the horizon, is a plan sometimes employed. These need not see one another. As long as their relative positions on the chart are known, and the true bearing of the zero employed has been established, the angles to the ship can be plotted. The smoke from the ship when she herself is below the horizon will often enable a valuable angle to be observed. A single theodolite line is often of great value, as it can be utilised in conjunction with angles, bearings, or observations from the ship herself.

It is scarcely necessary to say that an observer will be on the topgallant yard of the ship, as he may, from atmospheric or local causes, be able to see something on the land before the theodolite observers catch sight of the ship. Surveyor aloft.

True bearings come in useful again. The angular distance between the sun and the mountain, or other object, seen from aloft, will be taken by the observer aloft, while the sun's altitude is taken from deck for the azimuth.

Another method is to have one or two ships anchored as far from the land as they can fix, which observe, and are observed from, the sounding ship, as she runs in and out on her lines. A ship can easily, in light winds, anchor in 100 fathoms, and even in deeper water. Tenders.

When land stations are employed, heliostats are useful, as informing the running ship that she is seen from the station. A flash will tell the officer aloft that a sounding can be taken, with the certainty of an angle being got to the ship, for which, perhaps, she has been waiting. Use of heliostat.

If not able to return and pick up the land before nightfall, blue lights and rockets are useful, both from stations and moving ship. Position after dark.

A true bearing of a light, or mountain, if visible, as it often is a great distance on moonlight nights, can be obtained in the northern hemisphere very conveniently by the angular distance from the Pole star, as described on page 281. This angular distance can again be taken from aloft; but in the case of Polaris, we require no altitude. Use of stars.

If Polaris is not available, a time azimuth of a star near the prime vertical will give a good result. Should the resulting longitude differ much from the assumed, it may be necessary to re-calculate. Altitude azimuths cannot be much trusted in at night.

When objects are visible from deck at night, and we can rely on the compass, very good bearings can be taken with the standard, if the lighting arrangements are properly fitted. Compass bearings.

A ruled "Deck Book," as now supplied, is convenient for ship's sounding. In this everything taken from the ship Deck book.

should be recorded, as, rounds of angles when the ship is used as a station in main triangulation; elevations with sextant, and the corresponding fix; sketches of little bits of coast, &c.

SEARCHING FOR VIGIAS.

Difficulty of disproof.
In searching for a "vigia" it is difficult to say when its existence is to be considered as disproved. Although experience shows that nine out of ten of these bugbears and blots on the oceanic charts have been mistakenly placed there, from reports of floating whales, wrecks, and patches of *confervæ* taken for discoloured water over a bank, &c., still the apparently astounding manner in which coral banks rise from very deep water must always make us careful of assuming from a hasty search, that no shoal water exists near a given locality.

Area of search.
The area over which to search must always be large, as the reckoning of the reporting ship, especially as regards longitude, may often be considerably in error. In the vicinity of such reefs also, currents are generally accelerated, and altogether we must allow a large margin, in undertaking to search for a danger reported in a particular spot.

Small area very possible without much indication.
In clear bright weather, coral banks will show some miles with the sun in the right direction: but under other circumstances it is quite possible for a ship to pass within a mile of a bank with as little water as three fathoms on it, without its being detected

Assuming that coral reefs are built on submerged mountain-peaks, a little consideration will show that there is nothing extraordinary in a shoal near the surface standing in 2000 fathoms water, on a base of not more than three miles diameter.

The annexed sketch, Fig. 30, will show that we are not assuming an improbable steepness of side to the submerged island.

This allows us to pass little more than $1\frac{1}{2}$ miles from such a shoal, and still get a cast of 2000 fathoms, so that even a positive sounding of great depth will only cover a comparatively small area, and soundings of a hundred fathoms, no bottom, do not assure us of anything to a certainty, except that the reef does not exist within a few hundred yards of that cast.[*]

FIG. 30.

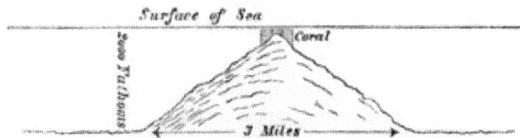

Nevertheless banks can frequently be diagnosed from a sounding, though deep, a little less than others around, and the only way to make certain that a bank does not exist is to follow up the direction of the slope with positive soundings. No bottoms are of little value.

Experience has shown that, in coral waters, the edge of a bank is a favourite spot for the growth of a small danger. This part of a bank should therefore be closely searched.

The difficulty of fixing the position at sea to within three miles or so, adds another element of uncertainty to our search, so that it is only by crossing and recrossing the area to be examined that we can at length say, *positively*, nothing is there.

Doubt as to our own position.

This is especially the case where the reported danger is out of the usual track of ships, as there is nothing improbable then in its having escaped notice up to that time. Where the locality is frequently passed over, there is more *primâ facie* reason for doubting the report, and in many instances, a cross-examination of the person making the report, will show how very slight is the ground for it. An actual cast of the lead seems a fact impossible to make a mistake about, but instances have occurred where this also has proved to be so, even with so-called "bottom" brought up. In cases, however, where a sounding *has* been obtained, we must conclude the report to

Credence in reports.

---

[*] Experience gained from recent extensive sounding operations appears to indicate that this angle of slope to great depths does not actually occur, and that a base of ten miles may be pretty safely assumed in the depth given.

be true, and a rigorous search must be made before the vigia can be obliterated.

Preliminary search. As a general rule, for the first commencement it is best to run lines east and west in or near the latitude reported, as this is more likely to be near the truth than the longitude.

When going to make an exhaustive search, the first day is perhaps best spent in doing this without getting more than may be one positive sounding, as we can cover more ground, and, if the danger exists, we have a good chance of finding it by sight, or by the soundings taken when the ship is running, as of course the deep-sea lead will be kept going constantly.

Lying-to near Vigias at night. It is rather unpleasant to be drifting about at night with reported reefs in the vicinity, and by no means a bad precaution is to ease a kedge anchor down to 100 fathoms or so, which may bring the ship up, or at any rate show, by drawing ahead, that bottom is reached, before she strikes on the reef.

At night the vicinity of a reef in open ocean may be indicated by fish, which invariably frequent these isolated spots, and here phosphorescence will help greatly in making their presence very apparent.

In daytime, birds, which generally congregate wherever fish are plentiful, may be an indication.

Decision rests with Hydrographer. In every case, of course, the surveyor transmits home a plan of his track and soundings, as it is at headquarters only that a decision on the matter can be arrived at.

Under the head of " Sea Observations " will be found hints as to early ascertaining of the ship's position, a most important matter on each morning.

Submarine sentry. The submarine sentry, now supplied to all surveying ships, should be constantly towing, set to about 30 fathoms. With kites fitted to float with the ship stopped, the chance of fouling the screw is much diminished, and the apparatus is of the greatest use in furnishing indications of small banks that may otherwise be missed, though it sometimes gives false alarms.

As the wire generally carries away close to the kite, it saves loss if a preventer of slack wire, secured to the tail of the kite, be spliced into the main wire about 4 fathoms up it.

CHAPTER IX.

TIDES.

ALL soundings in published charts are given for low water at ordinary spring-tides; we therefore want all the information we can get about the tides, and the very first thing to be done on arriving on the surveying ground is to commence observations on them. *(Tidal observations commenced at once.)*

There are very few parts of the world in which we have absolutely no knowledge of the tidal movement, so we have generally something to commence upon. That is, we usually know within an hour or two the time of high water at full and new moon, called H. W. F. and C. on the charts, as, except in estuaries or peculiarly shaped coasts, this will not differ greatly from places near at hand, and the same may be said for the range of the tide.

It will altogether depend upon our length of stay in any locality, as to what we can hope to find out about the tides. To get full information requires observation during months in succession, as in many parts the tides vary considerably at different times of year. The number of high and low tides in a day, in certain places, departs from the normal phase of from 6 to 7 hours for each rise or fall; in others, the tide will take longer to rise or fall, than *vice versá*, &c., &c. A long series of this kind is therefore very valuable, as the tidal theories are at present far from fulfilling all the requirements of observation all over the world, and good data are much wanted; but it is not often that the surveyor can obtain such a series. *(Different observations for different requirements.)*

It will be seen, then, that tidal observations for the practical reduction of soundings for purposes of navigation are one thing, and those for obtaining additional data for scientific investigation are another.

We shall mainly concern ourselves with the former, where much rougher observations are usually admissible; but here, again, it must depend upon the scale and nature of our chart what degree of nicety is requisite.

A few words on the Theory of the Tides will be given at the end of the chapter.

Tide Tables. The reader is referred to Dr. Whewell's Treatise on the Tides, published in the preliminary part of the Admiralty Tide Tables, for much information respecting their movement.

Local circumstances. A regular series of observations, even for our practical work, should be taken if possible; but in many cases the necessity for leaving tide-watchers encamped is inconvenient, and may be unhealthy, and we may have to be satisfied by obtaining what will be sufficient to enable us to construct the chart, which is our immediate business.

In other cases we may only be staying a few days at a place, as when making a plan of a small isolated harbour.

Observation indispensable. What we absolutely require in making a chart is to know the height of the water, whilst sounding is going on, above the level of low-water springs, which is called the "datum for reduction."

We shall also wish to ascertain, if possible, the "establishment," which is the time of high water at full and new moon, called in the charts, "High water at full and change;" the rise of spring-tides above our datum; and the range of the tides at neaps, and the time occupied by the rise and the fall of each tide, as these will give valuable information to the navigator.

We may here give definitions of some of the terms used in speaking of the tides.

Definitions. "Rise" of a tide is the height of the high-water level above the low spring datum.

"Range" is the difference between the height of high and

low-water levels of any one tide, without any reference to the datum.

The "semimenstrual inequality of heights" is the difference between the heights of spring and neap tides above mean water-level.

The "diurnal inequality of heights" is, in irregular tides, the difference between the height of high water of each successive tide.

The "age of the tide" is the interval between the time of new or full moon, and the time of the next spring-tide, and varies from $1\frac{1}{2}$ to 3 days.

The "lunitidal interval" is the time that elapses each day, between the transit of the moon over the meridian, and high water.

The "establishment" may be also defined as the lunitidal interval when the time of moon's mer. pass. is $0^h.\ 0^m.$ or $12^h.\ 00^m.$ This is called the "vulgar establishment."

The "mean establishment" is the mean of all the lunitidal intervals in a semilunation, and may differ considerably from the vulgar establishment. The latter is the high-water full and change given in the charts.

The "semimenstrual inequality of time" is the difference between the greatest and smallest lunitidal interval.

The "diurnal inequality of time" is, in irregular tides, the difference between the lunitidal intervals of each successive tide.

The time and height of the tide is ever changing, caused **Cause of varieties in Tides.** by the relative positions of the sun and moon, and these more or less regular variations are further affected by winds, and by the height of the barometer. The difference in level due to the latter may be taken roughly as a foot for every inch of the barometer above or below the mean barometer, high barometer causing lower tides.

The time of the moon's transit over the meridian gives us a **Observations referred to Moon's Transit.** rough measurement of the relative position of sun and moon in right ascension, and it is therefore to this meridian passage of the moon that we refer all calculations of the tides.

M

If the tides are regular, we shall find that on days on which the moon passes the meridian at the same time, the times and heights of high and low water will be the same.

This knowledge is very valuable in many surveys where from local causes we cannot always have a tide pole going, as from previous observation we can, when the tides have been found to be regular, construct a table founded on moon's meridian passage, from which we can take out a reduction for soundings, when working on a small scale.

When we arrive on our surveying ground, then, one of the first things to do will be to set up a tide pole, whatever is going to be the character of our observations.

Position for Tide Pole.

For this we want a sheltered spot, if we can find one, and also firm ground on which to place it, as nothing is more annoying than to find the pole down, especially when out of sight of the ship, when the tide-watchers, unassisted, generally succeed in putting it up again in a different position.

If a pier is available, there is nothing so simple and satisfactory as a plank secured to it, marked in feet and inches, the former being painted red, white, and blue alternately, with bold black figures.

Tide Poles.

If we have no pier, an ordinary spar, shod with an iron spike and painted as above, driven as far into the ground as possible and well stayed to heavy weights, anchors, rocks, or whatever we can get, will stand well, and generally answers our practical purposes. This may sometimes be so placed as to be read from the ship with a glass.

If however there is no shelter and much wash of the sea, and accurate observations are required, we must use a tube of some kind.

A square one of deals can be knocked up on board; but it must not be too small, as we shall want a slit down one side through which an indicator fixed to a rod carried by a float inside may work, and the water washing in by this slit will destroy the value of the tube, unless the area of it be large enough to make the water thus admitted too insignificant in quantity to disturb practically the surface of the water inside. Where there is not much range of tide, the slit can be dispensed with, and the rise and fall marked by an indicator protruding from the top of the tube (which in this case could be a boiler tube), and marking on a scale lashed so as to project above the tube. The water would be admitted by holes bored near the bottom of the tube, if it is to be placed on muddy ground.

A good portable automatic tide-gauge, suitable for all requirements, has not yet been made. A pneumatic gauge is now (1897) under trial, and promises well, but it cannot yet be said whether it will answer. *Automatic Tide-gauge.*

Whenever it can be done, a mark should be made on some fixed object near the tide pole, corresponding to some mark on the pole, which can then be replaced in the same position if it accidentally gets displaced. *Fixed mark for reference.*

Levels should also be carried to some permanent mark in the vicinity, and the difference of level between this mark and the datum given in the chart, with the object of enabling future surveys to be reduced to the same datum level. This is most important. When, as is the case of many civilised countries, there is a fixed plane of reference for land surveys, and level marks are available, the tidal datum should always be connected with such fixed plane.

The level of the water on the tide-gauge should be noted every hour of, if we are going to make a regular series, both night and day, if simply to get a datum for soundings, only of the day, except at springs, when it is as well to get the high and low water at night also, as night tides in some places and at some seasons are lower or higher than the day ones. *Time of Observations.*

It is not amiss in any case, when nothing is known of the tides, to observe for twenty-four hours, at half-hour intervals, as a commencement, as this will tell us whether the tides are regular or not, and we can take observations accordingly.

To get the time and height of high and low water accurately, observe every ten minutes, for half an hour or so, *High and Low water*

before and after high and low water, and calculate from these
records the exact time and height required.

This is best done by projecting graphically thus :—Divide
a line into equal parts to represent hours and minutes, and
from this, at the corresponding time, set off at right angles
distances, on any chosen scale, to represent the height of tide
registered at that time. These spots, joined by a curve, will
enable the time and height of high or low water to be arrived
at much nearer than by simple observation.

Thus, suppose we have noted—

					ft.	in.
X. 00 A.M.	..	..	..	..	12	3
10	..	..	..	..	12	8
20	..	..	..	..	13	0
30	..	..	..	..	13	1
40	..	..	..	..	13	1
50	..	..	..	..	12	10
..XI. 00	..	..	..	..	12	5

FIG 31.

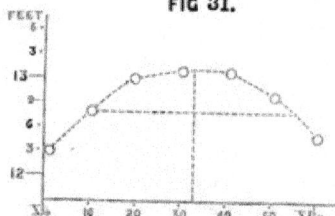

We project these as in accompanying Fig. 31, and by
drawing a horizontal line from the X. 10 position to the
opposite side of the curve, bisecting it, and letting fall a
perpendicular to the line of time, we find X. 32 as the time
of high water. The compasses, measuring the highest point
of the **curve, gives a** little **over 13 feet 1 inch as the** height
marked on the pole.

If we are at the place during the spring-tides, we can get a
fair low-water datum by observation, and all soundings will
be reduced to that, by the height marked on the pole above
this datum, at the time the soundings were taken each day

being subtracted from them. But it may happen that we arrive at the place a few days after a spring-tide, and leave again before the next one. The only thing to do is to note the high-water mark on the shore, and ascertain by measurement how far it is above the high tide of the day as marked also on the shore, subtract the same quantity from the low-water mark on the pole of that day, and call that the low-water spring datum, subtracting perhaps a foot or two extra, to be on the safe side.

Thus, suppose at high water our pole marks 13 feet 1 inch, and the high-water mark on the beach is 2 feet 6 inches above the level of the sea at that time; at low water the pole marks 5 feet 8 inches. This will give us 3 feet 2 inches as the probable low-water spring mark. If we reduce our soundings 2 feet below this to the 1-foot mark, we shall be pretty certain not to give too much water on shoal spots.

An approximation of this kind would be of course noted . on the chart when sent home, and also the manner in which the rise of spring-tides, which would be given as 14 feet, has been obtained.

In still rougher work, an approximation of the rise of the tide may be got by having a marked boat-hook held upright at the water-line at time of low water; the observer then places his eye at the high-water mark on the beach, and reads the mark on the boat-hook, where the horizon line cuts the latter, which will be the fall of the tide that day below high-water mark. If it is the high-water mark of the day that is so used, the result is the range of the tide for the day; and if the distance that the springs' mark is above the day high-tide mark can be measured, we can arrive at the full rise and fall, as in the last article. *Rougher approximation of Datum.*

This may be very useful in making a hurried plan of a bay, and thus the height of the water can be got by the officer putting in the coast-line from time to time during the day, without delaying him much, and to the great advantage of the correctness of the soundings being taken at the time.

The " vulgar establishment " is an exceedingly loose term, as given on the charts. As it is strictly only on days when the moon's mer. pass. is 12^h. or 0^h., that it can be directly observed, the surveyor is obliged to approximate to it in most cases. This perhaps matters the less from the fact that the establishment, even when correctly obtained, is seldom invariable.

The best way to approximate is to project the line of lunitidal intervals, and measure the length of the abscissæ from XII^h. and 0^h. for the vulgar establishment, meaning them if we get more than one.

FIG 32.

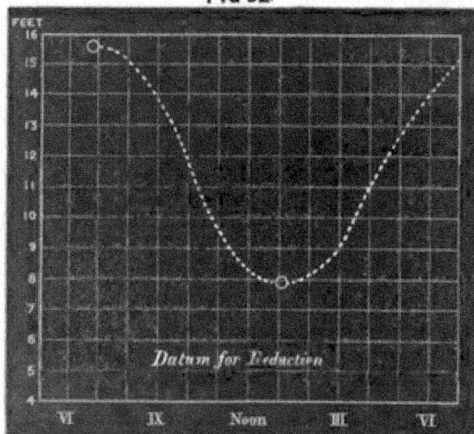

If the tides are regular, especially as regards the semimenstrual inequality, the establishment may be roughly determined by a method given in the article in the Tide Tables, from an observation of the tide at any period of the moon's transit, but which we shall not further discuss, as, in a case where it would be required, we should not know whether the tides are regular or not, and any assumption would probably end in very erroneous results.

In a case where only the high and low waters on any day are obtained, the height of the tide, at times between, is best to be got at by drawing a curve in the imagined course of the tide, after the manner of Fig. 32; the height on the pole at any time can then be taken from this, and a table of reduction formed for the day.

Thus, in Fig. 32, where we have only got the times and heights at high and low water on that day, viz. H. W. at VI. 50, mark on pole 15 feet 7 inches; L. W. at I. 10, mark on pole 8 feet.

On a piece of paper, either ruled in squares for the purpose, or on ready-printed squared paper, which is very useful to have by one for these like occasions, we draw a curve after the fashion shown. Then, supposing our datum for reduction to have been settled as 4 feet on the pole, our table of reduction for the day will stand as follows:—

VI	11 ft. or 2 fms.		I	4 ft. or ⅔ fms.
VII	11½ „ or 2 „		II	4½ „ or ¾ „
VIII	11 „ or 2 „		III	5½ „ or 1 „
IX	9¾ „ or 1⅝ „		IV	7 „ or 1¼ „
X	8 „ or 1⅓ „		V	8½ „ or 1½ „
XI	6 „ or 1 „		VI	10½ „ or 1¾ „
Noon	4½ „ or ¾ „		VII	11½ „ or 2 „

This will of course be, as all attempts at arriving at anything with insufficient data, only an approximation, but will probably be near enough for the purposes we want. If we intend great accuracy, we shall make arrangements to have the tide carefully observed throughout the day, whenever sounding is going on, and take every precaution not to be reduced to these straits.

Table of Reduction under certain circumstances.

When tides are found to be regular, a table of reduction may be formed from the observations during one or more complete lunations, by tabulating the tides according to the time of moon's upper transit. Such a table may be very useful when the *scale of the chart is small*; and sounding can be carried on under these circumstances, viz. regular tides, and scale of chart small, when no direct observations can be got.

In many places external circumstances control our wishes. For instance, it was found on the East Coast of Africa that if men were landed to make a regular series of day and night observations on the tides, fever generally ensued, and consequently the record was restricted to day tides. Again, the tide pole may have to be so placed on a shelving shore, or among reefs, that a boat would have to be used to go out at high water to read it, and this may not be convenient. Often, for considerable tracts among reefs, observations may be impossible, and a table deduced, as above suggested, from former observations may be then used.

Graphic Projection of Tidal Movement. In forming such a table, it is best to project the tidal curves as shown in Fig. 32, but by the hourly observations. It will then be seen whether the tides are regular, and days of similar time of upper transit of the moon can be compared, to see whether a table will give us the reduction near enough for practical purposes.

Surveying ships are now supplied with abstract forms for the projection of high and low waters in such a manner that the regularity or otherwise of a tide can at once be seen.

They provide for the record of the lunitidal interval, the moon's mer. passage : the declination of the sun and moon, apogee and perigee, and the mean time of the high water following the superior transit, and of the highest tide in the 24 hours.

A horizontal line in the upper part of the diagram is divided into hours with vertical lines drawn through them. These hours are those of the moon's superior transit. Directly under the position on the line of the local mean time of the moon's superior transit is plotted, as a dot, the high water next following that transit, at its proper position for height of the tide, as measured by the side scale in feet. When the next day's similar high water has been similarly plotted, the intermediate high water is plotted half-way between them at its proper height and the two low waters similarly interpolated.

The high waters following the superior transit are joined by a red line, and the intermediate high waters by a blue line.

Similarly with the low waters, taking care to note which are the low waters next following the superior transit. Each dot is joined to its next successive high or low water by a black line.

The lunitidal interval for each high water is plotted in the place provided, and the dots joined. By adding the moon's superior transit to the lunitidal interval as plotted, the mean time of each high water can, if necessary, be ascertained.

When the curves of the sun's and moon's declination, and of the moon's parallax are plotted, the general movement of the tide, and its relation to the positions of the sun and moon in declination can at once be seen.

The mean between each high and low-water height will give roughly the mean water-level.

To obtain the true mean water-level in a few days, other True Mean observations must be taken, and we subjoin an extract from Water- the "Instructions to H.M.S. *Challenger*," which contains full Level. directions, only adding that these observations must be made when the tide is moving normally, that is, when there are no strong winds to raise or depress the water-level.

" A good determination of the mean level-sea by the simple operation of taking means may be made, in less than two days, with even a moderate number of observations *properly distributed so as to subdivide both solar and lunar days into not less than three equal parts.* Suppose, for example, we choose 8-hour intervals, both solar and lunar. Take a lunar day at 24 hours 48 minutes solar time, which is near enough, and is convenient for division; and choosing any convenient hour for commencement, let the height of the water be observed at the following times, reckoned from the commencement:—

h. m.	h. m.	h. m.
0 0	8 0	16 0
8 16	16 16	24 16
16 32	24 32	32 32

"The observations may be regarded as forming three groups of three each, the member of each group being separated by

8 hours solar or lunar, while one group is separated from the next by 8 hours lunar or solar. In the mean of the 9 results the lunar and solar semidiurnal and diurnal inequalities are all four eliminated.

"Nine is the smallest number of observations which can form a complete series. If the solar day be divided into m and the lunar into n equal parts, where m and n must both be greater than 2, there will be mn observations in the series; and if either m or n be a multiple of 3, or of a larger number, the whole series may be divided into two or more series having no observation in common, and each complete in itself. The accuracy of the method can thus be tested, by comparing the means obtained from the separate sub-series of which the whole is made up.

"Should the ship's stay not permit of the employment of the above method, a very fair determination may be made in less than a day, by taking the mean of n observations taken at intervals of the nth part of a lunar day, n being greater than 2. Thus if $n = 3$, these observations require a total interval of time amounting to only 16 hours 32 minutes. The theoretical error of this method is very small, and the result thus obtained is decidedly to be preferred to the mere mean of the heights at high and low water.

"The mean level thus determined is subject to meteorological influences, and it would be desirable, should there be an opportunity, to redetermine it at the same place at a different time of year. Should a regular series of observations for a fortnight be instituted, it would be superfluous to make an independent determination of the mean sea-level by either of the above methods at the same time."

Mean Level as Datum. In some cases the mean level of the water may be made use of as a temporary datum for reducing the soundings.

If, for instance, we commence soundings in a place where we do not yet know the spring's range, but intend to get it accurately after some months' observations, we may find it convenient to reduce all soundings to the mean level, as found by meaning each day's high and low water. Then, when we have ascertained the level of low springs below this mean level, one uniform quantity will have to be subtracted from every sounding, which will save a good deal of complication and waiting, as the soundings may all be plotted without fear of mistakes in reducing them afterwards.

This mode will be mostly used for shallow channels, where a difference of a foot or two is an important matter, but it is liable to the error caused by variation in the height of the mean tide-level.

The direction and rate of the tidal streams and other currents must be observed.

This is best done under ordinary circumstances from the ship at anchor by means of a current log, which is simply a very large log-ship, and is worked in the ordinary manner, but with a longer interval of time. The line, which is small, is marked at every 10 feet, and is permitted to run out for an even number of minutes, varying according to the velocity of the current.

Then the rate per hour of the current = number of feet run out, divided by one hundred times the number of minutes.

Thus, if the log-ship is permitted to run for three minutes, and 220 feet of line pass out,—

Rate per hour $= \dfrac{220}{300} = 0 \cdot 73$ knots.

This current log should be hove at stated times, whenever the ship is on her surveying ground, and at anchor, and an entry made in the current log whether there is anything recorded or not, as negative results are in some ways as valuable as positive ones. Where the tidal range is great, and streams change their direction, these observations will be made at comparatively short intervals, in order to ascertain the movement of the water at different times of the tide. Where streams are strong, and of importance in navigation, assistants will be sent to heave the current log from a boat at anchor in different positions.

The current log can be kept by quartermasters, with supervision. A watch or clock with a seconds-hand is a requisite.

Tidal Streams and Surface Currents.

Current Log.

In the Current Book will be entered the position, time, direction of drift of the log-ship, number of minutes it was allowed to run out, and number of feet of line run out, wind and force. Blank columns for rate per hour, and time of tide, will be filled up afterwards by the officer discussing the currents.

Time of change of direction of Tidal Streams. The direction of the tidal stream will frequently change after high or low water, and when this occurs, we must endeavour to find out whether the change of stream occurs at a regular time of the tide, as this is an important point in the navigation of channels.

In channels connecting two open areas of sea, the general law is that the stream will run for some hours, often for three hours, after the tide, as indicated by the rise or fall on the shore, has turned. This makes it very confusing to speak of a stream as the flood or ebb stream, and the term east-going or south-going, or whatever the main direction of the stream may be, should always be used in preference; for in such a case the direction of a stream may be the same for the last three hours of the flood, and the first three hours of the ebb.

Theory of the Tides. The Theory of the Tides is one of the most complicated subjects that can be considered.

Recent investigations by Sir W. Thomson and Professor G. H. Darwin have shown that the tidal movement may be considered to be the resultant of as many as thirty-three separate tidal waves. Some are dependent upon the moon, others on the sun. Some occur once in the day, or are diurnal; others twice, or semi-diurnal; some have a period of a lunar month dependent on the moon's position in her orbit as regards the sun; others have six monthly periods dependent on the sun's declination. The moon's declination, a variable quantity with a long period of years, controls others. The position of the moon's node, and her varying distance from the earth, are responsible for considerable waves, the Perigee tide being always larger than one in Apogee.

The system of harmonic analysis has been adopted for

the clearing of these different waves, and is, when tides are very variable, the only method by which the calculation and forecasting of tides is possible.

It is not proposed to enter into this, which forms no part of the necessary work of a marine surveyor in the field, and is a subject in itself; but it may be mentioned that while in a few regions the time-honoured practice of calculating the time of the tide from the moon's meridian passage, and its height from the mean of tides observed in connection with that meridian passage may serve for practical purposes; in most parts of the world the movement is so complicated that for a satisfactory forecast the employment of harmonic analysis is necessary.

From what has been said it is evident that observation of the tide for a short period will, when tides are complicated, afford no means of predicting them. The movement may be wholly different when the sun is north of the equator and when it is south, and many other factors make it necessary to observe for at least a year to gather an idea of the behaviour of the tides at all seasons. General Remarks on Tides.

The variety in complicated tides is infinite. In some cases the water will rise to nearly the same level every tide, while the low water shows great differences; in others it is *vice versa*. In some cases the diurnal inequality, the difference in height of each successive tide, will be equally distributed between both high and low water levels; in others it affects only one, or may vary with the season.

One feature of a tide when there is much inequality is generally regular, *i.e.*, the succession of the inequality in height. At some places the higher high water is followed by a fall to the lower low water; the tide then rises to the lower high water, then falls to the higher low water, and finally completes its round by a rise to the higher high water again. At others this succession is reversed, but for most places the succession, whichever way it takes place, is the same throughout the year. Fig. 33 will show the two movements.

When there is great but regular inequality, the higher tide

will always be in the day time when the sun is on one side
of the equator, and in the night time when it is on the other.

When the diurnal tides are great, the inequality in height
will sometimes be such as to cause a mere stand in the tide
during either the rise or fall of one tide, giving the effect of
only one high and low water in the twenty-four hours.

As a general rule, it may be stated that in temperate
latitudes the highest tides take place at the equinoxes,
whereas in the tropics these occur at the solstices.

Curiously enough, and it has affected many of our notions
about tides, the tides about the British Isles are the most
simple that are anywhere found, that is to say, so far as the
individual movement of the tide at any one place is con-

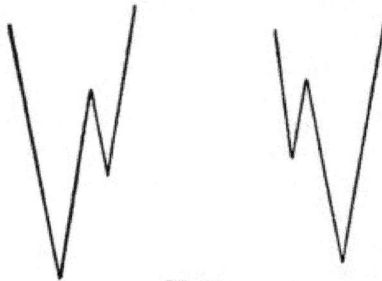

FIG. 33.

cerned. But they are largely affected by a further complication, known as "Interference." By this is meant the
appearance of another tidal wave, or perhaps more than one,
which affects the height and time of the resultant tide at any
place.

This is caused by either a tidal wave coming round from
an opposite direction, or by one reflected from another coast,
and it will be at once seen that if the crests of such tidal
waves coincide at any point with the crest of the primary
wave, the resultant tide will be higher, and if the crest of
one reaches a point at the same time as the hollow of another
the range of the tide may be, if the waves are of equal height,
nothing.

To this is to be attributed the variety of the height of the tide at different parts of a coast that appears fairly open.

Thus, in the English Channel, the height of the tide varies at different places on the same day from 5 feet to 25 feet; and the same phenomenon occurs in many parts of the world.

The tidal streams are not, however, directly affected in the same way, and there are many places where the rise of the tide is insignificant, but the tidal streams are very strong.

This is because the horizontal flow of the water is not determinable by the rise at the spot, but by the rise at other places, possibly at some distance, and by the fact that water once set in motion is not easily arrested.

The variations caused by Interference are wholly distinct from the differences in the height of tide and velocity of streams caused by the conformation of the land.

The vertical movement of the water in the deep open ocean is not great, probably not more than two feet, and the horizontal motion is practically nil, being a merely insignificant oscillation. It is only when the water shoals that the friction of the bottom, and the constriction caused by the water, which is in motion throughout its depth, being forced into a shorter column, cause the wave to become unnaturally heightened; and horizontal movements, which we know as tidal streams, are set up, by reason of water flowing from the higher to the lower level.

Thus, on banks in mid-ocean regular tidal streams are found, and could the height of the water be measured, it would be found to vary more than in the deep water around.

The opposition of a coast, and the shape of deep bays, gulfs and channels, accentuate these effects, and the height of the tide and velocity of the tidal streams varies in different places exceedingly. The oceanic tidal wave, thousands of miles in length, may have the distance between its crests shortened to hundreds, and the width of the portion that approaches the shore continuously narrowed as it passes up funnel-shaped passages.

Enough has been said to show that the tidal movements are exceedingly complicated, and though long-continued observation at one spot may enable predictions for that spot to be made, the variations are sometimes so great that at a short distance the phenomena are entirely different.

All prediction may be upset by what is known as the meteorological tide, that is the variation in the height caused by winds, and by the difference in pressure of the air on the surface of the water.

The inconsistencies in the tide thus caused affect the height of the water, more than the time of high or low water, and as they affect the mean level of the water, an unusually high tide induced by them does not mean an unusually low tide, but the reverse.

Wind will naturally have more effect when there is a funnel-shaped estuary than when it blows on to a straight, open coast, the heaped-up water at the wide mouth being forced higher and higher as it advances, for under such circumstances water will run up hill.

Mean Level. From the foregoing it is evident that the mean level of the water will considerably vary.

When steady winds blow at certain times of the year, the variation in mean level will be seasonal; in other places it may be constantly varying with the direction of the wind.

This variation in the mean level is important as regards navigation in some places. For instance, when a shallow flat exists which must be crossed to gain access to a harbour, and which at ordinary high water affords just sufficient depth, a change in the mean level may cause the high-water level to be one, two, or even three feet less than usual.

It is, therefore, most necessary for surveyors to acquaint themselves with the effect of wind at such places, and to record it.

Bore. In cases where the rise of tide is great, in a funnel-shaped estuary much encumbered by sandbanks, and where there is a continuous outflow caused by a large river, the phenomena of the "bore" appears.

This consists in the face of the rising tidal wave becoming so steep that it rushes up the estuary in the shape of a sudden wave, sometimes almost a wall of water, breaking as it advances, and the tide thus rises many feet in a few seconds, followed by a still rapid, but more gradual, further rise, so that the whole flood may only last one or two hours, the ebb prevailing for nine or ten hours.

The main factor in the production of a bore is the retardation of the lower part of the inflowing water by the friction of the bottom in shallow water, and by the action of the down-flowing river, so that a high tide rises quicker than it can flow forward, until its height and momentum enables it to overcome these obstacles by a final grand rush.

Bores are rare, but whenever encountered the surveyor should investigate the conditions, as but little is known of the details of most of them.

N

CHAPTER X.

TOPOGRAPHY.

THE sketching in of the topography, or detail of the land, is a point on which there is more variation, as to the manner in which it is done, than in any other of the steps of a survey.

It is the least necessary part of a chart, which is destined mainly to guide over the water and not on the land; but as we are guided over the water *by* the land, a perfect chart should have the features of the country correctly delineated, so as to assist the mariner in recognising the land by the mutual positions of peaks and other conspicuous objects. Furthermore, with our universal presence and interest all over the globe, it is impossible to say that an expedition may not want to start from some point on our chart, when information for a short distance inland will, in such a case, be most useful.

Width of Topography from Coast.

As a general rule, the land should be put in as far back from the shore as it is visible from the sea; but this is only a very general guide, and must depend upon the distance of the back ranges, and the size of our sheet of paper. When the most distant mountains are very far back, we cannot spare time to do more than fix their summits by angles, get their heights and the extent of the range, and the country between must be a perfect blank.

Rough Topography.

Often, in savage lands, the country will be too dense with jungle to be able to do much to the topography by walking over it, which is of course the only way to get it correctly mapped, and we must then be content to sketch what we can see from the sea, and from the coast. By making stations in the ship, drawing a sketch at each, and getting angles to all prominent parts, such as spurs of hills, valleys, ravines, smaller peaks, &c., which will be entered on the sketch, a very fair approximation of the position and shape of the more conspicuous elevations in the land, visible from seaward, will be made. The officer coast-lining will have got the entrances of all little streams fixed, and from the ship off shore we can recognise which ravines, or at any rate which of the larger ones, join on to these entrances. Topography put in in this way will present a somewhat detached appearance, and we can only fill up the hiatus by writing on the chart the general appearance of the land intervening between the hills, as far as we can see it from aloft, as, "rolling grassy plain," "densely wooded and undulating," &c. Sometimes, on a coast of this description, we can get back from time to time to an elevation we see from the ship to be partially clear, and a sketch from a position of that kind will materially improve our knowledge of the topography.

By referring to the sketch at page 84 it will be seen how, with similar views from different points, ravines and valleys may be cut in, and roughly drawn on the chart.

When, however, we can spare the time to perfect our chart, and the nature of the country permits it, we should walk all over it, and sketch the topography on the ground. *Regular Topography.* To do this, we must have as many conspicuous objects as possible fixed beforehand, and pricked on to a board, as for sounding or coast-line. Topography can be plotted afterwards, the same as can be done with coast-line or any other work, but it will be much more satisfactorily done if plotted at the time.

We then walk over our country, fixing ourselves with angles on commanding spots, plotting the stations by the station pointer or tracing-paper, and drawing lines from them to all things we want to plot, spurs of hills, houses, valleys, &c., &c., and sketching the details immediately

N 2

around us. To fix details for this purpose we shall often
have to content ourselves with two angles only, and as long as
we do not use such points to carry on our stations with, this
will be sufficient. A good deal of judgment is necessary in
selecting spots to make stations, which cannot come without
experience.

In placing the details on the paper on the rough board,
sketch in the line of a valley first by the stream at the
bottom, and then the adjacent hills or spurs.

A ten-foot or longer pole may be used with advantage in
sketching topography.

Difference of level can be at once obtained from a theodolite angle of elevation or depression by the formula :—

$$\text{Diff. of level in feet} = \frac{\text{angle in secs.} \times \text{dist. in miles.}}{34}$$

Contour-
ing.

Hills are best shown by contours. We do not of course
pretend that our contours are a fixed distance apart, but we
must endeavour to draw them approximately so, calling each
contour line, 25, 50, or 100 feet apart, as the scale may
require, and estimating the height of each spur, with the
assistance of a pocket barometer, if we have one, which will
give us roughly the height of each station above the sea, if
we read it when we land, and whenever we have occasion to
do so. Each contour must be continued on from one hill to
the other, or until it meets itself again round the hill; and
as their number and closeness together will roughly indicate
the height of the hill, we must be careful not to get more on
one side of a hill than another, or the value of this method
will be lost, and the contours will simply show the shape of
each spur, without reference to its relation in height, steepness, &c., to the next one, which is what we want to show as
well. These contours will perhaps not appear in the finished
chart, in which the mountains may be delineated in a different
manner, but they will form an excellent guide for the amount
of shade to be put on to the different hills and slopes, and it
is the readiest and quickest method of showing this at the
time.

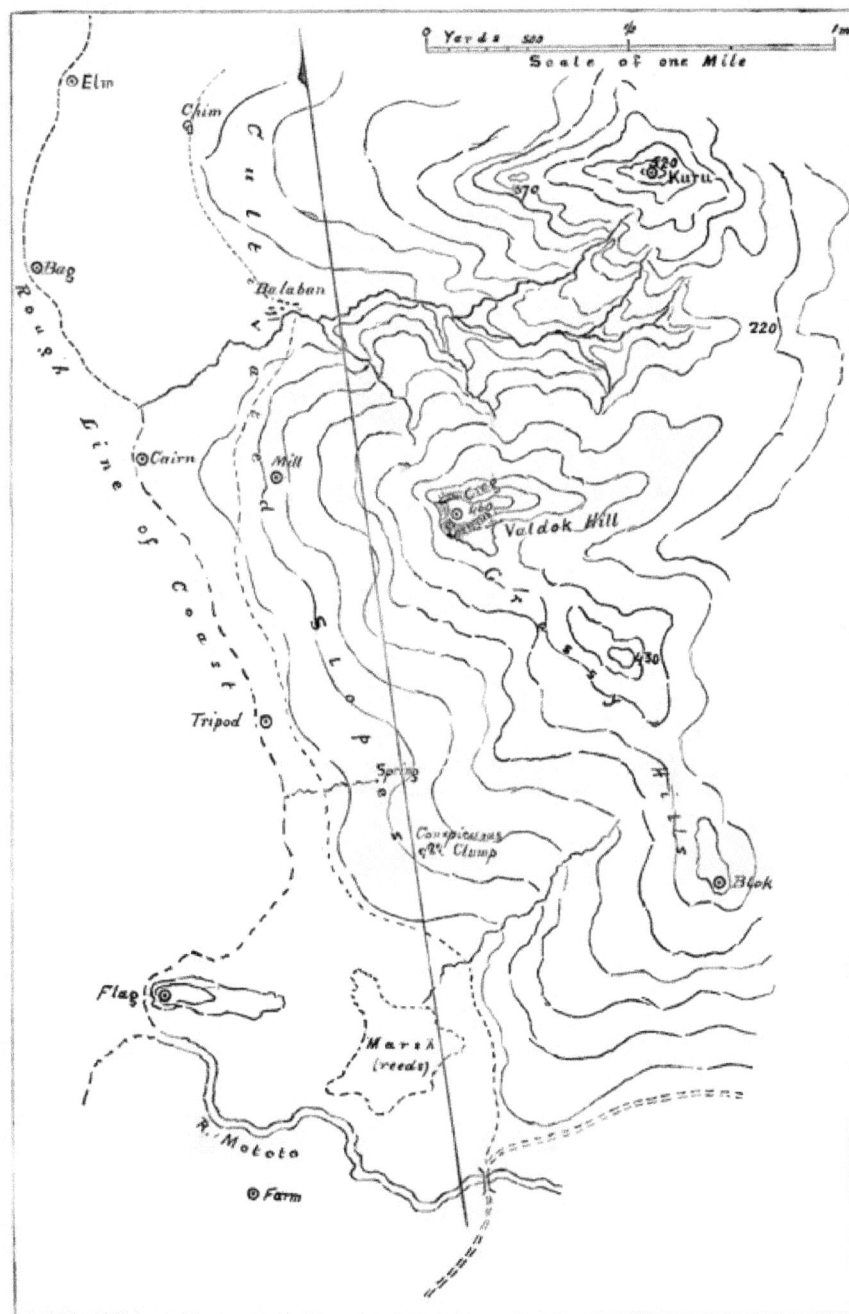

FIG. 34.

Red and blue Pencils.   Red and blue pencils are useful for topography. With the blue we show streams, and the red is used for marking roads. With only a black-lead pencil, the markings of these details are apt to get confused with the contour-lines to express the hills.

Pocket Sextant and Compass.   Much topography can be done with the pocket sextant and compass only, the latter being only used, however, when three objects to fix by cannot be got. The magnetic meridian, or several magnetic meridians, must be ruled on to the rough board, to permit the use of bearings. When the only objects available are much above us or below us, correct angles cannot be got with the sextant; and though we allow ourselves a certain amount of latitude in our angles for the purpose of topography, it will often be necessary to take a small theodolite for the purpose. A pocket sextant can be taken as well, and the theodolite, which requires more time to set up and arrange, only be used when the sextant angles will be too erroneous.

If we have a theodolite, we must take advantage of good opportunities to get a series of elevations and depressions for heights.

Difficulties with Sextant Angles.   In taking angles with a sextant to objects on different levels, try to find some natural mark which is exactly above or below, as the case may be, the object the farthest from your level, and nearly on a level with the other object, and take this instead of the object itself. But it must be noted that unless this second object is nearly at the level of the observer, the angle will still be incorrect.

Fig. 34 shows a rough field board, before any shading is placed on the sides of the valleys, which will be done with the brush before the work is considered completed.

CHAPTER XI.

HEIGHTS.

By Theodolite—By Sextant—Obtaining Distance from Elevation of a
known Height—Levelling.

FOR obtaining heights, we must mainly depend on angles of **Means used.**
elevation with sextant from afloat, and of elevation and
depression with theodolite from shore stations. The pocket
aneroid, though useful, as described under "Topography," to
get subsidiary heights, and assist in delineation of hills, is
not to be depended upon.

At all main stations, and, in fact, any station well fixed **Stations for obtaining Heights.**
and conveniently placed, angles of elevation and depression
to the objects whose heights we want, should be taken
throughout the course of the work. These are entered into
the "Height Book," and worked out when we can get the
distances, and occasion offers, the results being tabulated
and meaned.

Elevations and depressions can be taken from any station
whose height we shall eventually know; but it is evident
that any slight error in the true height of the observing
station will be carried on into all heights deduced from it,
and therefore it is well to get as many observations as we
can from stations at the water-level, or so placed that the
height above the water-level can be measured with a line.

In observing elevations and depressions with a theodolite, **Use of Theodolite.**
the instrument must be in fair adjustment, and carefully
levelled, and it is further necessary to take into account the
errors of level and collimation.

There are two ways of doing this. One is to take a series of observations with the telescope in its ordinary position, and then another with the telescope reversed, end for end, in the Y's, when the mean of these two observations for each object will be the correct amount of elevation or depression. This is the best way, and eliminates all error. It may, however, be sometimes convenient to proceed as follows :—

Ascertain the collimation error by directing the telescope on to an object in elevation, reading the vernier, then turning the telescope round until the level is uppermost, and again adjusting for the object and reading the vernier again. Half the difference between the readings is the collimation error, which, when the reading taken with the level uppermost is greatest, will be *added* to observations of elevations made with the telescope in its normal position, and *subtracted* from depressions. This collimation error is permanent for all positions of the horizontal arc.

For level error, at each observation of each separate object, the telescope must be brought horizontal by the level attached to it, and the vernier of the vertical arc read. Whatever it reads will be the level error.

The sign of the correction to be applied for this error is, for elevation, $+$, when the 0° of the arc is above the zero of the vernier when the tube is level, and $-$, when below. For depressions the signs will be reversed. Care must be taken that no mistakes are made as to these signs. For a tyro it is slightly confusing.

Both level and collimation error must be applied to each observation.

Sextant Elevations.

When the ship can be well fixed, sextant angles of elevation from her with a sea-horizon will be very good, as good, in fact, as elevations with a small theodolite, as they are free from all possible errors of levelling, &c., and a sextant measures angles to 10 seconds, whereas a small theodolite is only cut to minutes. Even when the ship is within the limits of the sea-horizon, the results will be good, providing the distance of the shore line is well known, and is not under half-a-mile. By observing from the lowest step of the accommodation-ladder, we can use a shore horizon at even less distances.

Sextant elevations, then, are very useful, but we do not generally get so many opportunities of obtaining series of heights by it, and when at any distance from the land, only the skyline of hills will be clearly seen, so that it is principally to the theodolite that we must look to give us a *sufficiency* of elevations.

Before dealing with the method of calculation of heights, we must refer to the effects of refraction. **Refraction.**

The apparent position of one object from another, as seen through our atmosphere, appears higher, whether we look up or down. The amount varies with the difference of densities of the various strata of the air, which are constantly changing.

All we can do is to take the mean refraction, and it has been found by experiment, that by taking $\frac{1}{12}$ of the distance, regarded as minutes and seconds of arc, and applying this to the observed angle of elevation, it will give us a fair mean result for the true angle of elevation, when this is small, as in all practical cases it is. It follows from this unknown amount of error in the coefficient of refraction that, when possible, objects should not be observed for elevation or depression at more than a few miles distance. We cannot always command the maintenance of this limit, any more than we can many other theoretical points in practical hydrographical work, but when circumstances are favourable, they must be regarded.

Looking upwards, or from a denser into a rarer medium, the effect of refraction is to increase the apparent elevation. This correction is therefore to be *subtracted* from elevations. As the effect, when looking downwards, is also to raise the object, or, in other words, to decrease the angle of depression, the correction for refraction must be *added* to angles of depression.

The angle of elevation measured by a theodolite, or the

Result of
Spherical
Form of
the Earth.
sextant angle when corrected for height of eye above the sea,
is the angle between the tangent to the earth's surface at the
observer's position, and the line drawn from him to the
object.   If the surface of the earth was a plane, all that
would be necessary to obtain the height would be to work
out in a right-angle triangle, Perp. = base × Tan angle of
elevation, after the latter had been first corrected for the
effects of refraction; but as the earth is a sphere, the tangent
to it, produced, will cut the line representing the height we
want, not at the point where it leaves the earth, but somewhere above that, depending upon the distance; the perpen-

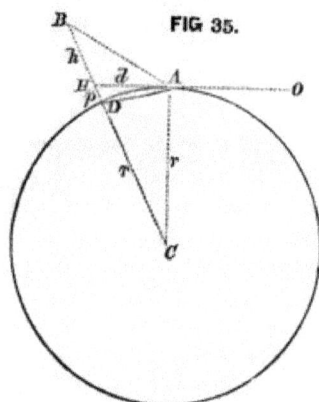

FIG. 35.

dicular, therefore, as worked out, will only give us a portion
of the height required, the other portion being that below
the tangent.

Explana-
tion of
"Dip."
Thus, in Fig. 35, A is the position of the observer, A H
the tangent to the earth's surface at his position, B a
mountain peak whose height, B D, we want to obtain.   The
angle of elevation measured by a theodolite is B A H, and
it is evident that the height we shall obtain by working out
the triangle will be B H, leaving H D to be found independently.   It will be seen that we are going to treat the
angle B H A as if it was a right angle, when it is evidently

more than 90° by the angle D C A at the centre of the earth; but our figure is much exaggerated to show things clearly, and in practice the distances we use to get elevations are so insignificant, comparatively to the diameter of the earth, and consequently the angle D C A so small, that we can neglect this quantity without introducing any error in the result. With a distance of 60 miles, when the angle is a degree, the discrepancy introduced into a height of 6000 feet is only 2 feet.

We require, then, to get H D to add on to B H in order to get the full height B D. This quantity, H D, is called "dip," an awkward nomenclature, as it is the same used at sea to express the angular quantity we apply to elevations taken with a sextant from a height, to reduce them to the tangent to the earth, whereas here it is used to express a linear quantity.

The problem can be solved in two ways. Either by finding H, D independently, or by adding the angle H A D to H A B to get the angle B A D, when a right-angled triangle gives us B D.

The latter method is the shorter and is now employed, and the former is therefore not described, but a table giving H D, the Dip, or the height of the part of the object observed obscured by the horizon, is given in Appendix Table O, as it may be sometimes useful to know how much of a mountain is below the horizon.

In the method now used, the angle H A D is found as follows :—

Angle H A B is the elevation, corrected for refraction, and the angle H A D (between the chord and tangent) is equal to half the angle at the centre, *i.e.*, half the distance in arc.

Suppose A B to be 60 miles,

$$\text{Then H A D} = + 30'$$
Correction for refraction ($\frac{1}{12}$ of the distance $= - 5'$
$-$ to elevation, $+$ to depression).

$$\text{Whole correction} = + 25'$$

No.	Place of observation	Object observed	Angle of Elev.	Angle of Dep.	Corr^n $\frac{1000}{6}$	Corrected Angle	Dist.	Diff. of height	Height of		Corrected Height	
									Object observed	Theod.	Place of observer.	Object observed.
1	Cork △ ..	Snow Hill △ ..	1·17·22	..	4·10	1·21·32	10·0	1441	..	15	..	1456
2	Ship Pos VI.	Fair Peak ..	1·14·20	..	3·1	1·17·21	7·24	990	..	0	..	990
3	Sok △ ..	Snow Hill △ ..	0·56·30	..	2·14	0·58·44	5·37	557	1463	5	901	..
4	„ „ ..	Crag △		1·01·30	2·37	0·58·53	6·28	653	257	5	905	..
5	„ „ ..	Low Hill		1·41·15	1·44	1·39·31	4.17	733	..	908	..	175
6												
7												
8												
9												
10												
11												
12												
13												
14												

Chart of

COMPUTATION.

1	2	3	4
Const. log. — 3·78359 Tangt. 8·37516 Log. dist. 1·00000 3·15874 1441 15 —— 1456	Const. log. — 3·78359 Tangt. 8·35240 Log. dist. 0·85974 2·99673 990	Const. log. — 3·78359 Tangt. 8·23265 Log. dist. 0·72997 2·74821 557·4 6 —— 562 1463 —— 901	Const. log. — 3·78359 Tangt. 8·23371 Log. dist. 0·79796 2·81526 653 257 —— 910 5 —— 905
5 Const. log. — 3·78359 Tangt. 8·46157 Log. dist. 0·62014 2·86530 733 908 —— 175	**6** Const. log. — 3·78359 Tangt. Log. dist. ——	**7** Const. log. — 3·78359 Tangt. Log. dist. ——	**8** Const. log. — 3·78359 Tangt. Log. dist. ——
9 Const. log. — 3·78359 Tangt. Log. dist. ——	**10** Const. log. — 3·78359 Tangt. Log. dist. ——	**11** Const. log. — 3·78359 Tangt. Log. dist. ——	**12** Const. log. — 3·78359 Tangt. Log. dist. ——
13 Const. log. — 3·78359 Tangt. Log. dist. ——	**14** Const. log. — 3·78359 Tangt. Log. dist. ——	**15** Const. log. — 3·78359 Tangt. Log. dist. ——	**16** Const. log. — 3·78359 Tangt. Log. dist. ——

Consequently, as $60' : 25'$

or $1' : 25''$ is a constant proportion for all angles.

Therefore the total correction, for Dip and Refraction, in seconds of arc, to the observed angle of elevation or depression is :—

$$\frac{\text{Distance in sea miles} \times 100}{4.}$$

This correction is to be added to angles of elevation, and subtracted from angles of depression.

Height Form bound into Book. A ruled form is supplied by the Hydrographic Office, which much facilitates the calculation of heights. This form, bound into a book, constitutes the Height Book. A specimen is given on last page, which nearly speaks for itself.

Entering and calculating Elevations. The angle observed to the object is entered under the head of either elevations or depressions, as the case may be; *as observed*, in the case of theodolite ; *minus the correction for height of eye*, if with the sextant ; and the distance in miles and decimals is entered under its head.

To get this distance, if we happen to have it worked out in the triangulation, we shall of course use the calculated distance ; but if not, which will be the case generally, we must measure it on our sheet, and enter the corresponding distance according to the scale.

In the column headed Corrⁿ, enter the correction for Dip and Refraction, obtained as above by multiplying the distance by 100 and dividing by 4.

We then work out the difference of height with these data on the opposite side of the page, the constant log being the log of feet in a mile, by which the distance must be multiplied to bring result out in feet. The log given is that for 6075 feet, the number of feet in a mile in Lat. 44°. Theoretically we should have a different log for different latitudes, but, as the utmost extent of error by neglect of this is 22 feet in a height of 6000 feet, we need not regard it. This difference of height is entered in its proper column.

Tables, computed by Commander Purey Cust, for taking

out the difference of height for any angle and distance, are now supplied. If these are at hand this computation is dispensed with.

The column for height of theodolite is a little confusing, as sometimes it will be merely the height of the theodolite-telescope above the ground, and sometimes the height of it above the sea-level, which we shall enter, according as we want the height of observer's position, or of object observed as will be presently explained.

We have now all the data necessary to obtain heights.

When we have accumulated enough observations, we set about getting out results.

There are four problems for obtaining heights, and the data **Height** we have for each observation will be combined according to **Problems.** what we want to arrive at.

These four problems are as follows :—

    1st. To find height of object observed, when height of observer is known, and the angle is one of *elevation.*

    2nd. Ditto, when angle is one of *depression.*

    3rd. To find height of observer, when height of object observed is known, and the angle is one of *elevation.*

    4th. Ditto, when angle is one of *depression.*

To understand the mode of combining the data, let us consider the Figs. 36 and 37.

In Fig. 36, which is the case where the angle observed is of *elevation,* and comprises Problems 1 and 3, we may have either X or Y known, and wish to obtain the other.

Suppose X to be known (Problem 1), to find Y, we have   **Height**

$$Y = X + (t + h) \quad\ldots\ldots\ldots \quad 1.$$

Formulæ.

If Y is known (Problem 3), to find X,

$$X = Y - (t + h) \quad\ldots\ldots\ldots \quad 3.$$

In Fig. 37, the case where the observed angle is one of *depression,* and comprises Problems 2 and 4,

Suppose X known (Problem 2), to find Y,

$$Y = (X + t) - h \quad\ldots\ldots\ldots \quad 2.$$

Suppose Y known (Problem 4), to find X,
$$X = Y + h - t \ldots \ldots \ldots 4.$$

These four formulæ, which it is also convenient to have written for reference in the Height Book, will enable us to solve any of the problems.

Column Height of Theod. When we are getting the height of the *Object observed*, we shall enter in the column of "Height of theodolite," $X + t$, or the height of theodolite above the sea; but when the observation is used to obtain the height of *Observer*, only t, the height of the theodolite above the ground, will be inserted.

Elevations from Sea-level first Meaned. We must commence by collecting results of elevations

Fig. 36.

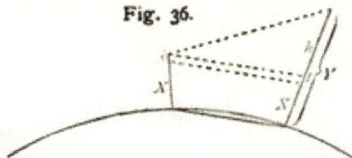

Fig. 37.

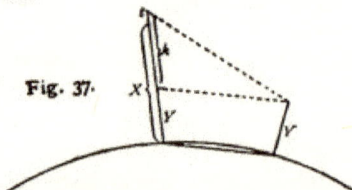

X = Height of observer's position.   h = Difference of height.
Y = Height of observed position.   t = Height of theodolite above ground.

from stations at the sea-level, or from stations whose height above the sea-level has been measured, which will give us the heights of objects observed; and also with depressions from stations to the sea-level, or to stations whose height above sea-level has been measured; which will give us the height of the observing-stations.

Absolute Heights. Heights so obtained are termed "absolute," as being calculated directly from the sea-level.

All such heights must be obtained first, then, meaning the heights of one station which has the most observations, or of which the results agree best, we can work out all other observations from that station to other objects. We then mean another, and so on, using our observations either to obtain height of observer or of observed object, as is most convenient, as we proceed.

These heights will be "dependent," as resting on the ascertained height of other stations. Dependent Heights.

No height can be considered as *exact*, that is not the result of both elevations and depressions, as no matter how nicely a set, of depressions say, comes out, they will all include the refraction error, for the refraction correction is only approximate. This is with reference to detailed surveys only.

Sextant angles of elevation must be corrected for the height of eye before being entered in the Height Book as angle observed. They are then treated in precisely the same manner as the theodolite elevations. Sextant Elevations.

The pocket aneroids should be tested up a known height, to get the value of each tenth, which will be from 92 to 100 feet for a tenth, each instrument varying slightly. As before mentioned, they are useless in getting accurate heights, but will give very good approximations up to about 4000 feet, if in good order and constantly worked; but their delicate chain-work is so liable to rust slightly at sea, that the links will frequently stick if the instrument is not carried up heights continually to work it. Placing under an air-pump will serve the same purpose. See Barometer, page 35. Aneroids.

It is useless to enter into intricate calculations of data obtained by so small a scaled instrument as a pocket aneroid; the impossibility of reading it exactly precludes any but approximate results, and a simple multiplication of the decimals of inches by the value of a tenth, as obtained above, is quite sufficient for the purposes for which we use the instrument, the differences being taken from the barometer as observed on board. Only for approximate Heights.

To obtain distance from an angle of elevation of a known height is like using a lever with the ends reversed, and is seldom had recourse to in surveying, as not being correct enough. Obtaining Distance from elevation of known Height

As it may be, however, sometimes useful, we give a formula.

$$\text{Distance in nautical miles} = \frac{34\,h}{E}$$

When h is height of mountain in feet

              E is the angle of elevation in seconds, reduced to water-level, and corrected by the addition of the dip, as explained in the rules for obtaining heights.

The same formula in rougher terms is

$$\text{Distance in nautical miles} = \frac{100\,h}{3\,E}$$

If the estimated distance should differ much from that given by the calculation, it should be recalculated with the correct allowance for dip and refraction.

An example is appended.

Given height of mountain observed = 2384 feet.

Elevation	..	..	0° 36′ 26″
Height of eye	..	..	16 ft.
Estimated distance	..		30 miles.

Obs. elevation	0° 36′ 26″	2384	..	3·377306
Height of eye	3 56	34	..	1·531479
	32 30			4·908785
	or 1950″	2700	..	3·431364

Corr. for Dip for 30′   1·477421

$$\text{or } \frac{100 \times 30}{4} \quad .. \quad 750 \qquad \text{Distance} = 30 \text{ miles.}$$

$$E = 2700$$

The rougher formula will give the distance as 29·4 miles.

LEVELLING.

Simple Levelling. Levelling with a staff is not very much required in marine surveying. Ascertaining the height above the sea of the fixed mark used for reference for the tidal datum is the purpose for which it is most used, but it is also required to find the height of the base of a lighthouse, &c.

This is called simple levelling, and gives us the height

between the two required points only, without any regard to distance.

A levelling staff and level is usually supplied to surveying ships; but a theodolite and marked boat-hook or pole will answer the purpose, if we have not got the regular apparatus.

Holding the staff at high-water mark, we place the instrument (level or theodolite) so far up the slope that we shall when it is carefully levelled by the level attached to the telescope, read off near the top of the staff. The reading of this, called the *back station*, being taken, the staff is taken above us, and planted so that we can read just above zero of the staff, which is now at the *fore station*. The theodolite is now moved up the hill until we shall again, when levelled, read near the top of the staff; this will be another observation of *back station*, and so on until the levelled telescope reads the staff on the spot whose height we want.

There is no necessity to keep in one line directly for the spot whose height we wish to measure; we shall do so if we can, as it is the shortest way, but in practice we are generally forced to zigzag.

The difference of the sums of the readings of back and fore stations will be the height required. Where former are the greatest, we are going up hill, and it is called *rise; vice versâ* is called a *fall.*

For HEIGHT OF —— LIGHT-HOUSE.

Back △		Fore △	
Reading of Staff.		Reading of Staff.	
Water-level	12·64	(1)	0·63
(1)	13·42	(2)	1·22
(2)	13·81	(3)	1·52
(3)	12·50	(4)	0·32
(4)	13·06	(5)	0·87
(5)	12·18	Base of L. H.	3·45
	77·61		8·01
	8·01		
	69·60		

Height of base of Light-house above High-water level .. 69·6 ft.

For our purposes, when the distance of the staff from the theodolite is not great, or when the distance of fore and back station from the theodolite are nearly the same, it will be sufficient to observe readings with the telescope in one position only; but when the rise of the hill is slight and distances increase, especially when the difference of distance between fore and back station is great, and we require accuracy, the telescope should be reversed in the Y's, and being again brought level by the bubble, readings should be taken a second time. The mean will be the true reading.

If the axis of the telescope and the attached level are perfectly parallel, and therefore in adjustment, it will be shown by the readings agreeing when reversed at the first station, and we shall know that we need not take this trouble; but it is necessary to ascertain this, as theodolites continually undergoing carriage by boat are liable to many accidents.

This method enables us, if necessary, to calculate the height of any of the stations where the pole is erected, but gives us no information as to the height of the spots where the theodolite stands. This can be obtained, if wished, by measuring the height of the axis of the telescope above the ground, when,

Height of theodolite position = height of back station + present reading of said back station − height of eye (back station being below us).

Distances measured will enable us to make a section of the ground traversed, but, as already remarked, this is not often required from the marine surveyor, and will not be enlarged upon here.

CHAPTER XII.

OBSERVATIONS FOR LATITUDE.

By Circum-meridian Altitudes of Stars—By Circum-meridian
Altitudes of Sun.

ASTRONOMICAL observations are largely used in all descrip- General
Remarks.
tions of marine surveying. In all but small plans the even-
tual scale of the chart is decided by the latitude and longitude
as obtained by observations of sun or stars, and we have seen
that true bearings often enter largely into the construction of
charts. In running surveys, or in searching for, or sounding
over shoals, in mid-ocean, everything depends on the positions
astronomically found, and every method of correctly finding
the latitude and longitude is in requisition. In considering
this subject, we will take first shore observations with
artificial horizons, where we require results as accurate as we
can obtain with the sextant, to which instrument remarks
will be confined, excepting so far as the theodolite is used
for true bearings ; and afterwards, sea observations.

In all observations of the heavenly bodies, instrumental Elimination of
errors, atmospheric effects, and personal differences, largely Errors.
influence the results. No matter how correctly we may take
the actual observations, unless we can eliminate these variable
quantities, the positions obtained will be in error.

On every occasion, therefore, where accuracy is aimed at,
the mode in which this elimination can be best carried out
must be considered. The general principle used in doing
this is to get two sets of observations for one result, in such
a manner that the errors of all kinds will act in opposite

directions in each set, and therefore disappear when the mean is taken. The precise way in which this is done will be described under each different observation.

LATITUDE BY CIRCUM-MERIDIAN ALTITUDES OF STARS.

Latitude by Observations "Absolute."
Determinations of latitude are more simple in one respect than those for longitude, as they are "absolute," that is to say, they depend solely upon themselves; whereas longitude has to be obtained by the difference of two sets of observations at two different places, and is further complicated by the eccentricities of the chronometers upon which, when there is no telegraph, we have to rely.

But more difficult.
But, on the other hand, the observations required for correct latitude are more difficult to take, as, to arrive at anything like an exact result, we must use stars, and each step of the observation of these in an artificial horizon is rendered less easy by the fact of their being made at night. It is much easier to become a good day observer than a good night one.

Elimination of Errors in observations for Latitude.
The errors to be eliminated as far as possible in observing for latitude are, firstly, errors of observation; secondly, instrumental errors, as centering error, index error of sextant, error caused by refraction in the rays passing through the glasses of the roof of the horizon, &c.; thirdly, atmospheric refraction, which varies much, and for which no known rule of correction thoroughly suffices; fourthly, personal errors, caused by each individual's mode of observing the contacts.

Errors of observation are eliminated by taking as many observations of altitude as we can, and we must therefore observe off the meridian, or what are known as circum-meridian altitudes; which consist in observing from a short time before the meridian passage to a short time after it, and adding a certain correction to each altitude to make it equal to the meridian altitude, and thus get a mean meridian altitude, which, if we can calculate the correction exactly, will be of much more value than actual observation on the meridian only.

There remain the other errors, some of which may be directly allowed for, but only approximately; others cannot be corrected at all, and the latitude resulting from observations of a single body, as *e.g.* the sun, will be therefore always more or less in error.

The only way satisfactorily to clear these errors is to observe stars, in pairs, of as equal altitude as can be found, one north, and one south, of the zenith. These errors will then act in opposite directions, as everything tending to increase or diminish the altitude on one side of the zenith, will act similarly on the other; but, in working out the latitude, the resulting error will increase the latitude in one case, and decrease it in the other, so that the mean of the latitudes obtained by each star of such a pair will approximate very closely to the correct one. *Pairs of Stars.*

To eliminate the artificial horizon roof error when observing pairs of stars, the roof must always be in the same position with respect to the observer, and therefore must be reversed when changing from face north to face south, and *vice versâ*. If observing a single object, as the sun, the roof must be reversed when halfway through the observation.

The use of a sextant stand, when once the observer has got thoroughly accustomed to it, is an immense assistance to good observations, as the images of the stars, instead of quavering and shaking with every slight motion of the hand of the observer, remain perfectly still, and can be made to pass over one another with great accuracy. *Sextant Stand.*

Certain preparations are necessary for *good* star observations, for all scurry that can be avoided, should be. *Preparation necessary.*

In the first place stars must be selected and arranged for observations according to their pairs.

If stars given in the 'Nautical Almanac' only, are used, the chances are very much against a sufficient number of pairs being obtainable, as only a small proportion of observable stars are there included, though the number has been lately much increased.

Star Cata-
logues.
A surveying vessel will have the Greenwich and Cape Observatory Catalogues of Stars, and out of these enough pairs can nearly always be picked to enable us to get a satisfactory latitude in one night, including stars down to the fourth magnitude, which can be easily observed on an average night by a practised observer with good instruments.

A special list of stars is supplied to all surveying ships, which includes all stars in the Greenwich and Cape Catalogues, and also a diagram with an ingenious method of pairing stars.*

Choosing
Stars.
The approximate altitude of each star must be calculated and inserted in a list in the angle book, together with the time of its meridian passage, its magnitude, the time that will be shown by the pocket chronometer that is to be used for taking time, whether it is N. or S. of the zenith; and each pair must be numbered. The nearer together in point of time the two stars of a pair can be placed, the greater will be the chances of the elimination of the refraction errors, as in a few hours temperature often varies much, dews form, and many differences may arise in the atmospherical conditions.

Stars over 60° of altitude are not usually good to observe; as, though a sextant will measure over 120°, the image of the star will not be sharp when reflected from the index glass at such a large angle, unless the glass be unusually good.

Altitudes of stars selected for pairs should not differ more than four degrees if possible. Generally pairs within this limit can be found.

If one star of a pair be lost, it is useless taking the other, unless a substitute for the one lost can be had. It is well, therefore, to be provided with spare stars for pairs, as this may often happen from clouds intervening, &c.

Care must be taken, in choosing pairs, to leave sufficient

---

* These are respectively compiled and devised by Lieut. H. B. T. Somerville and Commander Purey Cust, and much facilitate the selection of stars.

time between each meridian passage for the due observation of each star before and after culmination.

This will vary with the latitude and declination, as a star should not be observed so far from the meridian as to bring the mean of the altitudes observed less than a minute or so under the meridian altitude. Fifteen minutes, elapsing between each passage, will give plenty of time under most circumstances.

Time must also be allowed for changing the position from north to south, and *vice versâ;* but all this will vary with the quickness and experience of the observer. Beginners must be satisfied with a few stars, and must allow more time.

In preparing the ground we must look out for a spot *Preparing the ground.* whence we can see clear in the line of the meridian north and south, and one far enough from the beach to be beyond the distance where surf will shake our quicksilver. The latter point is sometimes, as for instance where jungle comes down to the very beach, difficult to find, but it is well worth looking for, and going inland a bit to get it; as otherwise good observations may be rendered impossible from the vibration set up. The more solid the ground the better, as it is astonishing what slight causes will suffice to set the surface of the mercury in motion. The use of the amalgamated trough, mentioned on page 13, will, however, enable observations to be obtained when impossible with the older form of horizon.

Wind is a frequent source of quaking mercury, and care should be taken to have the horizon trough firmly placed, and the roof so fitted that the wind cannot get under its lower edge.*

A screen of canvas to windward is sometimes a good thing, but on some ground this causes such vibration of the earth as to be worse than the free blast of the wind.

If more than one officer is to observe, a screen of some kind

---

* *Vide* Artificial Horizon, page 13.

should be put up north and south between the observers to
keep the lights out of one another's eyes.

In the tropics, and some other localities, mosquitoes must
not be forgotten.   In places where these plagues abound, it
is preferable to court the wind instead of shutting it out, in
order to free ourselves if possible of them.   Sandflies are
perhaps worse, as nothing will get rid of them, and many an
otherwise favourable opportunity of getting stars has been
spoilt by these wretched little insects.

The spot for the artificial horizon being settled on, and the
direction of the meridian taken with a compass, it is a good
plan to dig holes, if the nature of the soil permits, in which
to place the lantern used for reading off when not required, so
as to avoid unnecessary glare.*   The best place for these will
be on the left side of, and a little behind the observer's seat,
and two will be wanted for each observer, one for the north
stars' position, the other when facing south.   If the ground
will not admit of digging holes, buckets will answer the same
purpose well enough, but not so well.

If special lanterns can be got, these precautions will not
be necessary; but we are assuming observations with ordinary
ships' lanterns.

All these kinds of preparations should be made before
sunset, if possible; confusion will be sure to occur if things
are delayed till after dark.

For observations with a sextant stand, a small stool is
wanted, as described under "Sextant Stand," and another for
the observer's seat.   It much facilitates good observations for
the observer to be comfortable, especially when he is about to
observe for several hours consecutively.

Star Map.   A good star map is very useful to assist in recognising the
objects chosen.

Error of
chrono-
meter
require?.   It will have been necessary to obtain the error of the
chronometer on the day of our star observations (by single

---

* Some officers now fit their sextants with small electric lights as a
luxury.

altitude is sufficient), unless we have recently obtained error
at the same place, and have confidence that the chronometers
are going sufficiently well to give us the true time of place to,
say, two seconds.

Having thus made all preparations, and compared the
pocket watches with the standard chronometer, and another
as a check, we land to observe.   We may here remark that
we must again compare on returning on board.

Placing the artificial horizon north and south, we put on **Observing**
the roof, with the mark on it in the settled direction, according **Stars.**
as our first star is north or south of the zenith.

We then place the sextant on its stand, having first
. screwed in the inverting tube with the weakest-powered eye-
piece.   This should be adjusted to focus, as near as possible,
before screwing into the collar.

Place the sextant and stand on the stool, so that one of the
three legs which support the stand is at right angles to the
meridian, and on the right of the observer.

Set the vernier to a few minutes less than the estimated
double altitude of the star.

Move the stool with the instrument on it, so that, looking
over the tube, we can see the reflection of our star in the
artificial horizon.

Point the telescope at this, and set taut the screw that
fixes the handle of the sextant on to the bearing of the
stand, then working the sextant right and left in the stand
pivots, the other image of the star will soon dash across
the field.

Unless the star is moving rapidly in altitude there will be
no further need to move the sextant on its bearing, and care
should be taken that the screw is tight enough to prevent its
moving while turning the sextant up to read, or the telescope
will not point to the star, when directed to the horizon again,
which it should do at once, and so save time in redirecting.
Beginners often neglect this, failing to see the necessity for
it; and, losing time in looking again for the star after each
reading, miss one of the great advantages of a stand, viz. that

once fixed, the stars will always be in the field without any bother.

With faint stars, when there are several of nearly the same brightness close together, it is sometimes rather confusing, and a beginner will find it very difficult to be sure of his star; but comparison with the star atlas, and consideration of how the star wanted lies with regard to the others, will, after a little experience, clear up the difficulty.

If the star be a faint one, it is difficult to bring it down to meet its image in the horizon by taking the sextant off its stand; and if it is a bright one, there is no need to do so, as it cannot be mistaken if the estimated altitude is anywhere near the truth. There is then, in regularly pre-arranged observations, no necessity for doing this, and we shall trust entirely to the vernier being set to the calculated altitude for finding the star.*

Having brought the two images into proximity by hand, place the right hand on the screw at the end of the stand-leg that has been arranged at right angles, and the left hand on the tangent screw of the sextant, when, by working these two screws, the images of the stars can be made to pass over one another exactly, and the word "stop" given.

At this signal the attendant bluejacket will hold the lantern up for reading off. The light should be thrown on to the arc from a direction as nearly at right angles to the plane of the sextant as possible, to avoid parallax.

Resetting for each observation.
In taking the next observation, turn the tangent screw on or back alternately, before commencing to bring the images together again, so as to be entirely free from bias as to whether the star is rising or falling.

Time from Meridian.
The amount of time to observe before and after culmination varies with the position of the star and the latitude of the place as before mentioned; but, as a rule, commencing six

---

* Some observers find a small level fitted to the index bar, and another on the arm of the sextant stand, a great convenience in setting a sextant to the right star.

minutes before the calculated time of meridian passage, and continuing for a like time afterwards, will be ample, as we do not wish the correction eventually to be applied to the mean of the observed altitudes to bring it up to the meridian altitude to be more than one minute, if we can manage it.

Observing as close to the meridian as is recommended above, the decimal parts of seconds need not be recorded in taking the time. Decimals unnecessary.

The observed altitude of a heavenly body can be corrected to the meridian when the hour angle and the latitude are known; but if the hour angle is large, the calculation of the quantity to be added to the altitude is a complicated process, and it is only when the hour angle is small, and very little in error, that the formula assumes a simple and practical form. The error introduced by working with an assumed latitude also increases rapidly with the hour angle, so that we are confined in using this method to about twenty minutes of the time of meridian passage in ordinary latitudes; but in observations such as we are discussing now, which have for their object as correct a determination of the latitude as we can obtain, we must not observe more than about ten minutes from the meridian, and perhaps less. Reduction to Meridian.

The best method to use in reducing observations to the meridian is that known as Raper's, in which the principle is to add on to the observed altitude the amount necessary to make it equal to the meridian altitude, and then to calculate the latitude as in a meridian observation.

This amount is known as the "Reduction to the Meridian," and is so called because it is subtractive from the observed zenith distance.

The formula used by Raper is *—

$$\left.\begin{array}{l}\text{Reduction}\\\text{in secs. of arc}\end{array}\right\} = \text{Cos dec.} \times \text{Cos lat.} \times \text{Sec alt.} \times \frac{\text{Vers hour angle}}{\text{Sin } 1''}$$

Raper gives a table of $\dfrac{\text{Vers H A}}{\text{Sin } 1''}$ for every minute and

---

* For proof, see Appendix D.

second of hour angle up to thirty minutes from the meridian, which is very convenient, and is given in Appendix M. It saves a considerable amount of figures in the calculation, and thereby diminishes the chances of clerical errors; but the formula, as given above, can be worked out, if Raper's table is not at hand.

Sidereal Hour Angle. The hour angle as marked by a watch beating mean time, or nearly mean time, will not be strictly correct either for a star or the sun; as, for the former, it should be taken by a watch beating sidereal time, which gets over its 24 hours while a mean solar watch has only advanced $23^h 56^m 04^s$ nearly; and apparent solar time varies from day to day as the speed of the earth in her orbit varies; but, within the limits we observe, the difference in ordinary latitudes is scarcely perceptible, and if we observe the stars of a pair about the same distance from the meridian, any little discrepancy will disappear in the mean.

If it be desired to correct for the difference when observing stars, a constant log of 0·002000 added to the other parts of the equation will give a close approximation to the exact reduction.

Calculating the Reduction. In working out the circum-meridian observations of a star, it is not necessary to calculate the reduction for each individual observation. As $\frac{\text{Vers H A}}{\text{Sin } 1''}$ is the only variable quantity in the equation, and as it varies with the hour angle, by taking this out for each observation and meaning these quantities, we obtain a mean value of $\frac{\text{Vers H A}}{\text{Sin } 1''}$ which we insert in the equation, and add the mean reduction so found, to the mean of the altitudes. If working with Raper's tables, we take out the whole quantity $\frac{\text{Vers H A}}{\text{Sin } 1''}$; if without them, we look out the Vers H A only, and introduce the log Sin $1''$ into the calculation with the other logarithms.

Knowing then the error of the watch used on mean time of place, and the approximate latitude and longitude, the rule for

working out circum-meridian observations of stars will stand
thus :

1. Calculate Greenwich date.

2. Correct right ascension of mean sun (sidereal time of
'Nautical Almanac') for this Greenwich date, and subtract it
from the right ascension of the star, which will give the mean
time of star's meridian passage.

3. Apply to this the error of the watch, which will give the
time shown by the watch at the star's meridian passage.

4. Mean the observed altitudes, correct this mean for index
error, and divide it by two. To this apply refraction (corrected
for thermometer and barometer), which will give true altitude.

5. Write down the times of each observation in a column,
and taking the difference between each, and the time shown
by the watch at meridian passage, we get the hour angle at
each observation, which place in another parallel column.

6. Take out for each of these hour angles the quantity from
Raper's Reduction Table, or, if we have not got that, the
natural versine.

7. Add these quantities together, and divide the sum by the
number of observations, to get a mean.

8. Add together log cosine declination, log cosine estimated
latitude, log secant true altitude, and the logarithm of the
result of No. 7. If we are using versines, a constant log
$9 \cdot 316400$ is also to be added. (This is log cosec $1'' + 4$
$+ 0 \cdot 002000$.)

9. Look out the sum of these logs as a natural number, which
will be the number of seconds of reduction required.

10. Add this to the mean observed true altitude, which will
give the calculated meridian altitude, from which the latitude
is obtained in the usual manner.

The following is given as an example.

Rule for calculation of Reduction.

On July 11th, 1879.   At Buyuk Chekmejeh △ the following observations were taken.   Index error of sextant — 35″ Latitude (approx.) 40° 57′ 45″ N.   Longitude 28° 30′ E. Mean Time of Transit of star calculated, Xh. 12m.

Time by watch (Breguet 2086).			Altitude α Ophiuchi.		
h.	m.	s.	°	′	″
10	09	36	123	22	00
	10	11		22	40
	10	42		22	50
	11	14		23	15
	11	43		23	25
	12	09		23	35
	12	50		23	50
	13	14		23	50
	13	38		23	55
	14	06		24	00
	14	33		24	00
	14	55		23	50
	15	16		23	50
	15	37		23	45
	16	04		23	30
	16	24		23	15
	16	50		23	00
	17	20		22	40
	17	46		22	30
	18	15		21	55
Mean			123	23	16·7

The calculation will appear as follows :—

Watch Times.			Hour Angle.		Vers H. A. Sin. 1″
h.	m.	s.	m.	s.	
10	09	36	4	27	38·9
	10	11	3	52	29·4
	10	42	3	21	22·0
	11	14	2	49	15·6
	11	43	2	20	10·7
	12	09	1	54	7·1
	12	50	1	13	2·9
	13	14	0	49	1·3
	13	38	0	25	0·3
	14	06	0	03	0·0
	14	33	0	30	0·5
	14	55	0	42	1·0

Watch Times.			Hour Angle.		Vers H. A. $\frac{}{\text{Sin. } 1''}$
h.	m.	s.	m.	s.	
10	15	16	1	13	2·9
	15	37	1	34	4·8
	16	04	2	01	8·0
	16	24	2	21	10·8
	16	50	2	47	15·2
	17	20	3	17	21·2
	17	46	3	43	27·1
	18	15	4	12	34·6
					20) 254·3
			Mean	..	.. 12·71

		h.	m.					
Mn. Time of Transit	..	10	12	Cos. dec.	..	..	9·989329	
Long. in time ..	..	1	54	Cos. lat.	..	..	9·878027	
				Sec alt.	..	..	·323871	
G. M. T. ..	..	8	18	12·71	..	..	1·105510	
		h.	m.	s.			1·296737	
R. A. Mean ☉	..	7	16	05·9				
Corr. for 8h.	..		1	18·9	Reduction	..	19″·8	
„ „ 18m.	..			3·0				
		7	17	27·8			° ′ ″	
R. A. ✳ ..	..	17	29	22·6	Tr. alt.	..	61 40 51·4	
					Red ..	..	+19·8	
M. T. Transit	..	10	11	54·8			61 41 11·2	
Watch fast	..		2	08·1	Z. D. ..	..	28 18 48·8N.	
Time by watch at					Dec. ..	..	12 38 57·0N.	
Transit ..	..	10	14	02·9				
		°	′	″	Latitude	..	40 57 45·8N.	
Mean obs. alt.	..	123	23	16·7				
I. E. ..	..			−35·0				
		2) 123	22	41·7				
App. alt. ..	..	61	41	20·8				
Refraction	..			−29·4				
Tr. alt. ..	..	61	40	51·4				

When the pole **star** is observed, it must be worked out by **Pole Star.** the rule given in the Nautical Almanac, care being taken to take out all the quantities from the tables exactly by interpolation.

P

Moon not convenient.   The moon is but of little use for observations of any kind. Its rapid motion necessitates very careful corrections, which take more time than they are worth, and besides we have nothing to put against it to eliminate errors.

Results of Pairs.   The separate stars being worked out, we mean the result of each pair, and the mean of these again will give us the mean latitude.

Although from circumstances many more observations may be got of one star of a pair than of the other, no value can be assigned to one over the other, and the direct mean must be taken; but in meaning up the results of pairs, less value would be given to a pair in which the observations of one star are few in number, than to a pair where a proper number of observations of each star has been obtained. The necessity for assigning this value is increased where the observations are not only few but indifferent; but it would be a question whether it would not be better to omit such a pair from the final result altogether, and certainly it would be best to do so, if there are several other good pairs.

An example of the method of tabulating the different observations and pairs is given on next page.

The sets of stars given as an example were taken under favourable circumstances of sky and weather, and are not meant as a standard which we must expect always to get. Here no value is given, as each pair were nearly equally good, and the observations were nearly the same as to number.

It is, however, evident on an inspection of this set of observations, that the sextant had a centering error. It will be remarked that the latitude obtained from south stars diminishes as the altitude increases, and *vice versâ* with the north stars. This is wholly attributable to the centering error, and from these observations a very fair idea of that error for each altitude may be obtained, as explained on p. 7. It is therefore well, though the result will not be affected, to apply the centering errors if known.

In the example given only the direct mean was taken, though the second pair was open to suspicion, the seconds

LATITUDES BY CIRCUM-MERIDIAN STARS AT BUYUK CHEKMEJEH, JULY 11TH, 1879. ARRANGED ACCORDING TO ALTITUDES AND PAIRS.

SOUTH STARS.

Alt.	*	No. of Obsns.	Latitude.
°			° ′ ″
38¼	ζ Ophiuchi	20	40 57 52·6
45½	δ "	20	55·8
51½	δ Aquilæ	20	47·1
58½	κ Ophiuchi	16	44·2
61½	α Aquilæ	20	46·0
62½	ζ Aquilæ	20	39·2

NORTH STARS.

Alt.	*	No. of Obsns.	Latitude.	Latitude by means of each pair.	Remarks.
°			° ′ ″	° ′ ″	
40	Polaris	20	40 57 42·1	40 57 47·3	Taken consecutively.
44½	δ Ursæ Min.	20	49·1	52·4	
49	" "	20	48·7	47·9	
58	χ Draconis	23	51·4	47·8	Taken consecutively.
61	" "	20	53·2	49·6	
63½	δ "	17	54·4	46·8	
				6) 51·8	
	Mean Latitude ..			40 57 48·6 N.	

P 2

of both north and south stars being somewhat out of the
order observable in the remainder.

**Valuing
Results.**
If, however, the observations had been less uniform, and
there were not sufficient pairs manifestly better than others,
by which we could elect to stand, value would be assigned
by some such system as the following.

Assume a number as perfection, say 10, and give to each
pair its value in that scale. Then the sum of the products of
the value and the seconds of latitude by each pair, divided by
the sum of the values, will give the mean seconds of latitude.
The values should not be given by the results, or we shall
arrive at pretty much the same conclusion as we should by
assuming the latitude directly, but by the circumstances of
observation of each star, and the number of observations, remembering that reasons to doubt one star will as equally affect
the pair, as if both stars were bad.

Values must also be given when meaning the results by
different observers; but here again, if one observer is ad-
mittedly the best and his observations are also good, a more
probable result will be obtained by ignoring the latitudes of
the others altogether. They will not be totally lost, as experience in good observing is only to be obtained by plenty of
practice, and a young observer is more likely to take pains
when there is a chance, if his observations turn out well, of
having them included in the operations of the survey, than when
merely observing for what one may term barren practice.

**Difficulty
of valuing.**
This question of giving values is a very difficult one, and
many experienced observers prefer to take a direct mean both
for the result by each individual, and when finally meaning
the different observers' results, omitting those altogether
which seem least worthy of trust, and giving an equal value
to everything else. It is very much a case of judgment under
the circumstances; on some occasions one system will be best,
and at others the other, and we must so leave it, simply
remarking that whenever observations vary enough to make
the consideration of this question necessary, the final result
must always be regarded with suspicion.

LATITUDE BY CIRCUM-MERIDIAN STARS AT MESALE 1st AUG. 31ST, 1878 (RESULT BY VALUES).

SOUTH STARS.				NORTH STARS.				Latitude by mean of each Pair.	Value.	Product of Value and Secs. of Lat.	REMARKS.
Alt.	★	No. of Obs.	Latitude.	Alt.	★	No. of Obs.	Latitude.				
°			° ′ ″	°			° ′ ″	° ′ ″			
40	α Cygni	4	5 13 38·6	38	α Pavonis	22	5 14 33·1	5 14 05·8	8	46·4	Obs. of α Cygni, good.
44	Vega	25	37·6	51	α Indi	12	49·0	13·3	4	53·2	Two stars four hours apart. α Indi misty, Alts 7° different.
51	ε Cygni	3	37·5	51	β Sagittarii	19	45·7	11·5	5	57·5	ε Cygni only fair.
60	κ Pegasi	14	33·5	61	ε ,,	3	42·5	08·0	7	56·0	ε Sagittarii, pretty good.
						Sums			24	213·1	

$$\frac{213·1}{24} = 08''·8$$

(Observer B)   ..   ..   Mean Latitude 5° 14′ 08″·8 S.

Observer.	Latitude.	Value of each result.	Product of Value and Secs. of Lat.	
A	5 14 16·3	3	48·9	
B	08·8	5	44·0	Mean Latitude by 4 observers :—
C	03·0	5 ·	15·0	
D	12·0	6	72·0	5° 14′ 09″·4 S.
		19	179·9	= 09″·4
			19	

An example is given on the previous page of latitude arrived at by giving values.

Stars under Pole. In observing stars under the pole, we must not forget that the reduction will be subtractive from the observed altitude.

Planets. The larger planets are not good for observation, as they are so much bigger and brighter than the point which a star shows. On occasions, however, they must be used. The R.A. and Dec. must be calculated exactly.

Calculating apparent places of Stars from Catalogues. In using stars from the Greenwich or Cape Catalogues, it is necessary to calculate the apparent place of the star for the day. The method of doing this is given in the Nautical Almanac, in the explanation under the head of " Stars," where examples of a north and of a south star are shown worked out. Care must be taken to give the proper signs + or − to each logarithm.

Latitude by Circum-meridian Altitudes of Sun. Next to the observation of stars in pairs, the circum-meridian observation of the sun in the artificial horizon is the most correct and simple method we have of obtaining latitude ; but it is evident that we cannot use it when the altitude exceeds 65°, as a sextant will not measure the double angle. We must, in the case of the sun, be doubly careful in correcting the refraction, if we wish to get as near the truth as possible. There is nothing to be gained by observing both limbs of the sun, as the motion in altitude will be so small that it will not matter whether the images are opening or closing.

The roof of the horizon should be reversed at about noon and the sights worked out as two sets, roof one way and roof the other.

However careful we may be, we shall not expect our latitude by the sun only to be exact, and in many cases where we are going to be satisfied with this observation, it will not matter if the latitude be a quarter of a mile or so in error, and the reversal of the roof may often be dispensed with.

If however we know our centering error, and can depend upon the sextant, we can by its application get a vastly improved result, and none of these precautions should be omitted.

An observation of the sun cannot be meaned with an observation of a star the other side of the zenith, as all refraction errors, as well as errors introduced into the instrument by the heat of the sun, will be entirely different. *Sun and Stars can not be paired.*

Circum-meridian observations of the sun are worked out in precisely the same manner as those of a star, the only difference being in the ordinary corrections to declination, &c. We want the error of the watch on apparent time to calculate what it shows at apparent noon, the time the sun will be on the meridian. *Calculation of Sun Observations.*

An example is given on next page.

On November 15th, 1876, circum-meridian altitudes of the ☉ were observed with artifical horizon at Maghabiyeh I⁴. Approx. lat. 18° 15′ N., long. 40° 44′ E.   Bar. 30·00 ins. Ther. 80°.   Mean of observed altitudes Sun's Upper Limb, 106° 47′ 09″·1.

There is a source of error in obtaining an accurate latitude which must not be forgotten, especially when the scale of a chart depends on a difference of latitude, and that is the local attraction due to the irregular disposition of masses of land in the vicinity of an observation spot. A mountain mass on one side, or deep sea, will cause the local direction *Local Attraction.*

Time by Watch.	Hour Angle.	Vers. H. A. / Sin. 1″
h. m. s.	m. s.	
5 36 39	6 09	74·3
37 06	5 42	63·8
37 46	5 02	49·7
38 07	4 41	43·1
38 28	4 20	36·9
38 53	3 55	30·1
39 21	3 27	23·4
41 32	1 16	03·1
41 54	0 54	01·6
42 16	0 32	0·6
42 37	0 11	0·1
42 57	0 09	0·0
43 21	0 33	0·6
43 43	0 55	01·6
44 44	1 56	07·3
45 24	2 36	13·3
45 40	2 52	16·1
46 04	3 16	20·9
46 29	3 41	26·6
47 12	4 24	38·0
47 32	4 44	44·0
48 18	5 30	59·4
48 51	6 03	71·9

23) 626·4

Mean .. .. 27·23

		h. m. s.
App. noon..	.. ..	12 0 0
Watch slow	..	6 17 12
Watch at Transit	..	5 42 48

		o ′ ″
Obs. alt. ☉	..	106 47 09·1
I. E. ..	.. ..	−40·0
	2)	106 46 29·1
		53 23 14·5
S. D. ..	.. ..	−16 13
		53 07 01·5
Refraction..	.. ..	−35·0
True alt. ..	.. ..	53 06 26·5

Cos dec. ..	.. .. ..	9·976613
Cos lat. ..	.. .. ..	9·977586
Sec alt. ..	.. .. ..	0·221713
27·23 ..	.. .. ..	1·435044
		1·610960
Reduction ..	..	40″·8

		h. m. s.
App. noon ..	.. ..	0 0 0
Long. ..	.. ..	2 43 0
Gr. Date	.. ..	2 43 0
Dec. ap. noon. Gr.	..	18 39 47·8
		1 42·4
Dec. at Transit..		18 38 05·4 S.

		″
Var.	.. ..	37·80
		2·71
		378
		2646
		756
Corr. ..	..	102·43

		o ′ ″
Tr. alt. ..	..	53 06 26·5
Red. ..	..	+40·8
Mer. alt.	..	53 07 07·3
Z. D. ..	..	36 52 52·7
Dec. ..	..	18 38 05·4
Latitude..	..	18 14 47·3 N.

of gravity to slightly diverge from the vertical. The
surface of the mercury will not in such a case be truly
horizontal, and the altitude will be in error. This may often
largely account for differences between triangulation and
observation, and when the former is good, it may be in
certain cases desirable to rest on it rather than the scale by
observations. Formulæ have been drawn up for correction,
but they rest largely on assumptions of mass which cannot
be verified.

This attraction will also affect determination of time and
therefore longitude.

CHAPTER XIII.

OBSERVATIONS FOR ERROR OF CHRONOMETER.

General remarks on obtaining Longitude—Error by Equal Altitudes—
Error by two Stars at Equal Altitudes.

Shadwell's "Notes on Chronometers."

THE whole question of obtaining longitude by means of chronometers is so ably and exhaustively treated by Captain Shadwell in his "Notes on the Management of Chronometers," both as regards the treatment of the watches, the method of observation, and the various systems of obtaining meridian distances, that we refer the reader to that work for full information on the subject. Here we do not pretend to give more than the broader principles of the general question, but a work of this kind, intended for the perusal of young surveyors, would be incomplete without some reference to it.

Absolute Longitudes.

The methods of obtaining longitude, called "absolute methods," which give the longitude of the place as measured from the first meridian, directly and independently, such as observations of occultations of the stars by the moon, moon culminating stars, eclipses of Jupiter satellites, &c., are now rarely employed in nautical surveying, and may be said to be decidedly inferior in value to the results of good chronometric runs.

Similarly, altazimuths, portable transits, and other like astronomical instruments, are now seldom or never supplied to a surveying vessel. The sextant in a practised hand will give results equal to those obtainable by fixed instruments of small size, and has the great advantage of being more portable, and always ready.

To the sextant, telegraph, and chronometers, therefore, our remarks will be confined.

By the use of the two latter we obtain only the " difference of longitude," or "meridian distance," between two places, neither of which may have its meridian distance from the primary meridian of Greenwich determined; but by the accumulation of such observations, the absolute longitudes of certain places are from time to time decided.

These, then, become secondary meridians, on which the longitude of places in their vicinity depend.

When therefore a secondary meridian is changed in its value as regards its distance from the first meridian, all places whose longitude have been measured from it are changed also.

This is the work of the Hydrographic Office, which receives and collates all information. The nautical surveyor simply finds the difference of longitude, and transmits that information only.

A list of secondary meridians is given in the Instructions for Hydrographic Surveyors.

For our purposes we may look upon difference of longitude as divided into two main cases. The first, where the scale of a chart we are making depends on the astronomical observations for latitude and longitude at either extremity of our piece of coast. The second, when we wish to determine the relative positions of places more or less far apart, which are mainly required for the purposes of navigation.

In the first, we can nearly always use the system, hereafter described, of "travelling rates," which much adds to the accuracy of the result.

In the second, time, distance, and general circumstances often prevent our obtaining these, and compel us to use what we can get.

In obtaining the meridian distance between two places, either by means of a telegraph, or by carrying chronometers between them, the principle is the same, and is this, viz. that the difference of mean time of place at any moment at the two places is their difference of longitude in time.

[Marginal notes:] Remarks confined to Differential Longitude and Sextant. — Secondary Meridians. — Cases defined. — Principle of Differential Longitude.

If, therefore, we can find out that at the time that at a position A it is 9 o'clock, it is 8 o'clock at B, we know that the difference of longitude is equal to one hour of time, and that A is east of B. The telegraph enables us to do this in its simplest form, as, ascertaining the exact time at each end by astronomical observations, we can find out by an exchange of signals what is the difference of time.

Chronometric Difference of Longitude. In chronometric difference of longitude we have literally to carry the time from one place to the other. We ascertain the time at one place on a certain day, or, what is the same thing, we find out the Error of our chronometer on local time.

Supposing for the moment that our chronometer is keeping exact mean time, by carrying it to the other place and finding out its Error on local time there, we can deduce the difference of longitude by the difference of the two Errors. Thus, if our chronometer is four hours slow on mean time at A, and we find when we get to B that it is three hours slow, we know the difference of longitude is one hour, and that A is east of B. Unfortunately, chronometers do not keep mean time, and the problem is complicated by having to ascertain what time they *do* keep, or, in other words, what they gain or lose in each day, which is called the rate. If we can find this, we shall be able to get the difference of longitude just as accurately as if the chronometer was keeping mean time, as we can correct for this rate; but here, again, chronometers are not, and probably never will be, perfect instruments, and are liable to change of rate, and it by no means stands to reason that because a chronometer gains five seconds a day one week, it will do so the next, especially when the ship has been at anchor during one period, and under weigh for the other, and the temperature has not been invariable.

Chronometric runs are therefore liable to the errors arising from change of rate. To overcome this as far as possible, a number of chronometers are carried instead of one, and, if possible, what is called the travelling rate is obtained.

Travelling Rate. If the rate of chronometers is obtained at a station A, and we then go to another station, B, and obtain rate again there, and apply the mean of these rates as the assumed rate of the chronometers while being carried from A to B, we have no guarantee whatever that this assumption is correct, as the time employed in carrying the chronometers does not enter into the calculation at all, and they may have been going quite differently when the ship was at sea, with the vibration of the engines, motion of the ship, &c., to influence them, to what they were when the ship was at anchor, besides the important factor of change of temperature. If, however, we can return at once to A, and obtain the Error again, we can positively say that the chronometers have gained or lost so much between the first and second observations at A. Assuming this loss or gain to have been uniformly carried on throughout the interval, we shall have a travelling rate which will give a far nearer result than by using rates obtained at either end of our required base. By this means we only obtain one meridian distance for our double run forwards and backwards, but it will be of more value than two separate meridian distances obtained by fixed rates.

Even if we have to stop at B a few days, by observing *Modifica-* on arrival, and immediately before departure, we can eliminate *tion of Travelling* the gain or loss of the chronometers during the stay there, by *Rate.* subtracting it from the total gain or loss during the time of our absence from A, and dividing the remainder by the number of days actually travelling. We shall thus still get a fair travelling rate, if the chronometers are at all trustworthy as timekeepers.

This, then, is the system of travelling rates, which can be generally, and always should be, if possible, used in determining difference of longitude for the scale of a chart.

Whatever be the system of rates employed, good observations must of course be regarded as the foundation of all of them. We cannot control the irregularities of our chronometers, but we can, to a certain extent, make sure of getting fairly correct time by using the proper means.

To ascertain the Error of the chronometer as exactly as we *Elimina-* can with sextant and artificial horizon, we must endeavour to *tion of errors by*

Equal Al-
titudes.

get rid of the atmospheric and other errors, as we do in observations of stars for latitude, which in this case is attained by
observing at equal altitudes east and west of the meridian. It
will be evident that, whatever be the instrumental and other
errors, (excepting those of observation,) supposing them to
remain unaltered, the middle time between the observations
will be the same, as whatever tends to make the observed
altitude more or less in the forenoon, will act in the same
manner in the afternoon, and as we do not want to know
at all what that altitude is, but merely to ensure that it
is equal, A.M. and P.M., the amount of the errors is immaterial.

The method of equal altitudes, therefore, must be used
whenever we wish to get Error exactly. The Error of watch
obtained by single altitudes, called "absolute observations,"
will depend for its accuracy upon the corrections for each
source of error, which, as we have before stated, can only
be considered as approximate.

Superior
and In-
ferior
Transit.

Equal altitudes of the sun can be taken either in the forenoon
and afternoon of the same day, so as to find the Error at noon,
called Error at superior transit; or in the afternoon of one day
and the forenoon of the next, by which means we obtain the
Error at midnight, or at inferior transit. Theoretically, these
are equally correct, but in practice it is better to get Error at
noon, if we can, as the elapsed time being less, gives less latitude to the chronometers or hack watches for eccentricity.
The alternative is, however, very valuable, and saves many a
day, as when, for instance, we arrive at the place we wish to
observe at, an hour or two too late for forenoon sights. We
can then begin our set in the afternoon, and get away, if we
wish to do so, the next morning after forenoon sights, and
thus save several hours, a considerable consideration in
running meridian distances.

Principle
of Equal
Altitudes.

The principle of finding the Error of a timekeeper by observation of equal altitudes is, that the earth revolving at a
uniform rate, equal altitudes of a body, on either side of the
meridian, will be found at equal intervals from the time of
transit of that body over the meridian, and that therefore the

mean of the times of such equal altitudes will give the time at transit.

In the case of a body whose declination is practically invariable, as a star, this is strictly true, and the calculation of the Error of the watch is confined to taking the difference of the time shown by the watch, and the true calculated time of transit.

In the case of the sun, however, the declination is constantly changing; the altitudes are thereby affected, and an altitude equal to that observed before transit will be reached after transit, sooner or later, according to the direction of change in the declination. **Equation of Equal Altitudes.**

It is therefore necessary to make a calculation of the correction, resulting from this change of declination, to be applied to the middle time, to reduce it to apparent noon, which correction is termed the "equation of equal altitudes."

The observation of stars at equal altitudes will therefore be, theoretically, the best to use, as being the simplest, and they will indeed give as good results as those of the sun; but practically, the latter has generally been observed in marine surveying for the purpose of obtaining time. In many cases the inconvenience of landing, and carrying watches backwards and forwards for comparison, &c., by night, besides the increased difficulties of observing, and reading instruments by lamp-light, lead to the choice of day observations; but in places where clouds persistently veil the sun in forenoon or afternoon, the nights are often clear, and equal altitudes of stars become most valuable. **Stars and Sun compared.**

There are two other methods of obtaining error by stars, that are both good. The first, observing stars of corresponding altitude on either side of the meridian, working them out as absolute altitudes and meaning the results. The other is a method to which attention has been drawn by Captain A. M. Field, R.N., of observing two different stars of nearly the same declination at equal altitudes. This is published as a pamphlet by the Hydrographic Office, but is briefly described on p. 238. The great advantage of either of these **Other Methods by Stars.**

is that the complete observation is made in a very short time, thereby obviating the disadvantages of star observations, mentioned above.

Limita-tions of Observa-tions.   In taking the observations of equal altitudes in the artificial horizon, we are limited, as always, to altitudes between 20° and 60°, as the horizon will not permit us to observe a lower altitude than 20°, and the sextant will not measure much more than 120°. These restrictions will, however, only be inconveniences, as regards the sun, in extreme latitudes, as we must choose, as our time of observation, so as to minimise the effects of errors of observation, the period at which its motion in altitude is the greatest, *i.e.* when it is near the prime vertical, at which time, in all but high latitudes, the altitude will come between our limits. When the place of observation is near the equator, and the latitude and declination are nearly the same, we could observe up to a very short time of noon, the sun's motion in altitude being nearly uniform throughout the day; but we are in this case limited by the range of the sextant.

It is difficult to lay down any rule as to what is the smallest rate of motion in altitude we should observe at, as the greatest motion in altitude during the day varies so much with the latitude and declination. We can only say that we should, when we have any choice, not observe beyond an hour when the time of changing 10′ of double altitude exceeds 30 seconds.

Sets of Observa-tions.   Opinions have much differed on the number of consecutive observations that it is best to take to comprise in a set. The only theoretical limit is that the equation of equal altitudes should be practically the same throughout the set, as the variation in the time required by the sun to traverse the number of minutes of altitude between observations at the beginning and end of the set will not matter, as we do not care whether the mean of the times agrees exactly with the mean of the altitudes.

It seems well, therefore, to observe tolerably long sets, as errors of observation are thereby eliminated. The same result in the end will be attained by a large number of shorter sets ; but the value of each set is much enhanced if composed of a considerable number of observations, and it saves time and trouble in the calculations.

Too long sets are to be deprecated as wearying to the eye and hand, and the observations will therefore suffer from that cause, especially in hot countries, where the necessity for observing in the full glare of the sun makes it a trying operation.

We prefer to take eleven observations in a set. This allows the observer to commence his second set, of lower limbs, (A.M. observations) at exactly one degree more altitude than his first one, of upper limbs. It does not much matter, as each one has his own plans for these details, and soon falls into a regular method, which is the great thing to prevent mistakes.

The only point in fact of importance is, *always to observe in the same way.* Not only does it save time and errors, but it is necessary in combining observations, whether for rate or meridian distance, that they be as *similar in all respects* as we can get them. The whole system is a system of differences, and it is manifest that the result is the better the more like the observations are. It follows from this that the observers employed in any string of meridian distances should be the same, the instruments and watches the same, the temperature and time of observing the same, as far as possible. Also, supposing temperature to be the same, that a rate will be probably more correct if obtained by combining single altitudes of different days, both either A.M. or P.M., than by taking equal altitudes one day and single altitudes the other. *(margin: Observations must be similar.)*

If clouds prevent observation at *precisely* the same altitude, after transit, the mean of Error obtained by absolute sights, A.M. and P.M., at *nearly* the same altitudes, will be almost as good as equal altitude sights.

When the observers are good, the greatest error is frequently introduced in comparing the hack watches to be used for *(margin: Comparing watches.)*

taking time with the chronometers, and great pains should therefore be taken with this operation.

The watches used for taking time must be compared before, and after, both forenoon and afternoon sights, with the standard and another chronometer; and at noon, all the chronometers should be compared with the standard, and the hack watches with the same two chronometers.

In the case of stars, when return cannot be made to compare, the proper comparison for middle time must be deduced by interpolation from the comparisons before leaving and on returning.

Defects in Pocket Chronometers. Before saying more about comparing, we must remark that the seconds-hands of pocket chronometers are rarely placed symmetrically in the centre of the dial on which the seconds are marked.

These watches beat five times in two seconds, commencing with the even minute. The beat of the watch should therefore coincide exactly with every even second; the first beat from the minute being 0·4 sec., the second 0·8, the third 1·2, the fourth 1·6, the fifth at 2·0, and so on throughout the whole 60 seconds.

But it will be found, in nearly every watch, that the hand does not fall over the even second on some parts of the dial, although it may on others, and each watch must be examined by counting the beats from the even minute, to ascertain how the hand falls in different parts of the dial, or the time-taker will be at a loss to know what is the exact decimal which his watch is beating. For instance, supposing that at the 40 seconds' mark on the dial, the hand falls a little short of the mark one beat, and a little in advance at the next; unless he knows which of those beats is meant for the 40 seconds, he may be giving the time four-tenths of a second wrong. This, of course, refers both to comparing and taking time for the observations.

Method of Comparing. In comparing, which is perhaps best done by two persons, the "Stop" is given on an exact second of the box-chronometer (which beats half seconds), and the time by the pocket-

watch noted. A check is then taken, by comparing the
reverse way, calling the "stop" at an exact second by the
pocket-watch, and noting the seconds and parts of seconds
of the box-chronometer. As parts of seconds have to be
estimated on both watches, these two comparisons will frequently differ two-tenths of a second. This is as near as we
shall probably be able to arrive at the truth; but if the difference exceeds this, more checks should be taken, until we
are satisfied which was wrong.

No operation requires more care or more practice than
comparing, and while the simple method given above will
give good results when observers are experienced; varying the
method by stopping on odd seconds, will in the case of the
less experienced, prevent any bias on the part of the observer
taking the corresponding time.

In the same manner checks should be taken when comparing one box-chronometer with another.

If two pocket-watches are available, it will not be amiss
to use both, even when there is only one observer, as it
helps to eliminate errors of comparison. In this case half
the sets will be taken with one watch, and the other half
with the other.

Preparation of the ground is not necessary, as in the
case of observing stars, excepting so far as selecting the
spot, to ensure being able to see in both the A.M. and P.M.
directions.

Having compared the watches, we land to observe.

The watches to be used on the ground should always be
carried in their boxes, and great care must be taken not to
jerk them, and above all to avoid any circular motion. *Care in carrying hack watches.*

The method of observation for time differs from that
already described of stars for latitude, inasmuch as we observe at stated altitudes, generally at every 10', setting
the sextant for the purpose, and noting when the contact
takes place. In observing with the stand, therefore, we only
need to work the screw of the stand leg, to get the suns
vertically under one another. *Method of observa- tion.*

Observing
both limbs.

It is well to observe both upper and lower limbs, as though it will make no difference to the result, it is good to have constant practice at both opening and closing suns, and not have all one way in the forenoon and the other in the afternoon. If we begin by a set of upper limbs, and immediately after take a set of lower, as an invariable practice, there will be no confusion, and we shall soon naturally fall into the system.

It may be here noted that with the inverting tube, the movable sun (the sun reflected from index and horizon-glass) is *above* the other, when we are observing upper limb, and *below* when lower limb. Also that upper limbs in the forenoon are closing suns, and in the afternoon opening suns. It is necessarily *vice versâ* for lower limb.

Dark eye-
pieces to
be used.

Always use the dark eye-pieces, of which there should be several of different degrees of shade, as, if the brilliancy of the sun varies by passing clouds, no inherent error is introduced by changing these, which is the case with the hinged shades on the sextant. Moreover, the suns having been once equalised as to brilliancy with one eye-piece, by moving the up-and-down piece screw, they will remain equal, no matter what shade of eye-piece we use; but with the hinged shades, the position of up-and-down piece which equalises the suns as seen through one set of them, will be different to that required for others, besides the possibility of error thus introduced.

Suns
should not
be too
bright.

The suns should be as dark as possible. If too light shades are used, the irradiation spoils the sharpness of the limb.

Should be
as large as
possible.

Use the eye-piece with the greatest magnifying power, as it much facilitates correct contacts.

Setting the
vernier.

Great care must be taken in setting the vernier, and we must see that the tangent screw at the commencement of each set is run back to its full extent, so as to avoid risk of being "two blocks" in the middle of the set, and so probably lose an observation.

After bringing the zero of the vernier into what we believe

to be coincidence with the minute of arc required, glance right and left to see that the marks on vernier and arc are displaced in a symmetrical manner on either side. The eye will easier catch any inaccuracy in the setting by this means.

In setting, turn the tangent screw for the final adjustment the same way both forenoon and afternoon. Thus, if with altitudes increasing the tangent screw is turned to the right to attain coincidence, with altitudes diminishing the vernier must be set back below the required altitude, so as again to turn the screw to the right for final adjustment. This tends to eliminate error from slackness of screw.

Some observers, after giving a preliminary "Ready" at the commencement of each set, give no warning after, and simply "Stop" at each observation. With very careful time-takers this is sufficient, but experience of human nature leads us to say that it is better to call "Ready" about three seconds or so before each "Stop," and thereby avoid all chance of the time-taker having his eye and ear off the watch. *Warning calls.*

We should take more observations in our first half of equal altitudes than will be absolutely needed, so as to allow of some losses in the observations after transit, from obscurations of the sun. *Allowance for clouds after Transit.*

At the conclusion of observations it is always well to take the index error. It tells us whether our sextant is keeping a steady error, and also, by calculation of semi-diameter therefrom, whether the instrument is in adjustment for side error, and also, if we lose the other half of equal altitudes, and decide to work single altitudes instead, we shall have the index error observed at the time. *Index Error.*

After sights before transit, we must calculate the time the observations after transit will commence. By far the simplest plan, when engaged in observations, is to have an ordinary watch set to apparent time, which time the ship herself will in many cases be keeping, when by noting the time by this watch at the last observation, the time of commencement of *Calculating time for observations after Transit.*

the first observation after transit will be found by taking the time noted from twelve hours.

If we do not do this, and the ship be keeping mean time, we must find the mean time of the last observation by applying the approximate Error of the watch. Subtract this from twelve hours, and apply twice the equation of time, subtracting if apparent noon is before mean noon, and adding if *vice versâ*. This will give mean time of the first observation after transit, which can be re-transferred to the watch by the application of the Error.

We want the time by the ship's clock, to ensure leaving her at the right time, and the time by our watch, to avoid the chance of being too late on one hand, and scurry after reaching the observation spot, on the other.

Algebraically we can express this,

$$T = 12 - (t + e) + 2\,q$$

where T is mean time of first observation required.

  t is time by watch of last observation.

  e is Error of watch slow of mean time.

  q is equation of time.

Form in rough Sight Book.   The book in which the observations are registered should be ruled as in annexed specimen.

The first column is for the intervals between each observation of the first set of sights. The second, the time taken at those sights. The third, the double altitude. The fourth, the sum of the seconds of the two times. The fifth, the time at second set of sights; and the sixth for the intervals between the latter.

On next page is an example of a set of sights as written in the angle book.

Noting Intervals between sights.   When the time-taker is practised, it is well for him to note down the interval between the sights as he notes down the time, as it enables the observer at once to know, when the set is over, whether he has been getting good observations or not, as the intervals should theoretically be precisely the same.

♄ April 3rd, 1880, at Nagara Light-house △.
Time by Breguet (2086).

Interval in seconds.	Time by watch.	ʘ	Sum of secs.	Time by watch.	Interval in seconds.
	h. m. s.	° '	s.	h. m. s.	
	8 03 10·8	57 20	14·8	3 40 04·0	27·6
28·0	38·8	30	15·2	36·4	28·2
27·6	04 06·4	40	14·6	39 08·2	27·2
27·2	33·6	50	14·6	41·0	27·8
28·0	05 01·6	58 00	14·8	38 13·2	28·0
27·2	28·8	10	14·0	45·2	27·6
27·4	56·2	20	13·8	37 17·6	28·0
28·2	06 24·4	30	14·0	49·6	27·4
27·2	51·6	40	13·8	36 22·2	27·2
28·0	07 19·6	50	14·6	55·0	27·8
28·0	47·6	59 00	14·8	35 27·2	

$$11 \,)\, 5·0$$
$$14·45$$

Times at middle sight $\left\{ \begin{array}{l} \text{h. m. s.} \\ 8\ 05\ 28·8 \\ 15\ 37\ 45·2 \end{array} \right.$

$$2\,)\,23\ 43\ 14·0$$

Mean mid. time by watch 11 51 37·22

N.B.—The seconds of the result are obtained by halving the mean of Column 4.

We do not usually attempt to estimate time with a pocket-watch to less than two-tenths of a second, so we shall not find the intervals agreeing exactly, even supposing no errors of observation to exist. Taking everything together, if the difference in these intervals does not exceed one second and a half, we may consider that we have obtained very good observations. Another reason for noting these intervals is that we shall see if the sun's motion is becoming too slow.

In working out sights for equal altitudes, it is merely the mean of the middle times of each set that we wish to get, so that we need not mean up each column of times, but merely the sum of the seconds of each corresponding times, which we have in the fourth column of our sight book. Then taking

Method of Meaning.

the times of the middle observation, meaning those, and substituting, for the seconds of this mean, the mean of the seconds just found from the fourth column, we shall have the true mean middle time of our observations. Thus, in our example the mean of the fourth column is $14^s{\cdot}5$. We substitute this for the $14^s{\cdot}0$ obtained by adding the two times at the middle observation, and dividing by two, we get the mean middle time by the set as 11h. 51m. 37·22s.

In cases where the equation of equal altitudes is varying rapidly, we shall not find the middle times of two successive sets agreeing exactly, as they should differ by the amount of the variation of the equation of equal altitudes in the time.

Theoretically, this is an objection to long sets of observation, but practically, the errors of observation will exceed any little discrepancy introduced by assuming the equation to be uniformly variable during a set of 11 to 15.

Sights missed. If a contact is missed in either half set, it is no use to interpolate a time. The sight must be missed out of the double set altogether. It will not affect anything but the number of observations in the set, except when it happens to be the first or last of a set, when the set will become one with an even number of observations, and to get elapsed time we must, instead of the central observation, take the mean of the two times corresponding to the two central observations.

Personal Errors not eliminated by Equal Altitudes. Four sets of eleven observations each, half upper limb and half lower, ought to give accurate time; but it must be understood that equal altitudes do not eliminate personal errors, only instrumental and atmospherical ones. It is evident that, if an observer is, for instance, habitually too slow in recording his contacts, the resulting middle time will be so much after the true time of transit. Taking the case of an observer recording the time of contacts of opening limbs correctly, and of closing ones habitually too late; the middle time will still be in error. It is only when the observer records opening limbs in error one way, and closing in error in the contrary way, that these personal errors can be eliminated. This is of course not likely to happen with many people, and we must consider the time resulting from any observer's sights as being always in error by the amount of his personal equation.

In running a meridian distance, however, this personal error, supposing it to be tolerably constant, will disappear, as his time being equally in error, and in the same direction, (either too fast or too slow) at both places, the difference of time, on which alone longitude depends, will not be affected. From this it results that in running a meridian distance, the same observers must always be employed. *Dis-appears in Meridian Distance.*

The formula for finding the equation of equal altitudes is as follows :—

Equation $= A + B.$ *Formula for Equa-tion of Equal Altitudes.*

$$A \text{ (in seconds of time)} = \frac{1}{15} \times \frac{c}{2} \text{ Tan lat} \times \text{Cosec} \frac{\text{elapsed time.}}{2}$$

$$B (\quad \text{do.} \quad) = \frac{1}{15} \times \frac{c}{2} \text{Tan dec} \times \text{Cot} \frac{\text{elapsed time.}}{2}$$

where $\frac{c}{2}$ is half the change of declination in the elapsed time, or, as we use it in the computation, the change of declination in half the elapsed time.

The rules for noting the algebraic signs of A and B will be given hereafter.

In making the observations it is most convenient to ascertain the Error of the hack watch, and thence, by using the comparisons, to arrive at the Error of the standard. In the case where the watch has a large rate, as shown by the comparisons before and after sights, the elapsed time must be corrected for the amount gained or lost by the watch in the interval on mean time, which can be roughly calculated from the known rate of the standard. *Error obtained of hack watch.*

Having meaned the sights, and obtained the mean middle time for each set, and knowing the estimated latitude and longitude, the rule for working a set of equal altitudes at superior transit will stand thus :—

1. Ascertain elapsed time by subtracting the central time of observation before transit from the central time after transit, increased, if necessary, by 12 hours. Halve this, and if there *Practical Rule for Calcula-tion of Equation.*

are decimals, they can be rejected, the nearest even second
being used.

2. To 0 hours apply longitude to find Greenwich date of
apparent noon at place; this is first Greenwich date.

3. Find second Greenwich date, by subtracting the half
elapsed time from the first Greenwich date.

4. Correct declination at apparent noon in Nautical
Almanac for the second Greenwich date.

5. Correct equation of time at apparent noon for first
Greenwich date.

6. Multiply the variation in declination for one hour, by
the half elapsed time, to get $\frac{c}{2}$.

N.B.—The variation we want is that at Greenwich time of
local noon, we must therefore correct the variation given in
the Nautical Almanac for the longitude.

7. For **A**, add together the logarithms of $\frac{c}{2}$, the **tangent of**
the latitude, and **cosecant** of half elapsed time ; and for B
the logarithms of $\frac{c}{2}$, the tangent of the declination, and the
co-tangent of half elapsed time. Either subtract the log. of
15 from each of these sums, to reduce the results to time at
once, or take out the natural numbers of the sums as they
stand, and when A and B have been added or subtracted,
divide the result by 15, to reduce it to time.

N.B.—Tables are given in various works on nautical astronomy to facilitate the calculation of A and B; but as these
are only made out for every so many minutes of elapsed
time, interpolation is necessary when working with any
pretence to accuracy, and very little is gained by their use in
their present form.

8. To the mean middle time of the set, apply the equation
of equal altitudes with its proper sign (rule given below),
which will give the time shown by the watch at apparent
noon. N.B.—When working several sets, calculate them

simultaneously as far as this, and mean the results, thus getting the mean time shown by the watch at apparent noon.

9. Find the mean time of apparent noon, by applying the equation of time with its proper sign to 0 or 24 hours, and take the difference between this and mean time shown by the watch, for the Error of the latter, subtracting one from the other, according as it is intended to show the watch as fast or slow on mean time.

An universal system must be adopted of showing all chronometers and hack watches as fast or slow of the standard and on mean time, not some one way and some another, which leads to confusion. It does not much matter which is taken. The writer has always shown them as slow on mean time. Thus all chronometers are shown slow of the standard, and the standard and all others slow on mean time of place, or of Greenwich, as the case may be.

All watches to be either slow or fast of Mean Time.

The rule for giving A and B their proper algebraic signs is as follows :—

Signs of Equation.

At Superior Transit.

A is
+ if declination is decreasing and of same name.
+ if declination is increasing and of different name.
− if otherwise.

B is
+ if declination is increasing
− if declination is decreasing
} When elapsed time is *less* than 12 hours.

Reversed when elapsed time is greater than 12 hours.

At Inferior Transit.

A is reversed from what it would be at superior transit.
B is the same as at superior transit.

In working with inferior transit, whereby we find the Error at midnight, there is no difference in the rule, except that in calculating the change during the half-elapsed time, we use the variation of declination found by interpolation for the Greenwich time of local midnight.

Change of declination in Inferior Transit.

The next step is to calculate, from the comparisons taken with the standard before and after sights, a mean comparison

Calculating a Mean

Compa-
rison for
Haek
Watch.

to apply to the Error of watch found above, to arrive at the Error of the standard.

To do this, we take any sight, and by interpolation calculate the comparison at the A.M., and also the P.M. time corresponding. The mean of these two will give the comparison at noon. This should, if the watch has been going well, correspond very closely with the comparison actually taken at noon, and it will be satisfactory if it does so. If it does not, we cannot help it; but we shall know that the Error of the standard will be slightly incorrect from a jump in the watch, and shall be prepared to give the result a smaller value in consequence, in event of discrepancies with others.

Noon Com-
parison
not to be
used.

The mean comparison, as found above, must always be used, not the comparison *taken* at noon, which is done solely to ascertain how the watch has been going.

An example of the calculation follows.

AT MESALE l^d △   Aug. 31st, 1878.

SIGHTS OBTAINED FOR ERROR BY EQUAL ALTITUDES.   LAT. 5° 14 S.
LONG. 39° 40′ E.

	h. m. s.		h. m. s.
P.M. Time by watch ..	10 16 22·4	Long ..	−2 38 44
A.M. „ „ ..	3 55 50·6	½ El. T.	3 10 16
El. time ..	6 20 31·8	2nd G. date..	−5 49 0
½ El. time for use	3 10 16	31st.	

	o ′ ″		″		m. s.
Dec. ..	8 36 30	Var. ..	54·18	Eq. T. ..	0 12·09
Correction	5 14		5·8		2·03
Dec. ..	8 41 44		43344	Eq. T. ..	0 14·12
			27090		

Var ..	54·18	″	s.	Var ..	0·768
E.T.	3·17	314·244			2·64
2					

	37926			3072
	5418			4608
	16248			1536

$$\frac{c}{2} = 171\cdot6906''$$

2·02752 s.

Tan lat	..	8·961866		Tan dec		9·184541
Cosec $\frac{E.T.}{2}$		·131907		Cot $\frac{E.T.}{2}$ ..		9·961038
$\frac{c}{2}$	..	2·234770		$\frac{c}{2}$	..	2·234770
		1·328543				1·380349
15	..	1·176091		15	..	1·176091
		0·152452				0·204258
		A = − 1·421			B = − 1·600	
					A − 1·421	

Equation of equal alt. − 3·021 secs.

			h.	m.	s.
Mean mid. time	..	..	7	06	06·51
Eq. of Eq. alts		..		−	03·02
Time by watch at App. Noon		7	06	03·49

	h.	m.	s.			h.	m.	s.
Time by watch by 11 observations. ☉	7	06	03·41	⎫ Mean of two sets	7	06	03·45
11 ,, ☿			03·49	⎭			
11 ,, ☉			03·30	⎫ ,, ,,			03·44
11 ,, ☿			33·59	⎭			

Mean Time by watch .. 7 06 03·45
Mean time of App. Noon 12 00 14·12

Watch (Breguet) slow ... 4 54 10·67

To calculate the comparison between standard (A) and watch at noon, we have the following comparisons observed:—

		Before A.M. Sights.			Check.		After A.M. Sights.			Check.
		h.	m.	s.	secs.		h.	m.	s.	secs.
A.	..	4	16	55·0	02·6	..	5	37	10	19·2
Breguet	..	3	30	02·4	10·0	..	4	50	17	26·0
		0	46	52·6	52·6		0	46	53·0	53·2
			Mean 52ˢ·6					Mean 53ˢ·1		

		Noon.			Check.		Before P.M. Sights.			Checks.	
		h.	m.	s.	secs.		h.	m.	s.	secs.	secs.
A	..	7	50	50	06·2	..	9	47	25·0	38·8	50·0
Breguet	..	7	03	55·8	12·0	..	8	50	30·6	44·0	55·4
		0	46	54·2	54·2	..	0	46	54·4	54·8	54·6
			Mean 54ˢ·2					Mean 54ˢ·6			

		After P.M. Sights.			Check.
		h.	m.	s.	secs.
A	..	..	11	26	10·0	21·0
Breguet	..	..	10	39	14·8	25·8
			0	46	55·2	55·2
				Mean 55ˢ·2		

Taking any sight, say the middle times at the set we have
shown worked out, we find by interpolation that

$$\begin{array}{llll} & & \text{m.} & \text{s.} \\ \text{At} \quad 3\cdot56 \text{ by watch, comparison is} & 46 & 52\cdot74 \\ \text{At} \ 10\cdot16 \quad \text{,,} \qquad \text{,,} \qquad \text{,,} \qquad \text{,,} & 46 & 55\cdot07 \end{array}$$

and as these are equal times from noon, the noon-comparison
will be the mean of these, or $46^{m\cdot}\ 53^{s}\cdot90$. This differs $0^{s}\cdot3$
from the observed noon-comparison, which, supposing the
comparisons to have been carefully observed, means that
there has been a slight irregularity in the motion of the
watch, which must be remembered in comparing any meridian
distance founded on these sights with others; but in this case
it is so small as scarcely to be taken into consideration.

We now apply this mean comparison to Error of watch.—

	h.	m.	s.
Breguet slow ..	4	54	10·67
Comparison ..		46	53·90
Standard A slow on M.T. place	4	07	14·77

We next take the comparisons observed at noon between A
and all the other chronometers; and applying them to A's
Error, we get the Error of each.

ERROR BY EQUAL ALTITUDES OF TWO STARS ON OPPOSITE SIDES OF THE MERIDIAN, BY CAPTAIN A. M. FIELD, R.N.

The principle of this method depends upon the sidereal
time of passing the meridian of a place, by an imaginary star
having the mean right ascension of the two stars selected,
being compared with the time shown by a sidereal chronometer at that instant; the difference is its error on sidereal
time. A mean solar chronometer can be used equally well.
The sidereal time required is the mean of the right ascensions
of the two selected stars. The chronometer time at that
instant (mean or sidereal) is the mean of the times at which
the eastern and western stars had equal altitudes, with the
"equation of equal altitudes" applied with its proper sign.

When preparing a list of stars for observing, it will be necessary to first find the R. A. of the meridian for the times between which it is required to carry on the observations. As stars will generally be observed within 4^h of the meridian, the limits of R. A. of the stars falling within the required period may be obtained by subtracting 4^h from the first R. A. of the meridian, and adding 4^h to the last R. A. of the meridian as found above.

To arrange the stars in pairs, it is necessary to select two bright stars of nearly the same declination, not differing much from the latitude, but differing in R. A. by from 4^h to 8^h. The time at which they will be simultaneously of equal altitude will be about the time when the mean of their R. A.'s is on the meridian.

The time at which it will be necessary to begin observing will be governed by this, and the observations of one star should be completed shortly before that time, in order to allow an interval to prepare for observing the other.

It may be found that when two stars are of equal altitudes, that they are too near the meridian for obtaining the best results, or that the altitudes are too great for the sextant; in which case it would be necessary to commence observations of the eastern star as much as an hour earlier, and the western star will then have to be observed the same time later than when they would be simultaneously of equal altitude.

As a general rule, if the difference in right ascension is less than 6^h, the eastern star should be observed first; if it exceeds 6^h, then the western star; this is in order that the stars may be observed as favourably as possible with respect to the "prime vertical," but it will vary according to the latitude and declination. It will be noticed that if the observations are commenced with the eastern star, then they are, as a whole, taken further from the meridian than in the other case.

If the difference in R. A. exceeds 8^h, then there will probably be an interval between finishing the observations of one star and beginning those of the other, and part of the advantages of the method are lost; the same remark applies if the difference in R. A. is less than 4^h.

Having decided on which star to begin with, observe it continuously in the ordinary way, until the sidereal time is nearly equal to the mean R. A. of the two stars (the error of chronometer on sidereal time should be roughly known), when prepare to observe the other star, commencing at the same altitude as the last observation of the first star, and complete the series, which may be divided into sets in the usual way.

Owing to the more rapid change in "equation of equal altitudes" when there is a large difference in the declinations of the stars, it will be remarked that the "middle times" vary more rapidly than in the case of the sun; and the rapidity of this change increases as the observations get further away from the sidereal time at which the "imaginary star" passes the meridian.

If a mean solar chronometer be used the chronometric interval (corrected for rate) must be turned into a sidereal interval, and the resulting "error of chronometer" will be the error on sidereal time at that particular instant, from which the error on mean time can be readily deduced.

Equation of Equal Altitudes.

The rigorous expression, according to Chauvenet, is:—

$$\text{Sin } a = \cot \tfrac{1}{2} \text{ E.T. } \tan d. \tan \delta. \cos.a - \operatorname{cosec}.\tfrac{1}{2}\text{E.T.} \tan. l. \tan \delta.$$

where a = equation of equal altitudes $= \dfrac{h - h^1}{2}$.

$\tfrac{1}{2}$ E.T. $= \tfrac{1}{2}$ elapsed time $= \dfrac{h + h^1}{2}$

d = declination at upper meridian passage.

$d - \delta =$ „ „ observation E. of meridian.

$d + \delta =$ „ „ „ W. „

$\delta = \tfrac{1}{2}$ difference of the declination at the two times of observation.

h and $h^1 =$ Hour angles from noon at East and West observations respectively.

In the case of equal altitudes of two stars, one East and the other West of meridian, the above formula is strictly accurate, whatever may be the difference in the declinations; the half elapsed time being found as follows:—

$$\tfrac{1}{2}\ \text{E.T.} = \frac{h + h^1}{2}$$

$$= \frac{(R - S) + (S^1 - R^1)}{2}$$

$$= \frac{(R - R^1) + (S^1 - S)}{2}$$

Where, R and $R^1 =$ Right ascensions of star E. and star W.

S and $S^1 =$ Sidereal time of observation of star E. and star W. respectively.

$S^1 - S$, the difference of the sidereal times, may be taken as equal to the difference of times (as shown by a sidereal chronometer) of the observation of the two stars; the chronometer being of course corrected for rate.

If $R^1 > R$, add 24^h to R.

If the western star be observed first, in which case $S > S^1$, then S and S^1 are treated algebraically.

$d =$ the mean of the declinations of the two stars.

$\delta = \tfrac{1}{2}$ difference   „   „   „

In working the rigorous expression, an approximate value for a is first obtained, disregarding the term Cos. a, and then rework, using the value of a thus found.

Notes on the Computation.

The constant logarithm 4·138339 is an abridgment of Chauvenet's formula, by adopting it to circular measure, and is correct so long as the equation of equal altitude is small; certainly up to 4 minutes, no error is introduced by its use.

R

The cosine of a with an equation of equal altitude under 4 minutes changes so slowly that it is practically the same for each set unless separated by a very long interval.

The R. A. M. S. must be found for the Greenwich date corresponding to the chronometer middle time. In cases of a small equation of equal altitude and changing slowly, this is the same for each set ; but where it is large and therefore changing rapidly, as in the second example, the R. A. M. S. must be corrected for every set, but this merely involves applying the acceleration due to the difference of the chronometer middle time of each successive set to the R. A. M. S. for the first set.

A more or less accurate knowledge of the G. M. T. is therefore necessary, but this is inherent to the use of stars for time under all circumstances, and a second approximation must be made if the error on local time has been assumed more than three or four seconds in error.

It is worthy of remark that $\tan d \times \tan \delta \times 4\cdot138339$ and $\tan l \times \tan \delta \times 4\cdot138339$ form absolutely constant logarithms for the same stars, at the same place, for any night in the year, and it is only necessary to add the logarithm of $\frac{1}{2}$ E. T. to each to obtain the equation of equal altitude corresponding to that $\frac{1}{2}$ E. T.

The equation of equal altitude is always so large as to exclude all possibility of doubt as to which way to apply it, but the investigation gives the rule. Two examples are given; the first illustrates the case of two stars differing by $1° 40'$ in declination, and the second where they differ by $8° 25'$.

In the first instance the equation of equal altitudes hardly changes between the first observation and the last, and, therefore, the sums of the middle times do not vary. This is due to the difference in declination being small, and to the fact that the middle time of observation is only $10^m 16^s$ from the "time of crossing," or the time at which the stars had equal altitudes, the one rising and the other setting, and is represented by the "$\frac{1}{2}$ interval."

In the second instance the equation of equal altitude changes very rapidly; the difference in declination is large, and the "$\frac{1}{2}$ interval" is also somewhat large, viz., $44^m 20^s$; but the "$\frac{1}{2}$ interval" may, nevertheless, be extended to upwards of an hour if necessary, to wait for clouds, &c., without affecting the accuracy of the result.

The acceleration in the change of equation of equal altitudes may be considered as practically uniform for the short intervals of four or five minutes necessary to obtain sets of observations, and therefore, although it will be noticed that the sums of the middle times change rapidly, yet it will nevertheless be perfectly accurate to take the means and calculate the equation of equal altitude corresponding to the "$\frac{1}{2}$ interval" for the middle observation, no matter how rapidly the equations may change.

EXAMPLE 1.

On the evening of 28th November, 1892, at Hong Kong Dockyard, latitude 22° 16′ 55″ N., longitude 114° 9′ 49″ E., the following observations of Markab (West of meridian) and Aldebaran (East of meridian) were obtained:

Interval.	Markab (W.) Time by watch.	Double Altitude.	Sums.	Aldebaran (E.) Time by watch.	Interval.
″	h. m. s.	° ′	″	h. m. s.	″
	1 15 14·8	106 20	55·2	1 38 40·4	42·8
43·2	1 15 58·0	106 00	55·6	1 37 57·6	43·2
43·2	1 16 41·2	105 40	55·6	1 37 14·4	43·6
43·6	1 17 24·8	105 20	55·6	1 36 20·8	43·2
42·8	1 18 07·6	105 00	55·2	1 35 47·6	—
			5) 2·2		
			2)55·44		
			57·72		

	h. m. s.
Markab ..	1 16 41
Aldebaran ..	1 37 14
	2)2 53 55
Middle time ..	1 26 57
½ Interval ..	0 10 16

Mean middle time by watch = 1   26   57·72

½ Interval ..   ..   = 0   10   16

COMPUTATION.

		h. m. s.
½ Interval ..	=	0 10 16
Mean Middle Time ..	=	1 26 57·72
Equation of equal altitude ..	+	1 01·99
Time by watch of crossing ..		1 27 59·71

Tan d ..	..	9·441883
Tan δ ..	..	8·162336
Constant logarithm ..	..	4·236339
Constant logarithm (for all sets with these stars) ..		1·742758
Cot ½ E. T. ..	..	10·095962
Cos equation equal altitude ..	..	9·999996
		1·838716

	=	68·98
		130·97
Equation of equal altitude : ..	=	m. s. 61·99 or 1 01·99

		h. m. s.
R. A. Markab ..	=	22 59 24·93
R. A. Aldebaran ..		4 29 47·08
		3 29 12·01
Mean R. A. ..	=	1 44 36·005
½ Difference ..	=	2 46 11
½ Interval (Sidereal Time) ..		0 10 18
½ Elapsed Time ..	=	2 34 53

Tan l ..	..	9·441883
Tan δ ..	..	8·162336
Constant logarithm ..	..	4·136333
Constant logarithm for all sets with these stars ..		1·913425
Cosec ½ E. T. ..	..	10·293754
		2·117179

		secs.
	=	130·97

GREENWICH DATE.

		o ' "
Declination ..	: :	14 37 47 N.
Declination ..	: :	16 17 45 N.
		30 55 32
d	=	15 27 46
δ		0 49 69

		h. m. secs.
Middle time by watch ..	: :	1 26 57·7
Approximate error ..		7 44 57
		9 11 55
S. M. T. ..	=	7 36 39 E.
Longitude in time ..		
Greenwich date, 28th November		1 33 16

		h. m. secs.
Sidereal time ..	=	16 31 23·380
Acceleration, 1 h.		9·856
35 m.		5·750
16 s.		0·044
R. A. M. S.		16 31 39·030
Mean R. A. ..	=	1 44 36·045
S. M. T. of crossing ..	: :	9 12 56·975
Time by watch of crossing ..		1 27 59·710
Watch slow on M.T. at 9.13 p.m.		7 44 57·265

Example 2.

On the evening of 13th November, 1892, at Fullerton Battery, Singapore, latitude 1° 17′ 11″ N., longitude 103° 51′ 15″ E., the following observations of Markab (West of meridian) and γ Orionis (East of meridian) were obtained:—

Interval	Markab (W.) Time by Watch	Double Altitude	Sums	γ Orionis (E.) Time by Watch	Interval
″	h. m. s.	° ′	″	h. m. s.	″
—	2 41 12·4	106 00	20·0	4 14 07·6	41·6
43·6	2 41 56·0	105 40	22·0	4 13 26·0	40·0
42·4	2 42 38·4	105 20	24·4	4 12 46·0	42·0
44·0	2 43 22·4	105 00	26·4	4 12 04·0	40·4
43·2	2 44 05·6	104 40	29·2	4 11 23·6	40·0
43·6	2 44 49·2	104 20	32·8	4 10 43·6	40·0
43·2	2 45 32·4	104 00	36·0	4 10 03·6	—

7) 50·8

2) 27·26

43·63

Mean middle time by watch = 3h 27m 43·63s

½ Interval .. .. = 0 44 21

	h. m. s.
Markab ..	2 43 22
γ Orionis ..	4 12 04
	6 55 26
Middle Time ..	3 27 43
½ Interval ..	0 44 21

COMPUTATION.

	h. m. s.			h. m. s.				° ′ ″
† Interval .. =	0 44 21	R. A. Markab ..	22 59 25·100		Declination ..	14 37 47·2 N.
		R. A. γ Orionis	5 19 23·548		Declination ..	6 12 55·5 N.

Mean Middle Time .. = 3 27 43·63
Equation of equal altitude .. + 3 31·23

Mean R. A. .. 4 18 48·648
20 50 43·7

Time by watch of crossing .. 3 31 17·86

.. 2 09 24·324
d 10 25 21
δ 4 12 26

† Difference 3 09 59·2
† Interval (Sidereal Time) 0 44 28·5

GREENWICH DATE.

† Elapsed Time .. 2 25 30·7

			h. m. s.
Tan d ..	9·264681	Tan l ..	8·351616
Tan δ ..	8·866649	Tan δ ..	8·866649
Constant logarithm ..	4·138339	Constant logarithm ..	4·138339

Middle time by watch .. 3 27 43·6
Approximate error .. 7 05 14

S. M. T.
Longitude in time .. 10 32 58
6 05 25 E.

Constant logarithm for alt sets with { the ― stars .. 2·369669
Cot † E. T. .. 10·132731
Cos equal altitude .. 9·999917

Constant logarithm for alt sets with } .. 1·366604
Cosec † E. T. .. 10·220670

Greenwich date, 13th November 3 37 33

2·402350

1·583474

Sidereal time .. 16 32 16·020
Acceleration, 3 h. 29·569
37 m. 6·078
33 s. 6·091

secs.
252·85
38·32

secs.
38·32

R. A. M. S. .. 15 32 50·758
Mean R. A. .. 2 09 24·324

Equation of equal altitude .. 214·23 = m. s. 3 31·23

S. M. T. of crossing .. 10 36 33·666
Time by watch of crossing .. 3 31 17·860

Watch slow on M.T. at 10.37 p.m. 7 05 15·706

CHAPTER XIV.

MERIDIAN DISTANCES.

Telegraphic—Chronometric.

UNDER this head we shall consider all the methods, available for our purposes, of obtaining difference of longitude.

TELEGRAPHIC MERIDIAN DISTANCE.

Where a telegraph can be used, it is of course the best, and at the same time the simplest, means of obtaining difference of longitude.

This method consists in sending a current through the wire at a known local time from one place, the local time of arrival at the other place being noted. The difference of these is the difference of longitude.

Retarda-tion. In theory, the passage of the current through the wire is instantaneous; but in practice it takes an appreciable time, when the distance is considerable, and the electrical condition of the wire is not first-rate; and to eliminate this, and also to decrease errors of sending and receiving, we must send several sets of signals in both directions equally, the mean of which will give the true time.

A little consideration will show that if a signal is sent from A to B, a place to the westward, and it takes two seconds to traverse the wire, the time at B will have had those two seconds in which to catch up the A time, which is so much ahead; or, in other words, the difference of the two times as shown will be two seconds too little. Whereas, if the signal is sent from B to A, the watch at A, already ahead, will advance another two seconds before the signal arrives, and the difference of the two times will be two seconds too much. The mean of the two will therefore be correct, and half the difference of the two will give the " retardation of the wire," a matter, however, purely of curiosity as far as our results are concerned. In land lines the retardation is about $\frac{1}{10}$ of a second per thousand miles, with submarine cables it is larger and varies.

To eliminate all personal errors, the only satisfactory method is for the observers to change ends. When this is not possible, personal errors when obtainable should be applied. These can only be determined by a considerable series of observations for time, and the error of each observer or the person chosen as standard being recorded and meaned. The steadiness, and therefore the value, of the personal error can thus be judged.

In this case a number of chronometers is not necessary. All that is wanted is one good time-keeper. If, however, two watches are at hand, it is not amiss to ascertain the Errors of each separately, and use them both in transmitting the signals.

Only one watch necessary.

A box-chronometer is the best for sending and receiving signals by, and if practicable, it may be a good plan to land one, and let it stand in the telegraph office for a few days beforehand to settle down, comparing the watch actually used at sights with it, before and after observations.

Sights must be obtained on the day of sending the signals, and the latter should be transmitted at or about noon or midnight. Where the places are far apart in longitude, it can only be near noon at one place, and Error must be obtained at the other, either on the day before or after, as well, so as to be able to correct the Error to the time of interchange of signals. Using land lines, day-time is most favourable for exchange of signals.

Time to exchange signals.

Observa-
tion Spot.
If the observation spot can be at, or close to, the telegraph office, it is convenient, as the watch will not have to be carried about; but in many instances the local arrangements will not admit of this.

Galvano-
meter.
Telegraphic instruments differ very much ; but it does not much matter which are used, as long as they are similar at both ends. The deflection of an ordinary galvanometer-needle of Wheatstone's instrument, or of the Morse recorder, or of the more delicate mirror of long submarine cables, will all serve our purpose. Preference is given to one or the other by different observers. The writer prefers an instrument giving a sound, to the silent movement of the suspended mirror.

Each signal will consist of one deflection, and the key should be kept pressed down for about a second.

Pre-
arranged
method of
sending
signals.
In sending the signals, it must be clearly arranged beforehand what is going to be done.

A good plan is as follows :—

In commencing, give a warning, say of three rapid signals, at ten seconds before an even minute by the sender's watch. The first signal will then go at the even minute, and at every ten seconds another, *missing the fifty seconds, to mark the even minute,* for three minutes, ending with another even minute.

After an interval of three or four minutes, a similar set will be sent in the reverse direction.

If at the receiving ends the signals agree, this will be quite sufficient, unless we intend to use another watch.

An example of a telegraphic meridian distance is appended.

It will be seen that the resistance of the wire and instrumental retardation was less on one day than on the other, amounting at one time to nearly a tenth of a second, and at the other to only twenty-five thousandths.

April 3rd and 18th, 1880.

Observation spot at Constantinople was at **Leander's Tower**, observers having to go 1½ miles to Telegraph Office by caïque.

Observation spot at Dardanelles **at** Nagara Light-house, observers having to go three miles **to** Telegraph Station by steam pinnace.

Series.	Watch Times.		Local Times.		Meridian Distance.	Remarks.
	Sending.	Receiving.	Sending.	Receiving.		
	h. m. s.		h. m. s.			
Dardanelles to Constantinople.	11 31 00	Missed.	11 42 48·2	Missed.		Sent by Breguet 2084.
		h. m. s.		h. m. s.	h. m. s.	Received by Dent 6119.
	10	11 44 15·2	58·2	11 53 22·1	0 10 23·9	
	20	25·3	43 08·2	32·2	24·0	
	30	35·2	18.2	42·1	23·9	
	40	45·3	28·2	52·2	24·0	
	32 00	45 05·3	43 48·2	54 12·2	24·0	
	10	15·2	58·2	22·1	23·9	
	20	25·2	44 08·2	32·1	23·9	
	30	35·2	18·2	42·1	23·9	
	40	45·3	28·2	52·2	24·0	
	33 00	46 05·2	48·2	55 12·1	23·9	
	10	15·2	58·2	22·1	23·9	
	20	25·2	45 08·2	32·1	23·9	
	30	35·2	18·2	42·1	23·9	
	40	45·2	28·2	52·1	23·9	
	34 00	47 05·2	48·2	56 12·1	23·9	
				Mean ..	0 10 23·93	
Constantinople to Dardanelles.	11 49 00	11 35 55·2	11 58 06·9	11 47 43·4	0 10 23·5	Sent by Dent 6119.
	10	05·0	16 9	53·2	·7	Received by Breguet 2084.
	20	15·0	26·9	48 03·2	·7	
	30	25·0	36·9	13·2	·7	
	40	35·0	46·9	23·2	·7	
	50 00	54·9	59 06·9	43·1	·8	
	10	37 04·8	16·9	53·0	·9	
	20	14·8	26·9	49 03 0	·9	
	30	24·9	36·9	13·1	·8	
	40	35·0	46·9	23·2	·7	
	51 00	54·8	12 00 06·9	43·0	·9	
	10	04·9	16·9	53·1	·8	
	20	14·9	26·9	50 03·1	·8	
	30	Missed.	—	—	—	
	40	34·9	46·9	23·1	·8	
	52·00	55·0	01 06·9	43·1	·7	
			C. to D.	Mean ..	0 10 23·76	
			D. to C.	,, ..	23·93	
		April 3rd	Mean Mer.	Dist. ..	0 10 23·84	

April 18th, 1880.

Series	Watch Times.		Local Times.		Meridian Distance.	Remarks.
	Sending.	Receiving.	Sending.	Receiving.		
	h. m. s.	h. m. s.	h. m. s.	h. m. s.	h. m. s.	
Dardanelles to Constantinople.	11 41 00	11 51 50·8	11 50 06·4	12 00 30·1	0 10 23·7	Sent by Breguet 2084. Received by Dent 6119.
	10	52 00·8	16·4	40·1	·7	
	20	10·8	26·4	50·1	·7	
	30	20·8	36·4	01 00·1	·7	
	40	30·8	46·4	10·1	·7	
	42 00	50·8	51 06·4	30·1	·7	
	10	53 00·8	16·4	40·1	·7	
	20	10·8	26·4	50·1	·7	
	30	20 8	36·4	02 00·1	·7	
	40	30·8	46·4	10·1	·7	
	43 00	50·8	52 06·4	30·1	·7	
	10	54 00·8	16·4	40·1	·7	
	20	10·8	26·4	50·1	·7	
	30	20·8	36·4	03 00·1	·7	
	40	30·7	46·4	10·0	·6	
	44 00	50·8	53 06·4	30·1	·7	
					0 10 23·7	
Constantinople to Dardanelles.	11 57 00	11 46 09·2	12 05 39·3	11 55 15·6	0 10 23·7	Sent by Dent 6119. Received by Breguet 2084.
	10	Missed.	—	—	—	
	20	29·2	59·3	35·6	·7	
	30	39·2	06 09·3	45·6	·7	
	40	49·2	19·3	55·6	·7	
	58 00	47 09·2	39·3	56 15·6	·7	
	10	19·2	49·3	25·6	·7	
	20	29·2	59·3	35·6	·7	
	30	39·3	07 09·3	45·7	·6	
	40	49·2	19·3	55·6	·7	
	59 00	48 09·3	39·3	57 15·7	·6	
	10	19·3	49·3	25·7	·6	
	20	29·3	59·3	35·7	·6	
	30	39·4	03 09·3	45·8	·5	
	40	49·2	19·3	55·6	·7	
	12 00 00	49 09·3	39·3	58 15·7	·6	

C. to D. Mean ..	..	..	0 10 23·65
D. to C. „ ..	..	..	23·70
April 18th, Mean Mer. Dist.	..		0 10 23·67
„ 3rd, „ „	..		10 23·84
Final Mean Mer. Dist.	..		0 10 23·75

CHRONOMETRIC MERIDIAN DISTANCES.

When we have no telegraph we must have recourse to chronometers for conveying the time.

Having obtained sights at the two places whose meridian distance we require, we come to the consideration of the rate to be used.

If we have been able to run backwards and forwards, as recommended on page 220, we shall use a travelling rate.

The algebraic formula for finding the meridian distance by travelling rate, when we return at once to the original station, is as follows :— *Formula for Travelling Rate.*

$$M = \beta - a - n\frac{a' - a}{m + n},$$

where M = meridian distance,

a = error at place A, before starting,

$a' = \quad\quad$ „ $\quad\quad$, on returning,

$\beta = \quad\quad$ „ $\quad\quad$ B,

n = No. of days between first observations at A and those at B,

m = No. of days between observations at B and those at A, on returning.

Then $\dfrac{a - a}{m + n}$ = travelling rate.

This can be put another way, by example, as follows :—

Let us suppose we have obtained Error at Maziwi at noon, August 27th; we have been to Mesale, and there obtained Error at noon on the 31st; and then, returning to Maziwi, have obtained another Error there at midnight of September 1st–2nd. *Example of Travelling Rate.*

To find rate in this case, we simply divide the difference of the Errors ascertained on the 27th and 1st by 5½ (the interval between them). This rate, multiplied by 4 (the interval between sights at Maziwi on 27th and Mesale on 31st), will give the quantity to be applied to the error of the chronometer in question at Maziwi on the 27th, to give the error

on the 31st on mean time of Maziwi. The difference of this
and the error of the same chronometer on mean time of
Mesale, as ascertained on that day, will be the meridian
distance by that chronometer.

Form for Meridian Distance.
In working out a meridian distance with several chronometers, it is convenient to use a form, as shown in the
example of the above-cited instance (page 255).

Rejection of results.
Here so many of the chronometers agree closely, that
the result by D seems doubtful, and looking at the comparisons taken every day with the standard, we see it has been
going very irregularly; we therefore reject it. This should
not be done without some independent evidence of this kind;
and in a meridian distance, where the interval of time is
great, or where all the chronometers have been going but
fairly, as shown by the daily comparisons, it is very unsafe
to reject chronometers solely because they vary from a small
majority of the others.

Another case of Travelling Rates.
Supposing that we had had to stay at Mesale for a few
days before returning to Maziwi, we can still find a fair
travelling rate.

The formula for this is as follows:—

$$M = \beta - a - n\frac{(a^1 - a) - (\beta^1 - \beta)}{m+n},$$

where, the other letters representing the same values,

β^1 is the error at place B before leaving.

Here the travelling rate is,

$$\frac{(a' - a) - (\beta^1 - \beta)}{m+n}.$$

This can be exemplified thus. We obtain sights at Maziwi
on 27th and at Mesale on 31st as before. Then sights again
at Mesale on the 4th noon, and again at Maziwi on the 6th
noon.

From the difference of the errors on the 27th and 6th we
should deduct the difference of the errors on the 31st and 4th,
and divide the remainder by 6, the sum of the intervals from
the 27th to the 31st, and of the 4th to the 6th, or, in other

MERIDIAN DISTANCE BETWEEN MAZIWI Δ AND MESALE Δ, USING TRAVELLING RATES. CHRONS. SLOW ON M. T. PLACE.

	A	B	C	D	E	F	G	H
1878.	h. m. s.	h. m. s.	h. m. s.	h. m. s.	h. m. s.	h. m. s.	h. m. s.	h. m. s.
a ... Maziwi, Aug. 27	4 05 01·33	0 38 59·13	4 55 26·58	3 48 58·53	3 08 42·63	2 59 10·98	5 52 45·73	3 10 56·18
a' ... „ Sept. 1½	11·05	37 43·70	56 12·55	49 29·20	44·10	16·55	53 05·05	11 09·75
5½ days	09·72	1 15·43	45·97	30·67	01·47	05·57	19·32	13·57
$\frac{a'-a}{m+n}$ Daily rate	— 1·767	+ 13·714	— 8·368	— 5·576	— 0·267	— 1·012	— 3·512	— 2·467
$n\,\frac{a'-a}{m+n}$ 4 days rate	07·07	54·86	33·43	22·30	01·07	04·05	14·05	09·87
Maziwi, Aug. 31	4 05 08·40	0 38 04·27	4 56 00·01	3 49 20·83	3 08 43·70	2 59 15·03	5 52 59·78	3 11 06·05
Mesale „	4 07 16·77	0 40 13·02	4 58 08·47	3 51 30·07	3 10 51·97	3 01 23·42	5 55 08·37	3 13 14·52
Meridian dist...	2 08·37	2 08·75	2 08·46	2 08·24 Reject.	2 08·27	2 08·39	2 08·59	2 08·47

Mean Meridian Distance .. .. $2^m\ 08^s\cdot47$ E. of Maziwi.

Range .. .. $0^s\cdot48$.

words, the number of days actually travelling, which will give us the rate. We then proceed as before.

The travelling rate obtained in this instance will not be as good as in the former case, as the chronometers will have had two disturbances instead of one, and the rate they may settle into on starting the second time, after four days' quiet at anchor, may not be the same as before; but it will still be better than obtainable by any other method, and if circumstances of weather, sea, and temperature are nearly alike on both journeys, and the intervals are not long, we shall probably get a very good result.

Travelling rates, obtained thus, should always, as already remarked, be used when the scale of the chart depends on the observations.

The method is very simple, and, used for this purpose, none of the considerations of temperature, &c., hereafter mentioned need be thought of, as the time is short.

Other Rates. We now come to the consideration of the rates to be used on other occasions, especially when voyages are long, and circumstances change much during them.

This is a very wide subject, and besides the fact that it has already been fully discussed by Captain Shadwell, in his masterly treatise before referred to, neither space nor the intention of these pages permits our going very far into it, and we shall content ourselves with giving general descriptions of cases, together with formulæ for them, with just sufficient reasons to allow of their being understood.

The whole question rests on,—What makes chronometers vary?

Why Watches change their rates. The labours of many observers show us that the answer is :—

1. Imperfection in the workmanship of the watch.

2. Changes of temperature.

3. The quality of the oil in the pivots, and its age (*i.e.* the time elapsed since the watch was last cleaned).

4. Accidental shocks or vibrations imparted to the watch.

A supplementary question may be asked—Which of these

is the most important ? To which the general answer is that, according to circumstances, any one may be.

First, Imperfection of workmanship.

For this manifestly there is nothing to be done. A badly made chronometer will go so erratically that we shall soon lose confidence in it, and reject it from all results, returning it as soon as we can. There are, however, but few chronometers that pass through the hands of the Royal Observatory which will come under this head, and doubtless many a chronometer has been classed in this category from ignorance of the circumstances of its compensation, and its resulting variation under change of temperature. If on a voyage, *during which temperature is uniform*, a chronometer placed with others, under the same conditions of protection from injury, &c., goes erratically, while the others maintain their rates pretty steadily, we may fairly conclude it to be inferior.

The uniformity of rate of a chronometer while on shore, or when the ship is at rest, cannot be taken as a conclusive test.

Secondly. Change of temperature.

A chronometer is supposed to be compensated in such a manner that at two temperatures, a varying number of degrees apart, the rates will be equal. At all other temperatures the rates will vary, reaching a maximum at about the mean temperature between the other two.

Let us, for brevity, call this temperature of maximum rate T.

If we then examine the rates of a chronometer we should find a steady change of rate in one direction (nearly always in the direction of acceleration of gaining) from low temperatures to high, until we reach T, when the change of rate should vary in the opposite direction.

For every chronometer we shall have a different quantity for T, and different coefficients of change. Many chronometers are supposed to be compensated for $T = 60°$, the mean temperature generally experienced over the globe; but it would seem that makers cannot command the point T; any-

Marginal notes: Imperfection. Change of Temperature. Temperature of Maximum Rate.

way many have T over 90°, so that for such a watch, in practice, the direction of change is invariable, which will result in a great accumulation of difference of rate when passing through hot and cold climates, and where the coefficient is large, in great absolute change of rate.

Different conclusions. Different observers on the performances of chronometers have come to different conclusions on the subject of the law of change for a degree on all parts of the scale, which can only be accounted for by supposing that they have experimented on different classes of time-keepers.

Some have stated that they vary regularly, so as to have the same rate at an equal number of degrees above or below T, and have established the proportion of variation at the square of the difference of T, and the temperature required.

Other experiments have shown that the manner in which watches vary is not quite so regular as this, and that the coefficient of change is generally less at temperatures higher than T, than at those below.

No strict law. The fact is, that there is no invariable law on the subject, a watch being too complicated a machine to admit of any practical conclusion, unbased on actual experiment with each individual watch.

Practical experiments. Experiment, however, does give results that can be practically used, and tables of rates can be formed from observation of the watch at different fixed temperatures, which, with some watches, will undoubtedly give better results than by using invariable rates.

Liverpool Observatory. Tables of this kind are now furnished to ships sailing from Liverpool, whose chronometers are rated at the Bidston Observatory, the director of which, Mr. Hartnup, has studied the question for many years.

The rate of the watch to be used for determining the position at sea is then taken day by day from the table, according to the temperature experienced, and added to the accumulated rate since departure, obtained in a similar manner.

It seems pretty well established that the coefficient of

change for a degree remains the same, or nearly the same, for each individual watch, although the absolute rates of the watch (which depend upon many things) may vary.

The chronometers issued to H.M. ships have no such information sent with them, for this reason. *Chronometers in H.M. Ships.*

The timekeepers are carefully chosen from many sent to the Royal Observatory by different makers for trial, and only those whose compensation is such that they show very little change of rate at a great variation of temperature, or, in other words, whose compensation is as perfect as may be, are taken, the limit allowed being one and a quarter second of change of daily rate for forty-five degrees of temperature.

This reduces the variation of rate, arising from change of temperature probable in a voyage, to very small quantities, which would be lost in the variation arising from other causes, and it is not considered necessary under these circumstances to give data for allowing it.

Thirdly. The oil in the pivots. With good oil the inequality arising from age shows itself in the shape of a gradual and tolerably uniform acceleration of rate, generally in the direction of gaining, with a new chronometer, and when the instrument is older and all parts somewhat worn, in the contrary direction. It should be excessively small, and our opinion is that in the practical question of meridian distances, the labour of ascertaining it is not repaid by the result. It is difficult to separate the error due to this from that originating in-defective mechanism, and though formulæ have been elaborated for its detection, we do not propose to give them here. *Quality oil.*

Fourthly. Vibration and shocks. However well chronometers may be stowed, the jars from seas striking the ship, and other like accidents, must be communicated more or less to the chronometers. The vibration of the screw is in some vessels sufficient to pass through all the soft cushions in which they may lie, and must have its effect, more especially from the fact that the watches themselves are hanging in the metal gimbols, in which there must be play sufficient to *Vibrations and shocks.*

s 2

allow them to swing easily, and therefore enough to set up small shocks on any violent movements of the ship.*

---

* In connection with the observations of the Transit of Venus of 1874, Lord Lindsay conveyed nearly sixty chronometers to Mauritius. These were kindly permitted by him to be used in assisting to determine the meridian distance between Mauritius and Rodriguez, when they were shipped on board H.M.S. "*Shearwater*," under the author's command. As the results by these watches, both of the distance between Mauritius and Rodriguez, and Mauritius and Aden (between which latter places they were conveyed in the mail steamer), were remarkably good, and as the results by the "*Shearwater's*" chronometers which were admitted into the distance Mauritius to Rodriguez were not so satisfactory, a description of the manner in which Lord Lindsay's watches were stowed may not be out of place. We may add that the "*Shearwater*" had to beat up for eight days against a strong trade-wind on one occasion, and was a very lively ship.

The watches were taken out of their gimbols and placed in square boxes, which held nine of them each. The partitions of these boxes were thickly stuffed with very soft material (cotton wool) covered with satin, so that each watch lay in a bed of down which was made exactly to fit it.

Each box was fitted with a metal framework after the fashion of gimbols, the outer pivots of which fitted into carefully turned sockets, in two upright columns of wood, which were firmly screwed to the deck. Each pair of uprights carried three boxes of watches.

The effect of this was that any slight shocks to the boxes caused by seas striking the ship, or by longitudinal slipping of the pivots, were entirely deadened before reaching the watches themselves.

This mode of stowing necessitated taking the watch up bodily in the hand to wind, which at first sight seems dangerous, and undoubtedly does present more opportunity for accident than the ordinary method; but, as far as the author is aware, none took place during the five or six months the watches were thus treated, and the admirable agreement of the results seems to show that this system was unusually successful.

Whether it could be adopted on board men-of-war, especially small ones, which are usually employed in surveying duties, is another matter, as it certainly demands more space, both for the swinging of the box and to allow of free access for handling the watches.

It was very convenient for comparing, as one watch could be held to the ear while the eye took the time by the standard.

It seems probable, also, that the temperature would be more constant, from the fact of the watch being imbedded in thick soft material. The lid of each box was also stuffed softly, and, when in place, pressed on the glass of each watch, excluding all air.

In our opinion the variation of rate arising from these causes is, with the generality of Admiralty watches, the larger proportion of the total change.

The only notice that can be taken of variation of rate due to this, is to consider it as detracting from the general value of the meridian distance ; and the nature of the passage, whether rough or smooth, should therefore be noted in the returns.

Magnetism is another disturbing cause, to which irregularities of chronometers have been referred. As no trustworthy conclusion as to this has been arrived at, we do no more than mention it.

It follows as a matter of course, from the preceding observations, that not only will the rate of a chronometer as ascertained before leaving a port be different to that found on arrival at another port, but that the sea rate for the interval will probably be different from either of them.

We have now to consider the means at our disposal for approximating to the true rate under different circumstances.

The most satisfactory circumstance under which we can determine meridian distance (after those already described) is when, having left a port A, called at B (the position we want), where we have only stayed long enough to get Error, and eventually arrived at K without further stoppage, *the longitudes of A and K are sufficiently well known to take them as secondary meridians.* Interpolating between places whose longitude is established.

In this case, by applying the known difference of longitude between A and K to the observations at A, we find the Error on mean time at K at the epoch of starting from A. The difference between this and the Error ascertained on arrival at K, divided by the duration of the voyage, will give us a fair sea rate, which we shall assume to be uniform and invariable during the voyage. A simple application, then, of accumulated rate up to the time of observations at B, will give us the meridian distance from A to B, dependent upon A and K being in certain longitudes.

We can use the same means if we call at more places than

one on the way between A and K, but each stoppage will probably detract from the value of the sea rate.

We are here using the sea rate only, and therefore shall take the date of the last observations at departure, and first on arrival, as the epochs for calculation. If we have obtained rate on departure and arrival, we shall gain valuable information about our chronometers, as we shall be able to see how far they have obeyed any theory as to gradual or uniform change of rate, according to the ordinary assumption that the sea rate is the mean of the two harbour rates.

The value of a meridian distance by this method will, as always, be influenced by the conditions of temperature, fair passage, &c., which must therefore be taken into consideration and recorded.

It will be remarked that by this method, a large amount of time is saved, and opportunities otherwise wasted are utilised to their full extent. Instead of the necessity of waiting, certainly at A and K, and perhaps at B as well, for from five to eight days, a simple call of a few hours at each is sufficient to obtain an excellent result. Moreover, instead of involving the eccentricities of chronometers during the time in harbour at each end, we only include in the calculation the actual time while travelling at sea, and thereby save the irregularities of a good many extra days.

Shadwell's Treatment. Captain Shadwell, in treating of this case, does not use an invariable sea rate pure and simple, but supposes that the rate of departure has gradually and uniformly changed into the sea rate, which he considers as the rate on the middle day of the passage only. He therefore applies for his determination of B from A an intermediate rate between the sea rate and rate of departure; but our experience does not lead us to think that this is an advantage, although by doing the same to the sea rate and rate of arrival, he gets a second meridian distance from B to K, and takes the mean of the two as his result. Our opinion is that, temperature being left out of the question, a better result is likely by using a uniform sea rate.

The algebraic formula for meridian distance by above **Formula for Interpolation.** method of uniform sea rate, is,

$$M_1 = \lambda_1 - \lambda + \tau \frac{\lambda_2 - \lambda - M}{t}.$$

Where M is meridian distance between the terminal points A
and K of the voyage.

M_1 is meridian distance between port of departure,
and a port B touched at on the voyage.

λ is error at A on leaving.

λ_1 ,, B on touching.

λ_2 ,, K on arriving.

t is interval between observations at A and K.

τ ,, ,, ,, A and B.

In any case of a ship's calling at a place as an intermediate port on her voyage between two other places, it
may be well to send home, beside the meridian distance
obtained in the ordinary manner, the information which
would enable the office, *if or when it possesses the true
difference of longitude between the terminal ports,* to calculate the difference of longitude of the intermediate place by
the last formula.

This necessary information will be—

$$\lambda, \lambda_1, \lambda_2, t, \text{ and } \tau.$$

In transmitting this information we could, for the facilita- **Meaning the Errors.** tion of computation afterwards, give only the mean of the
Errors of all the chronometers, instead of the individual error
of each, or in other words, assume an imaginary watch, the
result of which will give the same meridian distance, as the
mean of the meridian distances by each chronometer; but the
adoption of this method will of course preclude any estima-
tion of the value of the distance by the concurrence of
individual results, and should be therefore only adopted
when we have reason to believe from inter-comparisons during
the voyage that the watches have been going well together.*

---

* This method of Interpolation | remarks on it must be taken as my
is not recognised as being as valu- | private opinion only.—W. J. L. W.
able as I believe it to be, and the |

There is another adaptation of the methods of sea rates as obtained by Error at two places whose longitude is known, which is often useful.

Adaptation of foregoing. If we obtain Error before leaving A, and after some days call at B, whose difference of longitude from A is known, and there obtain Error again, we get a very good sea rate for the subsequent part of our voyage, which we can utilise to determine the position of C, any third place at which we may hereafter soon call, with a probable better result than by means of harbour rates. If absolute altitudes are used, they must of course be both either A.M. or P.M.

This method is especially useful for navigational purposes. Suppose a ship to leave Portsmouth and to call at Gibraltar for a few hours only. Error can be obtained, and by means of the known difference of longitude a sea rate deduced, which will give a better landfall for Malta, than the harbour rates at Portsmouth.

Mean Harbour Rates. When our voyage is simply from one port to another, and we wish to find the meridian distance between them, we must depend mainly upon the harbour rates ascertained before departure and on arrival.

The ordinary and rougher method is to assume that the rate has changed uniformly from the rate of departure to that of arrival, and that therefore the mean of the two rates will represent the mean rate during the passage. We believe that (owing to the many causes of variation impossible to formulate) in most cases, and especially where temperature has been, in the chronometer room, fairly uniform, this method will give as good a result as any other; but where temperature has changed much, the result of long meridian distances with such rates will have but very little value, and that a correction for temperature will much improve the result, if we can apply it.

French naval officers have done much in working out this question, and Captain Shadwell gives their separate theories and formulæ. To our mind the method of M. Mouchez is the most practical; and not undertaking to enter into the question of acceleration, nor depending on observations on the watches while in the Observatory, it is more adapted to actual work.

Mouchez proceeds on the assumption, which is near enough to truth for the method, that the rate varies uniformly with the temperature ; but in working on this hypothesis, we must not forget that for each chronometer there is a point of temperature at which the rate is at a maximum, and that the sign of the variation will change as we pass it. *Mouchez's Rule.*

He ascertains by observations for rate at different temperatures, undertaken by the officers when the chronometers are embarked, the coefficient for temperature by simply dividing the difference of rate by the difference of mean temperatures during the intervals of rating.

This coefficient of change will remain constant for some period, though the actual rates themselves will alter from other causes ; nevertheless, the more these observations are multiplied the better, and the latest determinations will be used in practice.

In determining the sea rate for a meridian distance, he applies to the rate of departure the change of rate due to the difference between the mean temperature during rating, and the mean temperature during the passage, which gives one value for the sea rate. Doing the same for the rates of arrival, he gets another value for sea rate. The mean of these two he takes as the final mean sea rate to be used. One weak point here is that the mean temperature, T, of the compensation will not be indicated, unless many observations at different temperatures are made. It will therefore add considerably to the value of this method if we can find T.

It will be more satisfactory if we can get this from the Observatory ; but a formula for ascertaining it is given by Capt. Shadwell, from M. Lieussou, which we here append, but we apprehend that in practice not many opportunities will present themselves for making use of it. It depends on the results of four observations for rates, at equal intervals of time, and at different temperatures, a difficult condition *Lieussou's Formula for ascertaining T.*

to satisfy except with artificial aid for the temperature. M. Lieussou remarks, "that four rates and four temperatures, observed at intervals of ten days, determine the constants for each chronometer with a precision sufficiently remarkable." With the other constants we do not propose to deal, but solely to give his formula for ascertaining T, which is

$$T = \tfrac{1}{2} \cdot \frac{(m_1 - 2m_2 + m_3)\left(t_2^2 - 2t_3^2 + t_4^2\right) - (m_2 - 2m_3 + m_4)\left(t_1^2 - 2t_2^2 + t_3^2\right)}{(m_1 - 2m_2 + m_3)\left(t_2 - 2t_3 + t_4\right) - (m_2 - 2m_3 + m_4)\left(t_1 - 2t_2 + t_3\right)}.$$

Here T = mean temperature of compensation required.

m_1 m_2 m_3 m_4 are the four observed rates corresponding to t_1 t_2 t_3 t_4 the four temperatures.

The intervals between the sets of observations for rates should be between 10 and 30 days.

Hartnup's Formulæ. Mr. Hartnup's formulæ are somewhat different, and do not give exactly the same results with the same data.

He observes the rate at three different temperatures not less than 15° apart, but there must be an equal number of degrees between them.

The same remark already made as to M. Lieussou's method will apply here, viz. that in service afloat it will be difficult to fulfil the conditions of observation. His formulæ are as follows :—

$$C = \frac{2\,(d - d_1)}{p^2}$$

$$T = t_1 + \frac{d + d_1}{2\,C\,p}$$

$$R = r_1 - (T - t_1)\,\frac{d + d_1}{2\,p}$$

Where C is the coefficient of change of rate,

T is temperature of maximum rate,

R is rate at that temperature,

t_1 is the middle temperature,

r_1 is observed rate at temperature t_1,

d is difference of rate between that at lowest temperature and t_1,

d_1 is difference of rate between t_1, and that at highest temperature,

p is difference between highest and lowest temperatures observed at.

Then to find the rate in any required temperature.

If N = any number of degrees from T.

Rate at $T \pm N = R + C N^2$.

In using the rates of departure and arrival in calculating a meridian distance, the Errors at the last observation at departure and first at arrival should not be taken for the epochs of calculation, but the mean of the two should be used for the purpose, for it is at the mean date between the two observations for each rate at which the latter is actually fixed. Thus, if we observe at a place A on the 2nd and 8th, and again on arrival at B on the 20th and 27th, we should take the mean of the two Errors on 2nd and 8th and call it the Error at A on the 5th, and similarly at B on the 23rd·5, and use the interval between these two epochs for the multiplication of the mean rate. *Epochs of Calculation.*

The formula given by Tiarks, and generally adopted, for calculating the meridian distance between two places by rates at departure and arrival, without any consideration of temperature is *Tiarks' Formula.*

$$M = \lambda^1 - \left\{ \lambda + t \left(a + \frac{b}{2} \right) \right\}$$

Where M is meridian distance required,

λ the Error at mean epoch of departure,

λ^1 „ arrival,

t the interval between the two epochs,

a the rate at departure,

b the difference between rate at departure and arrival.

In calculating t, the difference of time, due to difference of longitude between the two places, must not be forgotten; but, being reduced to the decimal of a day, must be added or subtracted to the interval between the epochs, according as we have moved westward or eastward. *Calculating Interval.*

Thus, if our mean epoch at A is at noon on the 20th, and

at another place, B, 30 degrees to the westward, at noon on
the 30th, the interval of time for accumulated rate will not
be ten days, but ten plus the difference of longitude of the
two places, or $10^{d}\cdot08$; for the sun, having completed the ten
days by returning to the meridian of A, will take yet another
·08 of a day to be on the meridian of B.

Similarly, in calculating sea rate from observations at
different places where longitude is known, we must allow for
this difference of time.

Thus, having taken sights at A at noon on the 2nd, and at
B, 20 degrees eastward, on the 11th at noon, the interval with
which to divide the difference of Error at A (corrected for
difference of longitude) and Error at B, to ascertain the daily
rate, will be $8^{d}\cdot94$. as the sun will be on the meridian at B
·06 of a day earlier than at A.

Tiarks' Formula with temperature correction.

The same formula, when intending to correct for temperature, will stand thus :

$$M = \lambda^{1} - \left\{ \lambda + t\left(a + \frac{b}{2}\right) + t\left(\frac{\theta + \theta^{1}}{2} - \theta_{2}\right)y \right\}$$

Where, the other letters signifying as before,

θ is mean temperature during rating at departure,

θ^{1} „   „   arrival.

θ_{2} „   during the passage.

y is the coefficient for temperature found from previous
    observations.

Algebraic Signs.

In all cases of correction for temperature the algebraic sign
of y must be remembered, that is, it must be applied according to the observed effect in altering the rate.

The same remark applies to the algebraic signs of all
quantities in the formulæ.

Thus in the formula :

$$M = \lambda^{1} - \left\{ \lambda + t\left(a + \frac{b}{2}\right)\right\}$$

the signs which are here given, as throughout, for chronometers slow of mean time and losing rates, will only be true

under those circumstances with increasing losing rates and when moving eastward. A consideration of the facts, and obvious effects of the corrections, is perhaps the best course to take to determine these signs.

A meridian distance, founded only upon rates obtained at one end, without any further correction, cannot be considered as of any value whatever, unless the voyage be very short.

When using the combination of harbour rates at each end of a voyage, A to K, to determine the position of some inter- mediate place, B, we must, to be consistent, remember that we are assuming that the rate has gradually and uniformly changed from that of departure to that of arrival, and that the rate to be used for a portion of the voyage will not therefore be the same as that for the whole of it. *Tiarks' Formula for interpolation with harbour Rates.*

Tiarks, interpreted by Capt. Shadwell, gives us the follow- ing formula.

$$M_1 = \lambda_1 - \left\{ \lambda + \left(\tau a + \frac{\tau^2}{2t} b \right) \right\}$$

Where M_1 is meridian distance A to B.

λ_1 „ Error at B.

λ „ Error at mean epoch at A.

a „ rate of departure.

b „ difference of rates of departure and arrival.

t „ interval between mean epochs of rating at A and K.

τ „ interval between mean epoch at A and observa- tions at B.

It is to this case that our observations on page 263 refer, to the effect that the data for calculating the position of B, as interpolated between A and K, may be also transmitted home.

A very good way of measuring meridian distance for the scale of a chart, when the actual distance between the stations is not too far, is by rockets. Parties landed at either end of the base whose difference of longitude is to be measured, ascertain the Error of their pocket chronometers. The ship, midway between the two, fires rockets vertically, *Use of Rockets.*

and the bursting of these, an instantaneous phenomenon, is noted by the watches at either end.

An ordinary service signal rocket can be depended on to mount 1200 feet, and should reach 1600. The bursting, if it occurs, as it should, at the highest point, will therefore be visible nearly 40 miles on either side, which will permit a base of 75 miles to be measured under very favourable circumstances of dark night, and clear atmosphere, when the stations are east and west of one another.

Rockets will not often, however, be seen this full distance; the balls of fire, released on bursting, are scarcely bright enough; and supposing the observers to be at the sea-level, the burst of the rockets will only just be above the horizon, in which position atmospheric disturbances are greatest, and may disperse the rays of light before they can reach the observer. Ascending a hill, therefore, will greatly assist clear vision, and the use of a pair of field glasses will do wonders. Twenty-five miles, on either side, should be measured in this way without any great difficulty.

Transmitting Results. It is important, in transmitting to the Hydrographic Office the results of a Meridian Distance, that sufficient information is given to enable it to be valued and compared with others between the same places. The form appended on page 271, is that now employed.

No. 24.

RETURN OF MERIDIAN DISTANCE, H.M.S.   , 1874.

                                 Captain.

From *Seychelles*.   To *Zanzibar*.

Observation spot, Seychelles—Hondouls Jetty, Mahé .. Lat. 4 37 15 S.

      ,,   ,, Zanzibar—Old British Consulate Garden   ,, 6 09 45 S.

Rates used—Mean rates of departure and arrival.

Error at Seychelles on Jan. 13th by Eq. Alts. ☉

      ,,   ,,   ,,   ,,   ,, 18th ,,   ,,   ,,

      ,, ,, Zanzibar ,, Feb. 1st ,,   ,,   ,,

      ,,   ,,   ,,   ,,   ,, 9th ,,   ,,   ,,

Duration of passage, Jan. 18th, 6 P.M. to Jan. 30th, 4 P.M.

Epochs for calculating accumulated rate, Jan. 15½, Feb. 5th = 20·545 days.

Chrons.	By Observer I.			Date.	Mean Temp.	Date.	Mean Temp.	Remarks.
	Rate of Departure.	Rate of Arrival.	Meridian Distance.					
	s.	s.	h. m. s.	Jan. 13	° 80	29	° 80	Sea smooth dur-
A	−1·280	−1·544	1 05 04·54	14	81	30	81	ing passage.
B	−1·158	−1·211	05·40	15	82	31	79	Steaming 7 days.
				16	80	Feb		Sailing 5 days.
C	−1·888	−2·849	04 59·48	17	79	1	80	Head generally
				18	81	2	78	West.
D	+2·212	+2·026	05 05·03	19	78	3	80	C. H. & F. going
E	−2·068	−2·361	04·96	20	77	4	81	irregularly by
				21	78	5	80	intercompari-
F	−4·903	−5·?67	02·80	22	76	6	79	sons.
G	+4·832	+4·864	03·40	23	77	7	81	
H	−2·668	−5·261	None calculated.	24	78	8	80	
J				25	75	9	81	
K				26	77			
				27	79			
				28	80			

Chronometers rejected C. F. & H.   Number used, 5.

                                                         h. m. s.

Mean Meridian Distance by Observer 1 .. 1 05 04·59

   ,,   ,,   ,,   ,,   ,, 2 ..   05·43

   ,,   ,,   ,,   ,,   ,, 3

   ,,   ,,   ,,   ,,   ,, 4

                                                       h. m. s.

Final Mean Meridian distance by arithmetical mean   1 05 05·0 W.

   ,,   ,,   ,,   ,,   ,, values assigned

CHAPTER XV.

TRUE BEARING.

By Theodolite—By Sextant—Variation.

In nearly all descriptions of surveys true bearings will be used.

By theodolite and sextant. The most correct method, from a shore station, is to use the theodolite, which will alone give a very good result for azimuth; but it is better to get the altitudes with a sextant and artificial horizon, when two observers are available.

The theodolite in this case is only used for taking the horizontal angle between the sun and the zero.

Three methods. There are three principal methods in use for obtaining the azimuth. By observations at equal altitude A.M. and P.M., by observations A.M. and P.M. at nearly the same altitudes, or by single observations.

Single altitude generally sufficient. The former is theoretically the more correct, as many errors are eliminated; the second is nearly as good; but our experience is that with single observations taken with the sun near the prime vertical, with instruments in good order, the result is quite as near the truth as is generally requisite in marine surveys. When an extensive piece of coast is being surveyed, we shall, as before stated, depend mainly upon the astronomical positions for the scale and bearing of the chart, but nevertheless accuracy in obtaining the original bearing for working is necessary.

A very important point is careful levelling, which should be done with the telescope pointing in the direction of the sun, and the accuracy of the movement of the telescope in a vertical plane should be tested, as no method will eliminate error due to want of such accuracy.

In the first method, the sun will be observed at an even stated altitude, and the sextant will be set beforehand, the observer using it giving the "stop" to the theodolite observer.

In the others, the theodolite observer generally calls the "stop," and the sextant observer takes whatever altitude it happens to be.

To arrive at a satisfactory result in either case, it is neces- Changing degree of sary to take several sets, with a different degree of the arc Zero. pointed at the zero in each, so as to eliminate the errors of the horizontal arc of the instrument.

As it is the bearing of the sun's centre which we obtain Correcting to sun's by working out the azimuth, the aim of the theodolite centre. observations is to get the horizontal angle between that centre and our zero; but it is manifest that we cannot trust our eye to place the cross-wires of the telescope exactly on the centre of the sun, nor can we place the wires truly vertical and horizontal.

If we could do the latter, we could arrive at the angles to the centre by merely observing the sun in one quadrant, and applying the semi-diameter × sec. alt.; but we must not trust this, if we want fair accuracy.

In equal altitude observations, the method is to fix on an Method of bearing by altitude for both sextant and theodolite, and set the vertical equal arc of latter at it. In the forenoon, bring the sun so that it altitudes. is in the lower half of the field, and approaching the vertical wire. The theodolite observer then keeps the limb of the sun in contact with the vertical wire, and below the horizontal one. If the theodolite is truly levelled, he will not need to touch his vertical tangent screw, but if necessary he must do so, to keep the upper limb of the sun as nearly touching the horizontal wire as he can. When the upper limbs of the sun in the artificial horizon are in contact, the observer calls "stop," and the motion of both tangent screws of the theodolite ceases. The horizontal arc is then read.

T

Then, without moving the theodolite in altitude, the other limb of the sun is brought on the other side of the vertical wire, and the reading made when the artificial horizon observer gives "stop," on the lower limbs of the sun coming in contact.

The sun will thus have passed between opposite quadrants of the cross-wires, as in the diagram Fig. 38.

Similar observations are made at the same altitude in the afternoon, the lower limb coming first. Each set will thus consist of two observations A.M. and two P.M. In this method the time must be taken exactly, which is a draw-back, as it either requires three persons, or that one should take time as well as his observation. There is, however, no necessity to know the *local* time very exactly, all we want is the true elapsed time.

FIG 38.

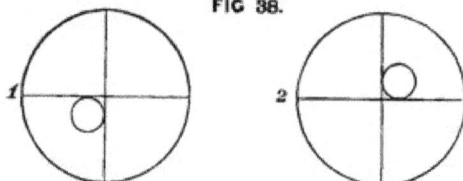

Calculating Bearing by Equal Altitudes.

To work out the equal altitude observation, the means of the times, and of the horizontal angles of A.M. and P.M. respectively, are taken.

If the sun had no motion in declination, the mean of A.M. and P.M. horizontal angle would be the angle on the horizontal arc corresponding to the true meridian, or, in other words, the bearing of the zero; but as this is not so, we must work a correction similar to the Equation of equal altitudes when obtaining time, to be applied to this mean of the angles.

The formula for this is—

$$\text{Correction} = \frac{c}{2} \text{ Cosec } \frac{\text{time elapsed}}{2} \text{ Sec lat.,}$$

where $\frac{c}{2}$ is half the change of declination in elapsed time.

This correction is additive to the angle when the sun is

moving from the nearest pole, and subtractive when moving towards it.

Let us take the following example—

		At Nut △. ⊕ Pagoda △ 360°.		
Alt.	Times.		Hor. Angle.	
		h. m. s.	° ′ ″	
39°	A.M.	8 20 14	15 05 30	Z. K.
	„	8 23 22	15 09 15	„
	P.M.	4 02 13	193 24 30	Z. K.
	„	4 05 20	193 28 45	„

Lat. 30° N. Declination corrected to Greenwich time of A.M. observation 18° 14′ N. Sun moving north.!

		h. m. s.			° ′ ″
Mean A.M. Times	..	8 21 48	A.M. angle	..	15 07 22
P.M. „	..	16 03 46	P.M. „	..	193 26 37
Elapsed Time	..	7 41 58			208 33 57
½ Elapsed Time	..	3 50 59	Mean angle	..	104 16 58

Var. of dec. in 1 hour $\quad$ 37″·44

$\qquad$ 3·85

$\dfrac{c}{2}$ .. $2·15866$

$\qquad$ 18720

Cosec $\dfrac{\text{E. T.}}{2}$ $\quad$ ·07279

$\qquad$ 29952

Sec lat .. $\quad$ ·06249

$\qquad$ 11232

$\qquad$ $2·29394$ .. $196″·7$

$\dfrac{c}{2} = 144·144$

Cor = 3′ 17″

		° ′ ″
Mean angle	..	104 16 58
	−	3 17
Angle of South Point	..	104 13 41
Or bearing of Pagoda	..	S. 104 13 41 E.

A number of similar sets, taken with different degrees as zero, will give a very correct result, and though all instrumental errors will not be eliminated, the majority of them will disappear.

In "single" observations, each set will consist of four _{Method by} contacts, in each of which the sun will be tangential to the _{Single Altitudes.} vertical wire in a different quadrant of the field.

T 2

The mean of these will then be the angle to the sun's centre, corresponding to the mean of the four altitudes.

When the altitude is being taken by a sextant, it will only be necessary for the theodolite observer to be *very* exact with the contact of the side-limb of the sun; but his upper or lower limb, as the case may be, should be as nearly touching the horizontal wire as possible, to insure the elimination of the wire error.

It is quite immaterial in which quadrant the observer commences; but whatever plan he adopts, he should always

FIG. 39.

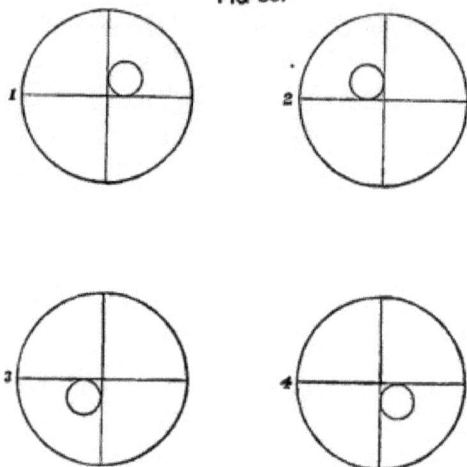

observe in the same manner, as it prevents confusion and mistakes. The sun will appear as in the diagrams in Fig. 39.

When taking the observation with the theodolite alone, it will of course be necessary to see that both the horizontal and vertical wires are truly tangential to the sun's limbs.

Six sets should give a very good bearing; but if the theodolite is a very small one, or is known to be badly graduated, more may be necessary.

Half the altitudes in the artificial horizon may be taken with upper limb and half with the lower; but this is not important, as if the observation be made when the sun is near the prime vertical, a small error in the altitude will but slightly affect the azimuth.

The azimuth of the sun having been obtained by the ordinary rule of nautical astronomy, the true bearing of the object is found by applying the mean of the theodolite angles of that set.

In finding a true bearing with sextant only, it will be more accurate if two observers are employed—one to take the altitude, the other to measure the angular distance at the same instant. **Method by Sextant.**

If only one observer is available, he must take altitude and angular distance alternately, taking care to end with the same

FIG 40.

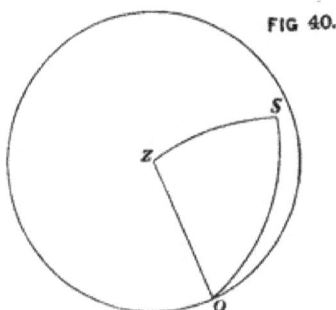

observation as that with which he begins, so that the mean of each kind will correspond as nearly as may be in time. Thus, if he begins with altitude, he must also end with altitude.

This method should not, however, ever be used when a theodolite is available, and is only adopted for true bearings from the ship, in an irregular survey.

In this instance we have to calculate the horizontal angle, which with the theodolite we obtained directly. **Calculation of Horizontal Angle.**

The object should be so chosen that the line joining it with the sun should not make a larger angle with the horizon than 20°, and the less the better, as any inaccuracies of observation will not then be much increased when the horizontal angle is deduced. If we take an object 90° or more from the sun, these conditions will be fulfilled, the sun being of course comparatively low, and near the prime vertical.

Two Cases.   There are two separate cases :—

First, when the object whose bearing is desired is on the horizon ; and secondly, when it has a sensible altitude, as a mountain top.

In the first we have to solve a quadrantal triangle as shown in Fig. 40.

Object on horizon.   In this Fig. Z is zenith, S is sun, and O the object on the horizon.

We have Z O = 90°. Z S the apparent zenith distance, and O S the observed angular distance, to find O Z S, the horizontal angle required, or

Cos horiz. angle = Cos ang. dist. × Sec. app. alt.

Example.

(Object on horizon, two observers with sextants and artificial horizon.)

On June 1st, 1881, at Ship IV. 7ʰ 24ᵐ A.M. mean time of place, observed altitude of ♎ 30° 13′ 50″, mean angular distance of ♎ to Pine △ on horizon 84° 26′ 20″, object right of ☉. Lat. 40° 26′ 15″ N. Long. 28° 00′ E.   Index Errors − 35″ and 0″.   H.E. 20 ft.

		h. m.						°	′	″
M. Time pl.	..	7·24		☉'s dec. 1st	..	22	6	59·6 N.		
Long. in time	..	1·52					2	08		
Gr. Date 31st	..	17·32		Corrected dec...		22	04	51·6		
„ 1st	..	−6·28		Pol. dis.	..	67	55	08		
		°	′	″						
Obd. alt.	..	30	13	50		Var.	..		20·0	
Index error..				−35					6·4	
		30	13	15					800	
H. E.	..		4	15					1200	
		30	09	00				128″·00		
S. D.	..	+15	48							
								°	′	″
App. alt.	..	30	24	48		Obs. Ang. dist.	..	84	26	20
Ref.	..		−1	31		S. D.	..	..	15	48
T. alt.	..	30	23	17		True Ang. dist.	..	84	42	08

		°	′	″		
Lat.	..	40	26	15	Sec.	·118550
Alt.	..	30	23	17	Sec.	·064181
		10	02	58		
P.D.		67	55	08		
		77	58	06	½ Hav.	4·798724
		57	52	10	½ Hav.	4·684677
						9·666132

Azimuth of sun	..	..		N. 85° 49′ 25″ E.
Cos. true Ang. dist...	8·965353			
Sec. app. alt.	·064294			
Cos. Hor. Ang.	9·029647	..	83° 51′ 14″ Hor. ang.	

Azimuth ☉	.	..	N.	85°	49′	25″	E.
Hor. angle	..	..		83	51	14	
			N.	169	40	39	E.
True bearing Pine △	..	..	S.	10	19	21	E.

In the second case, we have a spherical triangle with three **Object elevated.** sides known, as in Figure 41.

Here, Fig. 41, we have Z O, the zenith distance of the object,

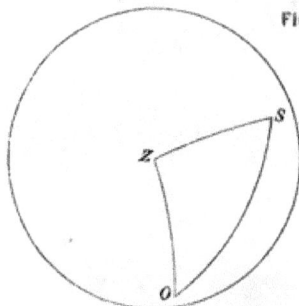

FIG 41.

Z S, and O S, as before, the apparent zenith distance of sun, and angular distance; to find O Z S, the horizontal angle required, which can be done by any of the applications of the formula

$$\text{Cos } O \, Z \, S = \frac{\text{Cos } O \, S - \text{Cos } Z \, S \, . \, \text{Cos } Z \, O}{\text{Sin } Z \, S \, . \, \text{Sin } Z \, O}.$$

Example.

(One observer with sextant, sea horizon, alternate observations, object elevated.)

At S. Ann's △, October 5th, 1881, Lat. 5° 10' S. Long. 57° 14' E., the following observations were taken for true bearing of Snow Peak. Height of eye 10 feet, object right of ⊙. M.T. place 8.00 A.M. I.E. – 50".

Alt. ⊙			Ang. Distance of Snow Peak ◖			Elevation of Snow Peak.				
°	′	″	°	′	″			°	′	″
30	06	10	94	14	40	On arc ..	1	26	10
	13	00		16	10	Off „ ..	1	24	30
	20	15		18	30				
	28	00		20	00		1	25	20
	36	10		21	20				
	42	50							

	h. m.						
4th M.T. pl...	20 00	⊙ dec. ..	4 53 44 S.	Var ..	57·7
Long. ..	3 49		7 30		7·8
Gr. date 4th	16 11	Corr. dec...	4 46 14		4616
„ „ 5th	– 7 49				4039
		P. D. ..	85 13 46		
				6)	450·06
					7′·30″

	° ′ ″			° ′ ″
Mean. obs. alt. ⊙ ..	30 24 24	Mean obs. ang. dist. ..	94 18 08
I. E.	– 50	I. E.	– 50
	30 23 34		94 17 18
H. E.	– 3 07	S. D.	+ 16 02
	30 20 27		94 33 20
S. D.	+ 16 02	Corr. ang. dist. ..	
App. alt. ..	30 36 29		
Ref.	– 1 30		
Tr. alt. ..	30 34 59		

	o ' "		
Lat. ..	5 10 00	Sec. ..	·001768
Alt. ..	30 34 59	Sec. ..	·065052
	25 24 59		
P.D. ..	85 13 46		

110 38 45 ½ Hav. .. 4·915068

59 48 47 ½ Hav. .. 4·697741

9·679629 S. 87° 30′ 11″ E. Azimuth⊙

	o ' "		
App. alt. ⊙ ..	30 36 29	Sec. ..	·065163
Alt. snow peak..	1 25 20	Sec. ..	·000134
	29 11 09		
Ang. dist.. ..	94 33 20		

123 44 29 ½ Hav. .. 4·045413

65 22 11 ½ Hav. .. 4·732408

9·743118

			o ' "
Horizontal angle ..	..		96 08 33
Azimuth ⊙	..	..	S. 87 30 11 E.
True bearing snow peak..			S. 8 38 22 W.

The Pole star may be used in the northern hemisphere to obtain true bearings at night. [margin: Use of Polaris.]

Circumstances under which this is useful are related at page 155, which see.

The Greenwich time must be known, and the angle between the Pole star and object whose bearing is required, must be large.

Measure the angle and take the time.

Ascertaining the sidereal time of observation as in ordinary Pole star calculation, add six hours to it for a second sidereal time.

Out of Table I. in Nautical Almanac, take the correction with first sidereal time, which, applied with the reverse sign to the latitude, will give the altitude at the time.

Take out a second correction with second sidereal time, which will be the rectangular deviation of Polaris from the meridian.

To calculate the horizontal angle answering to this, the formula is

Sin horizontal angle = Sin correction × Sec. alt.

which will give the true bearing of Polaris, east of meridian when first sidereal time is between 13 h. 20 m. and 1 h. 20 m., west when otherwise.

Example.

August 10th, 1881, Lat. 43° 30′ N., Long. 66° 30′ W., at 13.34 G.M.T. Observed angle from Polaris to Seal I^d. Light 80° 10′, right of Polaris.

	h. m.		° ′
G. M. T. ..	13 34	Cor for 1st Sid. T.	− 0 17
Long.	4 26	Latitude	43 30
M. T. ship ..	9 08	Altitude Polaris ..	43 13
Sid T. noon ..	9 16		
Acceler. ..	2		
1st S. T. obs.	18 26	Sin. Corr. for 2nd S. T.	8·35018
	+ 6	Sec. Alt.	·13741
2nd „ „ ..	0 26	Sine True B	8·48759

Corr. for 2nd S. T. .. 1° 17′　　Polaris .. N. 1° 45′ E.

Cos. ang. dist.		..	9·2324
Sec. Alt. ..	..	..	0·1374
Cos. hor. ang.	..	.. ..	9·3698
Hor. ang. ..	..	.. ..	76° 27′
Polaris	..	..	.. N. 1 45 E.
Seal I^d. L^t. ..	..		.. N. 78 12 E.

VARIATION.

Accurate variations are very useful in all parts of the world, as from them the lines of equal variation shown on

charts are drawn; but to enable them to be so used, they
must be trustworthy.

Variations obtained by swinging the ship carefully with a
smooth sea, and in water of, say, over 50 fathoms are most
useful, as fear of local attraction is thereby removed. Sea observations.

The bearing of the sun, or of an object sufficiently distant
to maintain the same direction whilst steaming round, and of
which the true bearing is obtained, should be observed on
evenly distributed points. There is no necessity to observe
more than on every other point, and good results will be
obtained from the cardinal and quadrantal points.

In the mean of the total errors the deviation will be
eliminated, and the result is the variation.

A full report of the observations should be transmitted
home.

Shore variations are also of value, when taken on ground
free from suspicion of local attraction, for the determination
of the true variation, and, in other cases, for the information
they afford on the amount of the local attraction as obtained
by comparison with the variation found from sea observations
in the vicinity, a point of much interest, and often of
practical importance. Shore observations for Variation.

The requirements for a good shore variation, that the
Hydrographic Office can put confidence in, are as follows :

1. The true bearing of different points (about six) as equally
distributed as possible round the circle whose centre is the
observation spot, must be well and accurately observed with
a theodolite.

2. Bearings of all these points must be taken by the
compass from the observation spot.

3. Different sets of observations must be made with different
pivots and with different cards.

4. The spot on which the observation is made should be
free from every suspicion of any iron in the vicinity, and the
nature of the rock, or whatever the formation may be near
the observation spot, should be mentioned in the return
transmitted home.

Points 1 and 2 are necessary precautions against the errors of the card caused either by bad graduation, or from accidental bending of the card. In ascertaining the true bearings, it will only be necessary to observe one object, when theodolite angles to the others will give their difference of bearing.

As regards No. 3, all compass cards have an error caused by inaccurate affixing of the magnetic needles, and it is necessary to multiply observations, and make certain the card is working properly.

Shore observations should be obtained at stations where the variation is already well known, when opportunity offers, as these will enable the Office to calculate the change of variation.

An example of observation for variation is appended.

Variation deduced at Office. Although the variation is here deduced to show the method, this would not be done in forwarding these observations to the Admiralty, as there are certain card-errors to be applied first.

VARIATIONS.

| Date. | Observation Spot, Nature of ground, &c. | Nature of Observation. | Objects. | True Bearing. | Magnetic Bearings. | | | | |
| | | | | | Standard Compass B 154. | | | Dover. H. O. | |
					Carl A.	J.	Spare A.	A.	B.
11 Dec. 1878.	West Base. Pasha Liman Iᵈ. Sea of Marmara. Alluvial soil. Lat. 40° 28' N. Long. 27° 34' E.	Single Observations Theodolite (6 in.) and Sextant. Artificial Horizon.	Marm. Mill .	N. 2 00 E.	N. 8 05 E.	N. 8 05 E.	N. 8 15 E.	N. 8 25 E.	N. 8 08 E.
			Chim . . .	56 16	62 10	62 20	62 18	62 40	62 35
			Slope Mill .	114 14	120 15	120 15	120 12	120 42	120 33
			Brush △ . .	179 54	186 20	186 10	186 18	186 10	186 15
			Rok △ . .	243 06	249 15	249 05	249 15	249 10	249 33
			Araplar Hill .	203 39	309 45	309 42	309 52	310 08	310 05
			Nest △ . .	342 37	348 50	348 40	348 42	349 05	348 50
			Mean . .	177 24	183 31 26	183 28 26	188 33 09	183 45 43	183 42 43
					177 24	177 24	177 24	177 24	177 24
			Rough Variation .. (without card errors)		W. 6 07 26	6 04 26	6 09 09	6 21 43	6 18 42

CHAPTER XVI.

SEA OBSERVATIONS.

Double Altitude—Sumner's Method—New Navigation—Short Equal
Altitude—Circum-meridian Altitudes of Sun.

As regards Surveying Operations. FOR surveying purposes, observations at sea are mainly required for fixing the ship's position when sounding banks, or
looking for vigias.

We cannot hope to attain to any very great accuracy, and
are much dependent on weather and the state of the sea and
clearness of the horizon. As longitude must depend entirely
on the chronometers, we must in cases where we require all
the accuracy we can get, as in fixing the position of banks
far away in mid-ocean, wait until we can again obtain Error
and rates to give the final positions; but with ordinarily good
chronometers our daily positions obtained whilst sounding
will be correct, comparatively one with the other, so that
we can at once plot and delineate the shape of the banks,
which is what we want at the time.

Refraction. In all observations at sea it must be remembered that the
horizon may be affected by abnormal refraction, and no
dependence can be placed on a latitude or longitude
deduced from altitude observed on one side alone. In
certain cases the error may amount to 2' or even 3'.

With a high sun at noon, when accuracy is aimed at, it is
well to observe the opposite side of the horizon; an
awkward observation at first, which practice will render
easy.

Position early in the day. One great object when sounding or looking for banks is to
obtain a position as early as possible in the day, after lying-

to probably all night, as in the vicinity of banks currents are nearly always set up, and in variable directions, so that we cannot at all depend upon dead reckoning, or upon finding ourselves where we laid-to the night before, and in many instances, unless we know in which direction to go, it is useless to move at all.

In such cases observations should be taken throughout the night; for, though they will give but approximate results, latitudes by pairs of stars north and south of zenith by the same observer should give under favourable circumstances a position within five miles of the correct latitude, which will tell us if we are drifting much in the line of the meridian, and also affords us an approximate latitude to work longitude by, in the morning. Star Latitudes.

The worst of it is, that circumstances apparently favourable are often not really so, as the great source of error in night observations is the impossibility of being certain of the horizon. A false horizon will frequently look so well defined as to mislead the best observer, and will of course throw out the resulting latitude greatly. Thus we can put no great faith in latitude by stars, and none whatever in a single observation, or even in a single pair, as the horizon in one direction may be true and in another false. It is only in a series of pairs of stars through several hours that we can have any confidence, as if the result of these agree fairly, or steadily show movement in one direction (the effect of a current) we may then feel pretty sure of our position as far as latitude goes.

Night observations at sea for longitude are not of much use; but, under unusually good circumstances of horizon, the mean of two star chronometers, one east, the other west of meridian, may be better than nothing. Star observations for longitude.

If, however, when the day has sufficiently broken to enable the horizon to be clearly seen, we can get observations of bright stars or planets on different bearings, we can obtain an excellent position by Sumner's method from which to start our day's work. Stars at daybreak.

Or, if we can only get one daybreak star, as soon as we

can get an observation of the sun, we can combine it with the daybreak observation.

Venus by day. The planet Venus can, when near quadrature, be observed all day. The altitude can be calculated, when she will easily be found in the field of the telescope. This observation is but too little used, as it is a most valuable one.

Different Methods of obtaining Position. As observations must be carried on throughout the day, in order to get as many positions as we can, we now come to the different methods of obtaining latitude and longitude other than by the ordinary means of longitude by chronometer and latitude at noon.

There are three methods of finding latitude and longitude at the same time, viz. by Ivory's rule for double altitude; by Sumner's method; and by a short equal altitude; and for latitude only, we have circum-meridian altitudes. These are all of service under different circumstances, which will be hereafter described.

DOUBLE ALTITUDE.

Ivory's Double Altitude. Ivory's rule for working a double altitude, with Riddle's extension, by which the longitude is also obtained, is too well known to require any special remarks.

Conditions for good Double Altitude. The condition requisite to make the position obtained by double altitude trustworthy, is mainly that the sun should change in azimuth a fair amount, otherwise one of the triangles will be so ill-conditioned that a small error in either altitude or time will have a great effect on the result.

In the generality of cases we have this condition by allowing about two hours to elapse between the observations, and we can therefore get a fair position by about half-past nine or ten o'clock.

In low latitudes, however, when the declination and latitude are nearly the same, the sun will rise so nearly on a circle of altitude as to change the azimuth very slowly, and we must wait till nearly noon before we can put any confidence in the observation. Unless then we lose our meridian or circum-meridian observation, a double altitude is under these circumstances of little use to us. It is, however, more to be trusted than a Sumner, when change of azimuth is small; but it may be broadly stated that unless we are *very* ignorant (from lack of previous observations or other causes) of our position in latitude, we cannot under these circumstances do much before noon with observations of the sun alone; but we can get a good position by combining a daybreak observation of a star with one of the sun, as soon as it has sufficiently risen, by Sumner's method.

SUMNER'S METHOD.

Sumner's method of obtaining the latitude and longitude at any one moment, which is but too little used in ordinary navigation, depends upon the fact that a heavenly body at any moment will be seen at an equal altitude from any part of a small circle of the earth, circumscribed round the spot where the body is vertical, with a radius equal to the zenith distance.

Knowing, therefore, the altitude of the sun, and the point on the earth whose zenith it is in, an observer will always be able to say that he is somewhere on the circumference of that circle. This alone would not give us much information, but it is seldom that we do not know our latitude to twenty or thirty miles. We shall know then that we are on that part of the circumference which includes these latitudes, and as when the sun is not very high the circle will be of large diameter, the portion of it within our limits may be, without much error, taken as a straight line.

In practice, then, having obtained an altitude of the sun, or any other body, we assume a latitude from our dead reckoning, and work out the longitude. By the aid of the azimuth tables we obtain the bearing of the body observed, and having plotted the position obtained, we draw a line at right angles to the bearing, which is the " position line." We now know that we are somewhere on that line. We must know the Greenwich time, and therefore our positions will be as much dependent on the chronometers as any ordinary longitude.

The position line can be also obtained by working with two assumed latitudes and joining the positions resulting, but the above is shorter.

Waiting until the earth has revolved a sufficient amount to alter the bearing of the sun, we repeat the operation, and obtain another line, the direction of which will differ from the former by the difference in the azimuths of the sun at the two observations.

If we have been motionless in the interval, the intersection of these two lines will give us our exact position (always dependent on the chronometers); but if we have moved, we must so far trust our dead reckoning as to transfer the first line in the direction, and for the distance, we have run, when the intersection of the second position of the first line with the second line will be our position at the second observation.

Plotting Sheet. H.M. ships are provided with a large sheet on which longitude is marked, leaving the navigator to complete the Mercator's projection by measuring the meridional parts for the latitude he is in.

On this, or a similar sheet graduated on board, the Sumner lines will be plotted, as it will both spare the charts, and from the increased scale provided, will give a better position than if the plotting was done on the usual small scale general chart of an ocean.

Limits of application. It is obvious that the value of a position will largely depend on the angle between the two lines, or, in other words, on the change in azimuth between the observations.

In low latitudes, therefore, when declination and latitude are nearly alike, we cannot use this method with the sun alone early in the day, as the sun will rise nearly vertically from the horizon, and we want a change in azimuth of at least 20° to give us a trustworthy position. The same circumstances will much detract from the value of a double altitude, as has been remarked, so that in such a case neither the one nor the other are much use as an absolute determination of position.

But Sumner's method has other resources. We can com- *Other combinations by Sumner's method.* bine lines obtained from two or more stars, or a line obtained from the sun with one from the moon or other heavenly body, as, for instance, a star obtained at daybreak when the horizon is sufficiently defined for us to trust it, or Venus in daylight, as before remarked. All we need is that the bearings of the two bodies differ sufficiently to give a good intersection.

By this means we can often get a good position early in the day, which we cannot get in any latitude with the sun alone, without a considerable interval of time elapsing.

This combination, therefore, of stars and the sun affords us the best and earliest opportunity of determining our position, and we should always endeavour to obtain it.

Should we be able to get a good meridian or circum-meridian altitude, of a star, we shall of course use the resulting latitude.

A sun-Sumner requires the same circumstances and obser- *Advantages over Double Altitude.* vations as a double altitude, but it has several advantages over the latter.

In the first place, the first half of the observation can be worked out at once; by which means we not only obtain the line on which we know we must be, and so have an approximation to our position at once, but also, having worked half of the calculation, it will not require many minutes after the second observation is taken, to complete it and obtain the true position.

Secondly, errors of calculation are less likely to be made in a Sumner, as it involves merely the ordinary "chronometer" problem.

Thirdly, the fact of obtaining a line of position is of great value in many cases, as we can always tell roughly in what direction to go to shorten our distance to any given point, unless it should fall on or near the line, and when searching

for a vigia this knowledge, early in the day after a night's
lying-to, will be invaluable.

Fourthly, we can repeat the observations a third time, and
so check our first position with but little labour of calculation
long before noon, especially in the case where we have combined a star with the sun, and are, perhaps, doubtful of the
star observation, either from faintness of the star or indis-
tinctness of the horizon.

Bearing of land and Sumner Line.   The true bearing of a distant mountain whose position is
known, will also give a position by combination with a Sumner

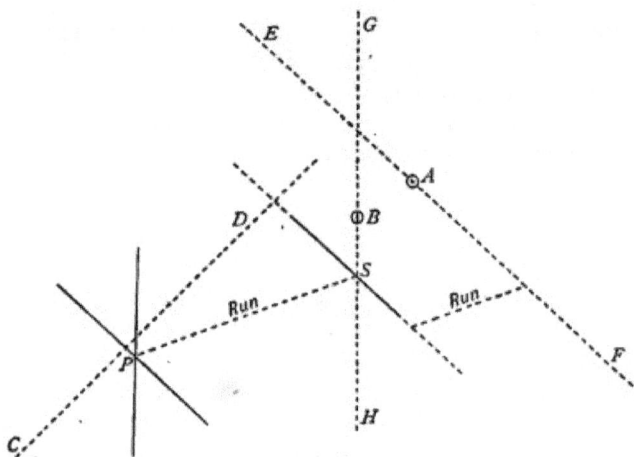

FIG. 42.

line, if its direction is such as to make a good cut with the
latter.

Example of Sumner.   In Fig. 42, let us suppose A to be the position found by
assuming a latitude and working out the altitude of a star
obtained at daybreak.   Drawing a line at right angles to the
bearing, we get our first Sumner line E F, and we know we are
somewhere on it.   Having run W. b. S. 6·2 miles, we get an
altitude of the sun; and assuming in this case a latitude a
little south, we get another position B, and draw another
position line G H.   To project the run, we draw a line in the

required direction, and for the distance run, from any part of the line A, and draw another line parallel to the line E F, through the end of the run line.   The position S, where this last intersects line G H, is the position of ship at second observation.

Running on in the same direction for 12 miles, we get another altitude of the sun, and another resulting Sumner line C D.   Transferring the two first lines by the run as before, we now have three lines intersecting, or nearly so, at P, and by their coincidence or not we can measure the accuracy of our former positions—to a certain extent, that is, for it must be remembered that as the intersection of our lines is governed by the run allowed, a current, or constant error in calculating the run, might give an apparently good position which may really be considerably in error, even when the third inter-section is obtained, with certain arrangements of the lines and the run.

Sumner's method is, in fact, the means by which all individual observations can be combined, and is from every point of view invaluable.

NEW NAVIGATION.

There is a method of obtaining position known as the " New Navigation," which is highly thought of by some surveyors. It is claimed that position obtained by this method from observations of a heavenly body between the azimuths of 20° and 70' is more accurate than by the ordinary "chrono-meter" method.

Having had no personal experience of this method we are unable to recommend it.

It is, therefore, considered sufficient to mention that it consists in assuming a latitude and longitude, and in calcu-lating a correction to this assumed position.   A line of position can then be drawn through the true position thus obtained, and the results combined by Sumner's method with other observations.   The calculation is given in the book quoted in the footnote.*

SHORT EQUAL ALTITUDE.

In low latitudes, where the motion of the sun in altitude is rapid nearly to the time of transit, a very good longitude may be obtained at noon, by a short equal altitude, taking observations about 20 minutes before and after noon.   The change of declination in this short interval will not affect the time, so that the middle time between the observations as shown by the watch, can be taken for the time by the watch at apparent noon.   All we have to do, therefore, is to take the difference between mean time of apparent noon and the Greenwich time, as shown by our chronometer, which gives us longitude directly.

CIRCUM-MERIDIAN ALTITUDES OF SUN.

These are of great value, as, when the observations are within the limits of time from noon, the resulting latitude is as correct as from a meridian observation, which may be lost from clouds.   They should be worked in precisely the same manner as the shore observations of the same description, and should be obtained as near noon as possible.   If more than four or five minutes have to be added to the observed altitude, they will not be of much value.

If Raper's most valuable book is at hand, a short and correct rule, in connection with two of his tables, will be found at page 232 of the thirteenth edition, which will give the reduction as nearly as requisite for sea work.

---

* Ex-Meridian Altitude Tables and New Navigation, by C. Brent, R.N., A. F. Walton, R.N., and G. Williams, R.N.   Geo. Philip & Son, London, 1886.

CHAPTER XVII.

Fair Chart—Reducing Plans—Delineation—Symbols—Colouring—
Graduation.

THE work is sent home to be published in several ways, **Trans-**
according to circumstances. **mission Home.**

When the detail, as it proceeds, is inked on the original
sheet itself, it may be necessary to transmit a portion home
before the survey is all complete, and a tracing is often used
for this purpose, as the original sheet, with the "points" still
accumulating, must be retained on board; but, if possible, it
is better to send work home on drawing-paper, which is not
liable to so many accidents from tearing, &c., can be more
fully worked up as regards detail, and can be better kept as
a record, though the originals will in the end be transmitted
to the Admiralty in any case.

When the detail is placed directly on the original sheet, it **Original**
is very difficult to keep it clean enough for everything to be **Chart.**
clear and distinct, as straight-edges, protractors, &c., will be
constantly placed on the chart over the completed part, and
lines must be often drawn over it. It *can* be kept clean
enough for transmission home as the finished chart, and by
doing so, all errors arising from imperfect transferring will be
avoided; but the surface of the paper must get so rubbed by
constant cleanings, that, if a large sheet, it is seldom satisfactory. Several hands may have been employed in it, and
the chart will then bear a piecemeal look. If this original
sheet is not sent home, a copy has to be made on another
sheet of paper, which will be the fair chart.

**Fair
Chart.**

The usual mode of making this is to place the new sheet under the old one, and prick the "points" through the latter, on to the former. A careful tracing having been made of the working sheet, it is placed on to the fair sheet, so that the points all correspond, and by means of transfer paper is traced on to the fair sheet, and inked in.

Great care is requisite, in transferring in this manner, that the tracing does not move from its proper position, and heavy weights must be used to prevent it from so doing. Errors have often crept in from careless transferring and want of proper examination and comparison afterwards.

In working with the method recommended by the writer, viz. each assistant's work plotted and inked on to his own separate board, and all then placed on one tracing, the final sheet can either be the original on which all the points have been plotted, if that has been kept clean enough ; or a sheet may be pricked through, as mentioned above, for the purpose ; but in either case only one complete chart will be made, the general tracing sufficing to show whether the work of different assistants has met, and what is wanted to complete.

This, or these (as in a large sheet there will be several tracings for different parts), will be the tracing used for making the final chart in this case.

These tracings should not be too large, as they are apt to get distorted. For fine work it is desirable to make small tracings on paper for the special work of transferring.

This chart will also be the work of one hand, who will, after transferring outline, soundings, &c., from the general tracings, have the original little bits before him while inking in ; these little bits having been taken off their boards, and so reduced, by having superfluous paper cut off, as to be handy to lay on the sheet.

**Original
Field
Sheets.**

By washing off the field boards, the paper will have become distorted and contracted, but not to a sufficient degree to interfere with the small detail of sinuosity of the coast, which is what we mainly want them for. Everything will have been traced on the general tracings before the paper has been removed, and care must be taken that this is so, as it cannot be done afterwards.

In whatever manner the final chart is sent in to the office, "Points" all "points" must be distinctly marked on it, especially main "to be points. These latter are often distinguished by the triangle which means theodolite station, and in surveys where the sextant has also been employed in triangulation, should cer tainly be so. The "points" are necessary to join one chart to another, and also, in case of future revision of the chart, they afford means to the reviser of measuring the accuracy of his predecessor's groundwork.

Plans sent home by officers in general service ships often lose much of their value from neglect of this. The existence of the "points," and their proper position, will at once give a confidence in the detail of the plan, that it is impossible to accord to the work of an officer, however zealous, of whom nothing is known as to his hydrographical capability, and who fails to give any indication in his chart of how it has been constructed.

REDUCING PLANS.

In a survey of an extensive nature, bays, harbours, &c., will often be done on a larger scale than the rest of the sheet. These must be either left blank on the coast sheet, or else re-duced from the large scale plans.

It may sometimes happen that a portion of an anchorage is surveyed on the small scale before it is decided to make a large plan of it, on discovering it to be worth while to do so. This must not appear, however, on the completed chart, it must be all reduced from the larger scale.

Instruments for reducing, *e.g.* eidographs, are not supplied, Reduction and the reduction is accomplished by "squaring." by Squaring.

This consists of ruling similar lines on both sheets, forming squares and diagonals all over the part to be reduced.

The two stations farthest apart on the plan, which must

also be plotted on the small scale chart, are joined by a line on
both sheets, the "directing line." Then, taking the smaller
first, divide this line into as many equal parts as is thought
necessary. These parts will be from a quarter to an eighth of
an inch long. Set off lines at right angles to the directing line
from each point measured, and then lines parallel to the
directing line, at the same distances apart as the others. The
portion of the sheet required is now covered with squares.
Rule also the diagonals. These will check the correctness of
the squares, as they should, of course, pass exactly through
each corner.

Now do the same for the large scale, making *an equal
number* of squares.

It will be seen that nothing is measured, everything being
done by subdivision of the directing line.

Great care is necessary to rule all these lines truly rectangular and equidistant.

Number the lines on each plan, to prevent mistakes, giving
the same number to similar lines. Letters may be put to one
set of lines, and numbers to those at right angles.

Then, taking proportional compasses, set to the difference
of the scale as ascertained by measuring the distances apart
of similar lines, the distance of each little detail of the plan
from the nearest lines, can be put down by the same distance
from the similar lines on the small scale.

Reducing is an operation demanding even more patience
and trouble than usual, and it is better to leave the space
blank than to reduce it carelessly.

DELINEATION, SYMBOLS, AND COLOURING.

The annexed specimen chart, taken from the 'Admiralty
Manual,' shows the method of delineation employed in fair
chart work.

The following symbols are in use in surveying, in field
books, and rough charts.

Can. △ Conical. △ with staff & ball △ with bell or whistle. △ gaslight.

Zolite Station

Admiralty Summit

River

Sand Hills

Fort (according to shape)

Marsh

Figures on the land show heights in feet above high water Springs.

The days of the week are thus symbolised by the astronomical signs of the planets.

Sunday	..	.. Sun's Day	..	.. Sun	.. ☉
Monday	..	.. Moon's Day	..	.. Moon	.. ☽
Tuesday	..	.. Teut's Day	..	.. Mars	.. ♂
Wednesday	..	.. Woden's Day		.. Mercury	☿
Thursday	..	.. Thor's Day	..	.. Jupiter	♃
Friday	..	.. Friga's Day	..	.. Venus..	♀
Saturday	..	.. Saturn's Day		.. Saturn	♄

The following signs are useful in the field books.

Objects in line, called transit `	ϕ
Station, where angles are taken	△
Zero, from which angles are measured	⊕
Single altitude Sun's lower limb	☒
,, ,, ,, upper ,, ...	☒
Double ,, ,, lower limb in artificial horizon	☒
,, ,, ,, upper ,, ,, ,,	☒
Sun's right limb	☒
,, left ,, ...	☒
Sun's centre..	⊕ ⊖
Right extreme, or tangent, as of an island ..	→
Left ,, ,, ,, ,, ...	←
Zero correct ,, ,, ,, ...	Z. K.
Windmill ,, ,, ,, ,, ...	☒
Water-level	w. l.
Whitewash	w. w.

Some charts are worked up by indian-ink alone in all parts; in others, colour is used to assist the delineation of the different parts, indian-ink being always used over the colour, in exactly the same manner as if there was no groundwork.

A wash of some colour on the land helps to throw it up very much; but care is very necessary in giving this edging that it be not too deep, and that too much water is not used, or the paper will distort, and the tracing will not fit. Also in drying, that it does so gradually and generally, not allowing a streak of sunlight, for example, to fall across one part of the sheet.

If using colour, the following tints should be used for the different parts.

Towns and Buildings	Carmine.
Hills	Payne's Gray.
Cliffs	Black.
Roads	Burnt Sienna.
Rivers and Lakes	Prussian Blue.
Sand, Sand Banks, Sand Hills or Sandy Islets	Gamboge, dots black, Carmine dots for low water round edge.
Shingle	Raw Sienna.
Coral	Carmine and Burnt Sienna.
Low Water Rocks	Burnt Sienna.
Mud	Payne's Gray, edge of fine black dots.
Mangroves, Cultivated Ground, Grass, Meadows, Trees.. ..	Prussian Green.
Swamps and Marshy Land ..	Prussian Blue.

The three and five fathom lines should be coloured with cobalt; the former with a light tint all over the space included between it and low water, and the latter with a narrow edge inside the fathom line.

N.B.—To make Indian-ink perfectly black, mix a little indigo with it.

When the country is mountainous, no general wash, but only a local green in the valleys, and on flat ground, has a good effect.

Hills.   As hills in most large scale charts are now engraved in contours, it is best to use this system in the fair chart.

Shading of indian-ink, put on with a brush, is done quickly, and shows up very well.

Simple contour lines will enable the chart to be engraved almost as well as the other modes, but does not look so well.

In charts issued by the British Admiralty, the shading is put on hills as though it were a raised map, with the light coming from the north-west.

Names.   In inserting the names, care should be taken that no letters are upside down. Thus, it is often necessary to write a name in nearly a meridional direction, and it will depend upon whether the trend of the name is east or west of the meridian, whether it is written from south to north or north to south.

Thus in the two instances given here, Fig. 43, if the names had been written in the opposite direction, part of them would have been upside down. All names should be readable by turning the head, without the necessity of moving the chart.*

All names of capes, &c., should be as much on the land as possible. The soundings being the most important part of a chart, they should·be kept as clear and distinct as practicable.

Different characters should be used for the names of different classes of objects. Thus, one style for bays, another for points, another for shoals, and so on.

The scale of the chart is got from the longest calculated distance on it. This will, in cases of plans, generally be the **Side for scale.** same as that we originally plotted from, in which case we

FIG. 43.

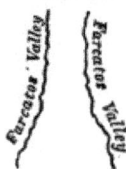

already know our scale. But if we were obliged to plot from a short side, and have since obtained data which will enable us to calculate a longer distance, we must measure the distance between the two points on our chart, and dividing this number of inches and decimals by the distance as calculated, we shall get the true scale.

It is well to indicate the two stations from which the scale is derived, by drawing a red line between them, and writing, either against it, or elsewhere on the chart, the calculated distance and bearing. If a long distance, this last should be the Mercatorial bearing.

In the case of extended surveys, or when there is no regular triangulation, the scale will depend upon the distance obtained between two stations by astronomical observations. This

---

* Vide " Instructions for Hydrographic Surveyors " for useful hints.

distance being calculated, the scale will be obtained as before.

Soundings thick.   The soundings in the chart sent home should be as thick as possible, without sacrificing legibility. There is always a great temptation to thin them out, so as to look better; but that is the work of the Office, and will probably have to be done again there in any case, as the scale on which the chart is published is usually smaller than that on which it is constructed, and if so, will not permit all soundings in the original to appear.

Natural Scale.   The natural scale, or the proportion which our chart lineally bears to the actual size of the portion of the globe it represents, is obtained by dividing the number of inches corresponding to one mile on our chart, obtained as above, by the number of inches in the nautical mile at the latitude. It is given in the form of a fraction, whose numerator is one.

Thus, supposing our scale is found to be 1·8 inches to a mile, in latitude 3°, we divide 72552 (the number of inches in a mile) by 1·8.

This gives $\dfrac{1}{40306}$ as the natural scale.

This natural scale should be noted on all sheets that are not graduated.

When the chart includes a considerable extent of coast-line that is intended to form part of a navigational sheet, it will have eventually to be redrawn on Mercator's projection, as it is on that projection all charts are published.

To do this, the sheet must be graduated, *i.e.* have the meridians and parallels placed upon it, as it is by means of them that a chart on one projection is redrawn on another.

GRADUATION OF THE SHEET.

Gnomonic Projection.   We have before said that a chart constructed by drawing right lines from one object to another, when graduated, has to be considered as being on the Gnomonic projection, and the

general features of this projection have been explained.* It
now remains to consider how to graduate such a chart.

A sheet may be graduated either before or after the chart
is drawn on it. The methods are substantially the same, and
will differ only in some preparatory work necessary in the
latter case.

We will first consider the case of graduation after the chart
is complete, and to do this we must suppose our observations
to be obtained, and that we know the latitudes and longitudes
of two stations on our chart as far apart as possible in
opposite corners of the chart.

FIG. 44.

We require, first of all, the reciprocal bearings of each
from the other, and the distance between them.

In Fig. 44 let A and B be two stations whose latitudes
and longitudes we have obtained; P is the pole. Add
to the diff. long. the spheroidal correction, and use this
corrected diff. long. in the calculations. Calculate by
spherical trigonometry the bearing of each station from
the other.

We have P B, P A the co-latitudes, and B P A the diff.
longitude. P B A and B A P are the angles required. The
latter subtracted from 180° will give us B A Q, and the

---

difference between P B A and B A Q is the convergency.
Find also the distance A B, to get the scale.

Now in Fig. 45 let A B be these same stations plotted on
our chart. Required to graduate it.

Join A B, and from A and B lay off (by chords) the reciprocal bearings of one another, ascertained as above, as A N,
B M, which will be meridians passing through those points.

From A and B measure, on the meridians, A H, B E, the
distance, according to scale, to the nearest even minute of
latitude (as 1', 5', 10', &c., as convenient).

At H and E lay off short perpendiculars to the meridians,

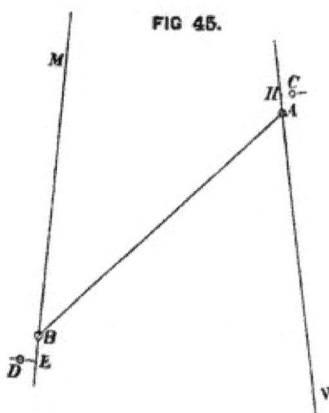

FIG 45.

and on these measure the distances H C, E D, the lengths of
departure, according to scale, to the nearest even minutes of
longitude that may be convenient.

In high latitudes and large scales, if the even meridian
required is many miles distant, error will be introduced by
this latter operation. It will only be correct for short distances, as the curve of the parallel, on which we ought to
measure this departure, will not coincide with the perpendicular to the meridian for more than a mile or two in such
a case.

We have now C and D, two stations on even meridians and

even parallels, which we shall take as our points for graduation.   This is exactly the case when we wish to graduate the
sheet first, so that henceforward the methods are identical.
In the case of after-graduation, when these even points have

FIG. 46.

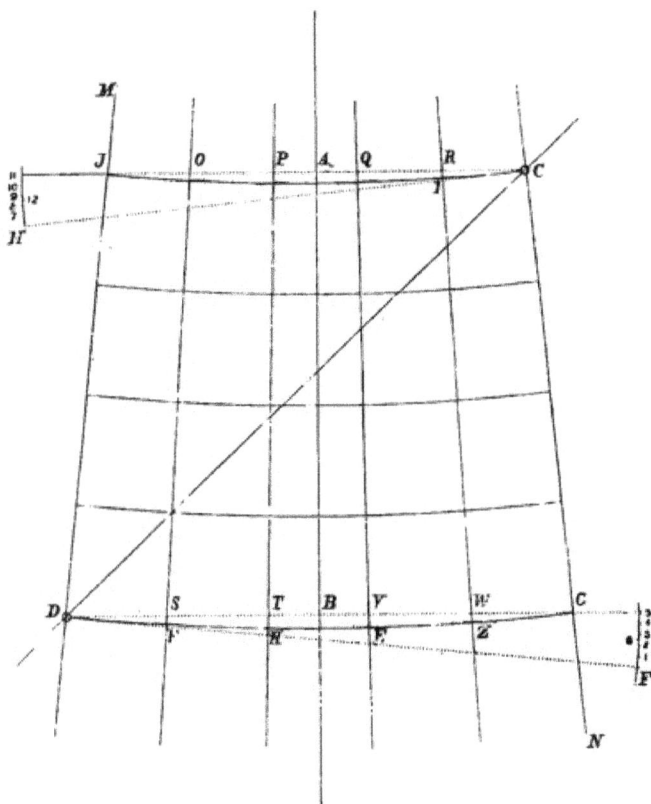

been obtained, we can rub out on our chart all lines already
ruled, to prevent confusion, and we will take a new figure for
the similar purpose of facilitating comprehension.

In Fig. 46 let C and D be the positions for graduation.

Calculate spherically, as before, the bearings of C and D from one another, and lay off the meridians C N, D M.

From C and D lay off the perpendiculars C H, D F, and from these perpendiculars lay off, on the side of the pole, half the convergency calculated for the difference of longitude in the latitude of C and D respectively, as C 11, D 5, cutting the opposite meridians respectively in J and G. Then J will be on the same parallel as C, and G as D, and J D, C G, should be equal.*

To get the central meridian of the chart, bisect J C and D G in A and B, and join them.

Then J G joined should intersect C D in the central meridian. This is a capital check for our correctness so far.

To get other meridians, divide J C and D G as many times as there are meridians required, and join them, as O S, P T, Q V, &c.

To get the parallels, which it will be remembered are curves, divide the half convergency chord, already measured, into as many parts as we have meridians.

In our figure we want five meridians from D to G, therefore we divide the chord into five parts, as 1, 2, 3, 4. Draw a small portion of D 4, cutting R W in Z. Z will then be the position on R W of the parallel of D G. By similarly drawing D 3 to cut Q V in E, D 2 to cut P T in H, and D 1 to cut O S in F, we obtain a series of points on the meridians, which, connected together, will form the curve of the parallel D G required. In high latitudes we want more meridians, to draw the curve exactly, than in low, and we must therefore be guided by circumstances as to the number of them.

Similarly, we obtain the curve of the parallel J C.

To draw more parallels, divide each meridian between the parallels obtained into as many parts as required, and join them.

This process demands considerable care and accuracy in

---

* See Appendix B.

drawing every line, and should be checked wherever practicable.

The margin of the chart is marked by subdividing the distance between each parallel or degree to the unit required.

There are other ways of drawing this graduation, all founded on the same principle. As this is, in the writer's opinion, the best of them, it is here given.

Finally, every original chart must have a memoir written on it, giving a brief description of how the chart has been constructed, the base used, observations for latitude and longitude, &c., &c., enabling the authorities at home to put the proper value on the work. **Memoir of Construction.**

It is scarcely necessary to describe the construction of a Mercator's Chart, as every naval officer learns it as part of his education. **Transferring to Mercator's Projection.**

To redraw a survey on Mercator's projection, similar meridians and parallels must be drawn on both charts, and enough of them to make the parallelograms formed by them small enough to reduce the discrepancy between the shape of any parallelograms on either chart to as little as possible. The soundings, coast-line, &c., in each parallelogram of the gnomonic chart are then transferred to the same parallelogram of the Mercator, by the latitude and longitude of each detail.

CHAPTER XVIII.

DEEP SEA SOUNDINGS.[*]

Wire Sounding—Dredging.

In the first edition of this work the method of deep sea sounding with a hempen line was alone described. Hemp has now been entirely superseded by wire, and therefore the machines employed in wire sounding and the methods of using wire will alone be treated of.

Advantage of Wire. Besides the advantage of weight, greater compactness of the apparatus, the celerity with which the weight descends, and the greater speed at which it can be reeled in, wire, from its small size, and the smoothness of its surface, enables in many cases soundings of greater accuracy to be obtained. In sounding in a surface current with hemp, the line was carried along with the current, and it was impossible to keep the ship over the lead. The result was that when the lead reached the bottom the ship was a long way astern, and an empirical correction had to be made to arrive at the vertical depth. With the fine wire now used the friction is so slight that the ship can be kept over the lead without the wire getting under the bottom, and the length of wire out is the depth.

Lucas Machine. The machine used in surveying vessels for wire deep sea sounding is that devised by Mr. Lucas, of the Telegraphic Construction and Maintenance Company, and has undergone several modifications.

---

[*] This chapter is entirely from information supplied by Captain A. M. Field, R.N., supplemented by notes from Captain W. U. Moore, R.N.

The large machine now supplied holds over 5000 fathoms
of 20-gauge wire, and is very compact.   It is fitted with two
brakes: one a screw brake for holding the reel when
required, the other an automatic brake for stopping the reel
when the weights strike the bottom.   A guider for the
purpose of winding the wire uniformly on to the reel is also
attached, and is worked by a small handle.

After leaving the reel the wire passes over a registering

FIG. 47.

Automatic Sounding Machine to carry 6000 fathoms of Wire.

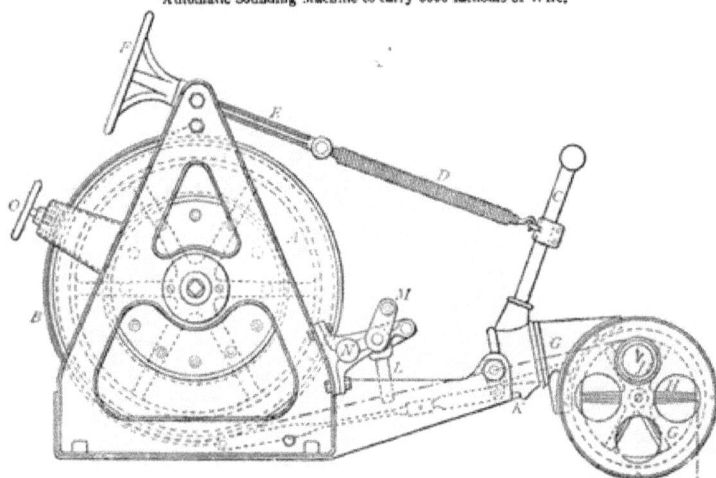

References.		
A Reel or drum.	E Regulating screw.	K Stop.
B Brake.	F Hand wheel.	L Wire guiding rollers.
C Brake lever.	G Swivelling frame.	M Handle for working roller.
D Springs.	H Measuring wheel.	N Bolt.
	J Indicator.	O Screw brake.

wheel, the dial of which indicates the amount of wire run
out, no matter how little or how much wire is on the reel.

A machine of smaller size, but very similar in type, is
supplied for use in boats for soundings of, say, more than 15
fathoms, and is also useful from the ship for serial temperatures and other purposes.

The larger machine is represented in Fig. 47, but further
detailed description will not be given, as the type may be
further altered.

Wire.   The wire used is galvanised steel wire of 20-gauge. It is supplied on drums in 5000 fathoms lengths, which are sometimes in one piece, but often have a splice in them. The drums are in hermetically-sealed tins. Though galvanised, the wire requires looking after. The galvanising process is not perfect, and it may be thin in some places, and even actually bare spots may occur. The wire should therefore be passed through an oily rag as often as possible, and oily cloths kept on the machine to protect the outer layers from damp air. After a long sounding cruise it is probably safer to condemn the wire on the machine.

The wire when new has a breaking strain of 240 lbs.

Smaller wire of 21-gauge has also been supplied, for the purpose of allowing a sufficient length to be on the reel for very deep soundings, but with the larger machines now supplied, will probably be discontinued. Its breaking strain is 190 lbs.

Splices.   Splices are made about five feet in length, one wire being laid round the other in a long spiral of about one turn per inch. The ends are soldered, and a seizing of fine wire laid over the end and for two or three inches up the splice. No end must project. Solder is then applied along the whole length of the splice. A third seizing can be placed in the centre.

Splices are the weakest points of the wire. They should be frequently examined, and their positions noted, so that, both in running out and heaving in, they may be eased round the wheels.

Sounding Rods and Sinkers.   For depths of 1000 fathoms and under, the lead can be recovered, and no detaching rod is necessary. A lead of 40 or 50 lbs. weight is suitable.

For greater depths two kinds of rods for slipping sinkers are supplied, the "Baillie" and the "Driver." Both are fitted with tubes to bring up a specimen of the bottom, and the same sinkers fit them both.

The sinkers are conical and cylindrical iron weights of 25 and 20 lbs., with cylindrical holes cast in them, through which the rod is passed. The sinkers are attached to the slipping arrangement of the rod by wire or cod line, the length of which should be so adjusted that as much of the rod as possible should project under the weight, in order to permit the rod to penetrate well into the mud. A rope grummet or iron ring fits round the bottom of the lower weight, to which is attached the suspending wire.

For water under 2000 fathoms, two conical weights are sufficient. In deeper water, a third cylindrical weight should be put between them.

It is important to have a piece of hemp-line, some ten **Splicing** fathoms long, interposed between the end of the wire and **wire into hemp.** the lead or rod. This is for the purpose of preventing the wire from kinking when the lead strikes. A piece of sheet lead about 1 lb. in weight wrapped round the hemp just below the junction, keeps the wire taut, while the hemp slacks. To splice the hemp to the wire, lay the wire in the lay of the hemp for about six feet, putting on a good racking seizing of well-waxed twine at about every foot. Test this splice well.

Before splicing, the wire must be led from the reel of the machine, between the jaws of the guiding lever, through the hollow spindle of the swivelling frame, and over the registering wheel.

The wire must be carefully transferred from the drum on **Winding** which it is supplied, to the reel of the machine, by mounting **the Wire.** the former on a temporary spindle, and fitting a brake, by which the wire can be kept taut. Winding must be even, the wire passing through a piece of greased canvas in a man's hand.

Small brass screw stoppers are provided for holding the **Wire** wire, if necessary, during a sounding. These should be fitted **Stoppers.** with a hempen tail to make fast to a cleat or other fixture.

For the greater depths it is usual to sound from forward, **Method of** but some officers have successfully accomplished it from aft **working** in fine weather. A projecting platform is fitted on the fore- **Machine.**

castle, to which the machine is bolted so as to plumb the water, being pointed in a direction slightly on the bow. An endless hemp swifter of 2-inch rope connects the deck engine and sounding machine. This is led through blocks to the forecastle, and so to the machine. One or two turns are taken round the drum of the deck engine, and the bight passes through a leading block with a jigger attached, which is placed abaft the deck engine. By means of the jigger the swifter can be kept to the requisite amount of tautness. The details of this arrangement will of course vary in different ships, and with individual tastes. Specially made sister blocks for guiding the swifter are now supplied.

As the wire runs out, the regulating screw of the brake must be gradually screwed up, so as to increase the power of the brake in proportion to the amount of wire out. The regulating screw is marked for each 500 fathoms. In fairly smooth water the brake will at once act when the weights strike the bottom, and the reel stops.

When sounding in depths of less than 3000 fathoms it is best to use only one spring, but beyond that depth two springs are required. The marks on the regulating screw are only intended as a guide; the real test is that the brake is just on the balance so as to act when the strain lessens, which may be known by the swivelling frame being just lifted off the stop. As the wire weighs $7\frac{1}{2}$ lbs. for each 500 fathoms, the 500-fathom mark on the screw should be at the position in which the screw has to be to sustain a weight of $7\frac{1}{2}$ lbs.; the 1000-fathom mark 15 lbs., and so on. This can be tested and the marks verified.

A spring balance is supplied for attachment to the brake lever when heaving in, by which the amount of strain can be seen, and the deck engine worked accordingly.

Signals.   It is necessary to establish some system of signals by which the officer on the forecastle, who is carrying out the sounding, can control the helm, main engines, and deck engine, both by day and night.

The signals given in Fig. 48 have been used with success for helm for day work :—

By night.

Green light for starboard helm.

Red light for port helm.

No light for amidships.

If lights are waved, hard over.

Fig. 43.

Helm.   *Red Flag.*

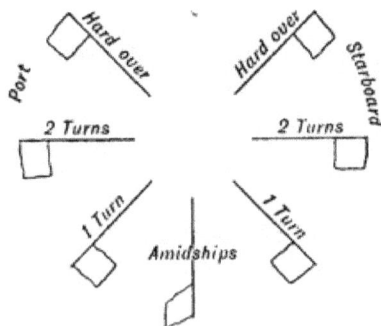

Fig. 49 gives signals for Main Engines :—

Fig. 49.

Main Engines.

Yellow Flag.

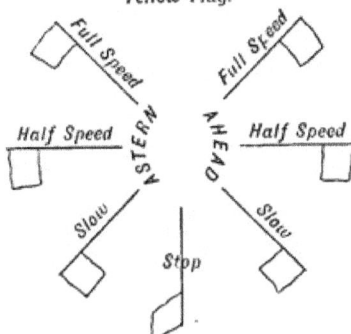

By night, a white light in starboard fore rigging for ahead,
and a white light in port fore rigging for astern; the height
indicates the speed.

For Deck Engine.

Blue flag held **vertically** downwards—Ordinary speed for
heaving in.

Blue flag held horizontally—Slow.

Blue flag held over man's head—Stop.

By night, pass the word or arrange a bell near the deck engine.

Preparing for a Sounding. See head rails cleared away as necessary.

Have ready wire stoppers, weights, sounding rod, grummet
or ring, and sling, oil can and spanners, dish and spoon for
collecting bottom specimen.

See that sounding machine is properly placed; and that
the swifter runs fair, and is put round both deck engine and
sounding machine wheel in the right directions; jigger in
place, but not taut. Place indicator at zero.

Hook the springs downward to brake lever, and see
regulating screw set to zero, and screw brake screwed down.
Wire guiding rollers must be turned back, by taking out
the bolt, and slewing the rollers clear, so as to allow the
wire to run clear. All parts of machine should be well
oiled, and the winch handles unshipped.

See that the wire is evenly wound, and taut.

Letting go. Ease down the weights, using a stick with hook at the end
to prevent a jerk, as the strain comes on the machine.

Attend the screw brake, and ease down gently and carefully to the first 100 fathoms or so, according to the weather,
after which the screw brake is no longer necessary, and may
be lifted clear of the rim of the reel.

As the wire runs out, screw up the regulating screw of the
brake.

When the bottom is reached the springs come into action,
the reel will stop, and the depth can be read on the indicator.
The length of stray hempen line, less the drift to the water's
edge, must be added.

When bottom is struck, ship one of the winch handles; **Heaving** then press the brake lever outwards to free the reel, and reel **up.** up some 10 or 20 fathoms as quickly as possible to get the rod off the bottom. If the sinkers have not detached, the effort required to reel up by hand will indicate that they are still on, and the operation of letting go and heaving up by hand must be repeated until it is certain that the weights are off, using the screw brake each time for letting go.

When the weights are gone, screw down the screw brake, and holding the brake lever by hand, run out the regulating screw, disconnect the springs and hook the spring balance, stopping the legs of the springs loosely to the balance to keep them out of the way.

See the guiding lever in gear.

Bring to and set up the swifter, which may be ready in position round the grooved wheel, the latter being secured to the shaft by means of the T head screw when required to connect.

Unship the winch handle, and heave in by the deck engine, taking care not to heave in so fast as to bring a greater strain, as shown by the spring balance, than 100 or 130 lbs. In smooth water and depth less than 2000 fathoms the strain may be as little as 70 or 80 lbs. with the wire coming in at a good pace.

When the indicator shows 30 fathoms, stop the deck engine, and heave the remainder in by hand.

When the hemp appears at the surface, heave very slowly, and when up to the machine secure the reel by screwing down the screw brake, and lift the rod in by hand.

It is very important to guide the wire on to the reel carefully and evenly by the guiding apparatus. Wire badly wound is sure to develop slack turns on running out, and will probably kink and snap.

When the rod is up disconnect the guiding gear, and get everything ready for another cast.

During the whole of the operation the ship must be carefully looked after, in order to keep the wire up and down, or as nearly up and down as possible.

Management of Ship. Though perfectly simple to effect after a little practice, inexperienced officers frequently part the wire or get erroneous soundings, and the following notes may be of service until experience is gained.

Sounding from forward, when there is little or no current, all sails are furled except the spanker, which should be set with the sheet to windward.

Keep the wind slightly on that bow on which the sounding machine is fixed; and never let the ship get more than two points off the wind unless a weather tide necessitates it.

Always endeavour to keep the position by small changes of speed and helm, avoiding high speed. This demands the closest watch, and the moment the wire is seen to be getting out of the vertical apply the brake and steam her up with the helm in the necessary position. Bear in mind that the helm is of little use, even with slight headway, unless the screw is working.

If you have failed to catch her in time, act decisively with engines and helm before matters get too bad, using high speed if necessary. A few quick revolutions of the screw with helm hard over is the best plan, checking the headway thus unavoidably given by a back turn when she has sufficient swing on, remembering that going astern always causes the ship's head to fall off from the wind, but to a less extent if the wind is on that side to which the ship's head naturally turns in a calm on reversing the engines.

The worst position is when the wire gets under the bottom, as it may catch the copper, and everything must be done to prevent this occurring. It will generally happen from allowing the wind to get on the wrong bow. In such a case it will probably be necessary to steam her rapidly round till the wind is on the right bow, and then let her drop down until the wire is again clear.

Should the wire foul the copper, or, as it may do, the gangway wire used for serial temperatures, it is sometimes possible to clear it by getting out the lower boom and using long hook ropes of ordinary sounding line with weighted smooth hooks.

A surface current across the wind complicates matters considerably. In order to get an up and down sounding the ship must be moved against the current, and the wind must then be more or less on the beam; with a weatherly current the wind may be nearly aft. Under such conditions it requires a practised eye and hand to manage the ship.

It must be understood that unless the wire is nearly up and down, it may be very difficult to say when the bottom is reached, and the depth, as given by the wire out, will not be accurate. If wire is run out after the lead has reached the bottom, kinks will result, and the wire will part.

It not infrequently happens that the wire parts for no apparent cause, but this may be due to a kink in the wire from a previous sounding.

Unless conditions of current, as above mentioned, necessitate it, it is generally fatal to let the ship fall off broadside to the wind. Should she be allowed to get round with the wind aft, there is probably no remedy but to heave in again, and commence afresh.

Care is necessary in heaving in the last 50 fathoms, so as to stop the deck engine in time.

The time interval with wire, when not pitching heavily, **Time** up to depths of between 2000 and 3000 fathoms, is about **Occupied.** one minute per 100 fathoms. Reeling in may be accomplished at nearly the same rate.

A sounding of 1000 fathoms may be obtained in 25 minutes from the time the weight is lowered to the time the order is given to put the ship on her course. 2000 fathoms will require 45 minutes, and 3000 fathoms 75 minutes.

Though the time of running out each hundred fathoms is no longer required, as with hemp, for ascertaining when the sinkers strike the bottom, it is well to take the intervals, as they assist in the regulation of the brake.

If a second wire machine is available (a boat's machine will **Serial** do), serial temperatures can be conveniently taken from the **Temperatures.**

gangway whilst the sounding is being obtained forward, thus gaining time.

A 30 lb. sinker is attached to the end of the wire, and the thermometers are secured to the wire by the metal clips at the back of the cases, at the required distances. See that the indices are down before attaching the thermometers.

There is a certain amount of extra risk in thus working from the gangway while the other wire is over, as the two wires may foul deep down, when the fact of the thermometers acting as toggles may make them difficult to clear. The time saved, however, justifies it in fine weather, and when experience in sounding is gained. To avoid heavy loss, however, not more than four thermometers should be on the wire.

Sounding on Voyages. Deep sea soundings on every voyage are now a recognised part of a surveying ship's routine. It is only in this way that depths so useful for submarine cable, as well as for scientific purposes, can be accumulated without the expenditure of time involved in special sounding cruises, though those are occasionally necessary. As a rough rule a sounding after daybreak, and before sunset, should be obtained daily, when observations can be got.

Dredging. Connected with deep sea sounding, though not such a common part of a surveyor's duty, is dredging, on which a few words may be useful.

The dredge consists of a strong iron frame, the sides forming lips, which are connected at each end by an iron bar, and are chamfered off to fairly fine edges. These edges slightly incline outwards, as seen in the sketch. On the iron bars arms are fitted, and to the eye at the extremity of one of them the dredging hawser is bent, the eye of the other arm being seized to it to form a span of such a strength that the seizing will carry away if the dredge catches.

Attached by wire seizings to holes in the lower part of each lip is a stout canvas bag, perforated with holes in its lower portion to permit the water to flow through.

A stout iron bar, to which three long swabs are secured,

is suspended by ropes from the iron end bars immediately below the canvas bag.

Dredges are of various dimensions, but a convenient size is as follows, as illustrated by the sketch Fig. 50:—

Fig. 50.

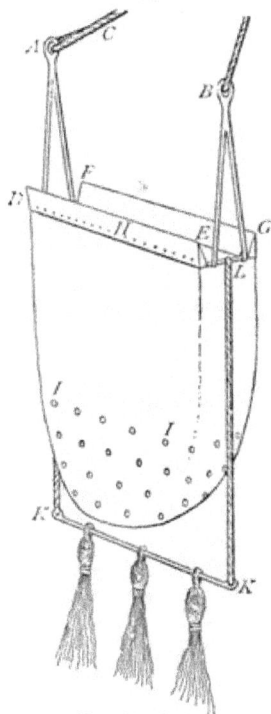

The Dredge.

A and B, arms, 2½ feet long.

C, hawser.

D E and F G, lips of the dredge, 2½ feet long.

H, holes to which bag is laced.

I I, perforations in the canvas bag.

K K, swab bar.

L, bar of dredge mouth, 6 inches.

The hawser is weighted with about 60 lbs., at 5 fathoms from the dredge.

On a sandy bottom, a net bag is substituted for the canvas bag, which gets full of sand.

On a rough bottom an iron triangle carrying swabs only is used, the arms being stopped lightly together so as to carry away, if caught ; or even one large swab at the end of a rope, weighted two feet above it to keep it down. These are especially useful on coral banks, where a regular dredge may very likely be lost.

To dredge, turn the ship away from the wind or current, and drop the dredge from aft with slight headway on, taking care that the bag and swabs do not capsize and foul the mouth of the dredge. Ease out the hawser to about three times the depth of water, and let the ship drift for about 20 or 30 minutes. The dredge can be hoisted in by a burton from the mizzen gaff.

CHAPTER XIX.

MISCELLANEOUS.

Distortion of Printed Charts—Observations on Under-Currents—Exploring
a River—Swinging Ship.

In printing charts from an engraved plate, the paper has to Distortion in Printed Charts.
be damped. This results in distortion on the sheet drying,
and angles laid off on a published sheet will never be found
to agree exactly, especially if the sheet is large. This must
always be borne in mind, in trying angles on a published
chart.

For this reason, when a published plan is to be examined,
a "dry proof" is supplied to the surveyor from the Admiralty.
This is an impression "pulled," as it is termed, on to a dry
sheet. It is much fainter than a damp-pulled copy, and
would not do for ordinary use; but being an exact facsimile
of the copper plate, all angles, bearings, &c., should agree
precisely, if the original survey is correct.

This fact of the distortion of published charts is not generally known, and many reports of so-called inaccuracies have
been made in ignorance of it. The amount of it varies with
the goodness of the paper, and the trouble bestowed by the
printer in damping his paper uniformly. It is a fact much
to be deplored, and the man who invents a means of obviating
it, will bestow a great boon on cartography.

Y

OBSERVATIONS ON UNDER-CURRENTS.

Though not in the ordinary run of surveying operations, a slight description of the method of discovering the direction and approximate rate of under-currents may be useful.

To ascertain these satisfactorily, special gear is necessary.

General Principle.
The general principle is to expose a large surface to the action of the under-current, and to support this in the water by a floating buoy which will present as small a surface as possible to the action of the surface stream.

The experiments must be carried on from boats, and therefore the gear must be as light as possible, for easy handling.

A series of observations on the under-currents in the Bosporus and Dardanelles resulted in the author's adopting the following :—*

Apparatus used.
A light, flat wooden board, 6 feet square, with a wing 2 feet in length, at right angles to the rest of the frame, was used as the submerged drag. (See Sketch, Fig. 51.)

To the extremities of the wing the sling, *a a*, was made fast, and to this sling the supporting line to the buoy was bent, at such a point as kept the surface of the drag vertical when the strain came on.

It weighed 70 lbs. in air, and took 120 lbs. of lead to sink it satisfactorily. These leads were made fast with a little drift, and another line, *c*, was bent, both to them and to the lifting sling, *b b*, so that the weight of the leads could be taken off the drag, when pulled up to the surface, before finally hoisting it into the boat.

An iron buoy, 1 foot in diameter and 5 feet long, supported this structure well when the surface current was not very strong, and only presented an area of less than one square foot to pull through the water.

When the surface current was swift, other buoys had to be added, attached in line to the upper end of the first, for under

---

* Observation of Currents in Dardanelles and Bosporus.

FIG. 51.

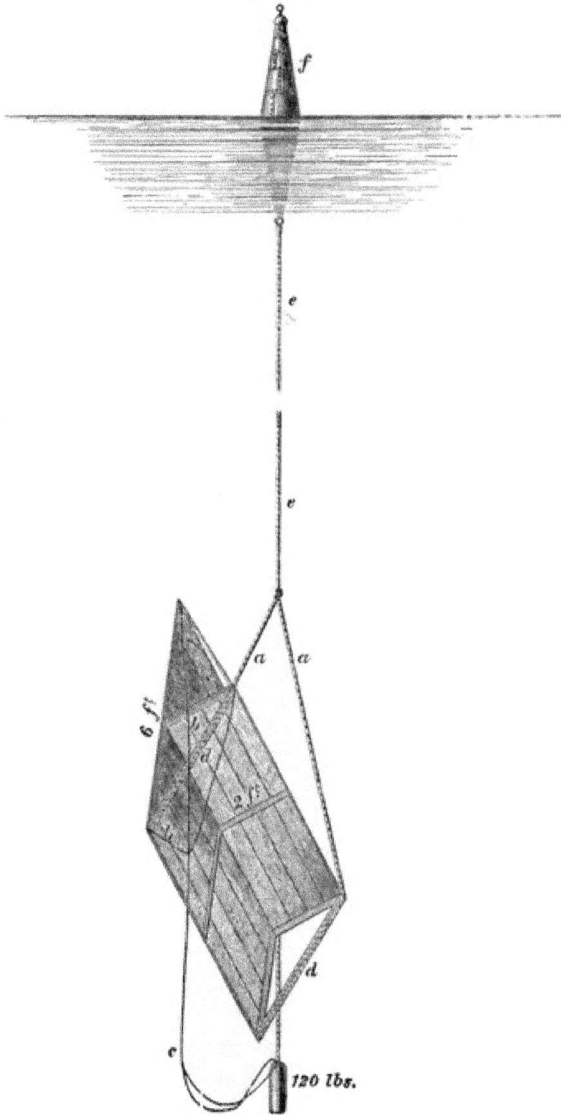

these circu~   ~o dragged under water,
an? ·   . ~ollowed. Several disappeared
        . ⌄ reappear when the apparatus got into
        .⌄ı, some for good and all.
        ⌄o ascertain the movement of the floating buoy, and there-
~ug Rate fore the direction in which the drag was carried by the under-
and Direc- current, a "fix" was taken to shore objects, and plotted on a
tion. large scale sheet of points, when the drag was let go free from

FIG. 52.

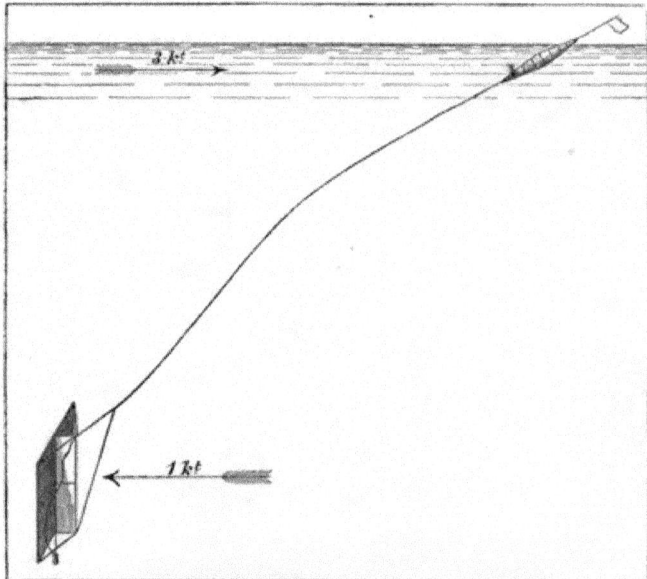

the boat.   Subsequent fixes and times taken enabled the
course and distance of the buoy in the intervals to be recorded on this sheet.

Surface A small buoy, weighted so as to float awash, put into the
Current. water at the same spot and time, and followed by another
boat, afforded means of ascertaining the surface current.

Defects. This arrangement worked very satisfactorily altogether, but
there are several defects in it.

The depth of the submerged drag will not be the length of the line allowed, but some unknown quantity less, as will be seen by accompanying sketch, Fig. 52.

This must be estimated.

The force expended in dragging the buoy through the surface water, and overcoming the friction of the suspending line, is also an unknown quantity, but will always have the effect of retarding the motion of the submerged drag. The rate therefore recorded, by the movement of the surface buoy, will always be *less* than the true rate of the under-current. *Rate less than true Rate.*

We do not imagine that the apparatus described above may not be much improved upon, but we give it as a starting-point for any officer employed in future investigations of a similar character.

Several instrument makers now turn out "Current Meters" of various forms. Doubtless these could be, with a little ingenuity, adapted to sea work, at least to show the true *rate* of an under-current. *Current Meter.*

A deep sea current meter devised by Lieutenant Pillsbury, U.S.N., has, with several modifications, been (1897) under trial, but is not yet brought into general use.

The observations are made more complete by ascertaining the temperature and density of the water, at the depths experimented on. *Temperature and Density.*

EXPLORING A RIVER.

Narrow rivers, navigable for boats, will generally be suffi- ciently laid down on a marine chart by a sketch survey, made from the boat (a steam pinnace, if possible), while passing up and down. Patent log and compass will be the instruments mainly used for putting down the direction and length of each reach; though if we have objects that we can use for a sextant fix, we shall of course use them in preference, at any rate from time to time. We must endeavour in every case to get a good fix at our furthest point, and the course of the river, as mapped by patent log and compass, will then be squared *Running Survey.*

in on that, and the fixed points at the entrance and any other
fixes we may have got.   Any elevated points near, which we
can ascend and fix, and from them get angles to bends and
reaches of the river, will much assist us, especially when,
which is so often the case, the river is thickly lined with trees
and jungle.

The patent log will be fitted, as already described, with the
dial on the gunwale and the fan towing astern.   Theodolite
legs standing in the stern-sheets make an excellent stand for
a prismatic compass, and enable us to get a better bearing
than by holding it in the hand.

It is in rounding the bends that the greatest error in mapping a river is introduced, as the distance run over while
gradually altering course must be estimated by eye, which
requires considerable experience at judging distances.

Current.   Current must be taken into consideration, and may be
obtained, if time allows, by anchoring the boat for half an
hour, and reading the patent log, or in shorter time, by
heaving the current log.

In a river where the tide extends some distance up, and
where the land is low and jungly, as in so many mangrove
rivers, our difficulties are much increased, as the velocity of
the current will be constantly varying, and we cannot hope
to obtain any sextant fixes to check our position.   In cases of
this kind, if it is desired to have any degree of accuracy in
the sketch, the only way is to run over the work again,
making an independent map, and squaring in afterwards a
mean of the two.

Plotting in Boat.   It is best always to plot as we go.   Mistakes are thus
rendered less likely, and the vexed question of the bends
can best be solved by placing their shape on the paper at
once.   We can also look at our work on the way down again,
and correct little inaccuracies more readily.

Survey on larger scale.   If it is desired to make a large scale plan of a river of
greater width, the best method is to employ several boats at
once, four if possible, which will triangulate their way up,
two on either side.   Starting from two fixed points at the

mouth of the river, two boats will remain there while the other two go up to convenient positions, whence they can see the boats remaining at the first points. Angles will then be taken from all, to everything conspicuous, and to one another, and the lower boats will, leaving marks at their old stations, move up to two new positions above the other boats, when the angles will be repeated, and so on, the lower boats moving on each time.

The shore line can either be sketched by the boats as they go up, or done afterwards more correctly when the marks are all up and fixed. Soundings, in the same way, can either be taken from the boats as they move from station to station, in which case they would cross over each time so as to get a diagonal line across the channel, or can be more regularly taken afterwards, as the circumstances of the case may require.

Everything must be plotted afterwards, and communication between the boats as they pass one another, when names can be given and objects pointed out for mutual observation, will greatly facilitate the comprehension of one another's angles, when putting down the points.

SWINGING SHIP.

Though the compass is but little employed in surveying, it is occasionally unavoidably brought into use.

As deviation varies, with lapse of time and change of latitude, it must be constantly ascertained by swinging ship.

The methods in use in swinging ship are well known, but perhaps a repetition may not be thrown away.

They are two in number.

One, by observing the compass-bearing of a distant object whose true magnetic bearing is known.

The other, by reciprocal bearings of the compass on board, and another on shore.

The first is the best and simplest when the magnetic **By distant Object.** bearing of the distant object can be well determined.

The object should be, at the least, six miles distant, and the more the better.

Its bearing can be obtained from observations on shore from a spot in line with the ship, or by true bearing with known variation applied.

Objects for this purpose are sometimes indicated on the charts or in the sailing directions, and the bearings given.

The deviation, for each position of ship's head, is then the difference between this fixed magnetic bearing and the observed bearing by compass.

The ship can either be hauled round with hawsers, at anchor, or, if the object be far enough off, can be steamed round a circle small enough to make no difference in the bearing.

If steamed round, it is well to repeat the operation, turning in the opposite direction, as the compass may partake of the swing of the ship, which will introduce error. The mean of the two will then be the bearing to use.

By reciprocal Bearings.   The second method is perhaps the one generally employed, and is very convenient with a theodolite at hand.

An officer is landed with azimuth compass and theodolite.

He obtains the bearing with the compass of some well-defined object, and setting up his theodolite, takes it for his zero.

In arranging the theodolite on zero, it saves calculation to point the degree and minute of the magnetic bearing to the zero instead of 360°. Thus, if the zero bears by compass S. 44° 20′ E. (supposed to be unaffected), set the vernier to 134° 20′. The angles read to the ship will then be the angle east of the magnetic north.

A flag on a long staff is held behind the theodolite, when all is ready.

The ship, under steam, and with a flag placed exactly over the standard compass, steams slowly round, hoisting a large flag close up to the mast-head just before the ship's head comes to each point, which is dipped at the moment of observation, when the bearing of the shore station is taken.

The flag on shore is dipped, to show that the angle of the

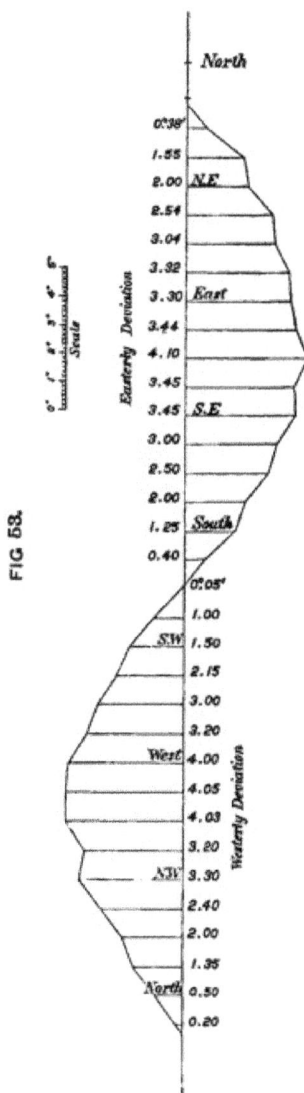

FIG 53.

flag over the compass has been obtained by the theodolite, and is again shown as a response, when the flag is mast-headed for the next observation.

The time of each observation is taken by previously compared watches.

In this case, too, the ship should be swung in the opposite direction, if it is deemed necessary.

The difference of the reciprocal bearings is the deviation at each observation.

If more than one observation at any or all points has been obtained, the results are meaned for the accepted deviation.

It is usual to observe at every point of the compass, for the ship's head, but in some vessels it may be necessary to subdivide this.

The readiest way of examining the results of our observations is by use of the Graphic method.   (Fig. 53.)

Drawing a long line, measure off equal parts along it, for the points of ship's head, and at each point on this normal lay off, at right angles, a line equal to the degrees and minutes of the deviation, on any scale we choose—easterly deviation to the right, westerly to the left of the normal.

Lines drawn through the extremities of these abscissæ will denote the curve of deviation observed.

By the irregularities of this curve, we can judge of the correctness of the observations very fairly ; and for our final table of deviation, we can draw a mean curve, if there are many irregularities, and measure to that for the amount of deviation for each point.

The valuable results for variation obtainable from swinging have already been mentioned at p. 283.

APPENDIX.

FIG. 54.

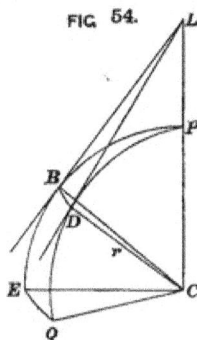

Here C is the centre of the earth, P is the pole, E P, Q P, two meridians a known distance apart. B L, E L, are two tangents to the meridians, at the middle latitude known, in the same plane as the meridian, and meeting one another and the axis of the earth C P, produced, in L.

Then B L D is the Convergency required, and D L C is the middle latitude, and B C D the departure. D C is a radius of the earth = r.

Now as B D is small, it can be taken as a straight line without sensible error.

We can also assume B L D and B C D to be right-angled triangles.

Then B D = D L × Tan B L D.

Similarly B D = r × Tan B C D.

Equating, we have D L × Tan B L D = r × Tan B C D.

But   D L = r × Cot D L C;

∴ r × Cot D L C × Tan B L D = r × Tan B C D,

or Tan B L D = Tan B C D × Tan D L C,

or Tan Convergency = Tan dep × Tan Mid Lat.,

and when Convergency is very small, we can say

Convergency = Dep × Tan Mid Lat.

B.—*In Graduating a Chart on the Gnomonic Projection.*

To show that the angle of half convergency laid off from the rectangle intersects the opposite meridian on the parallel, and also that the further subdivisions of the convergency intersect their respective meridians on the same parallel.

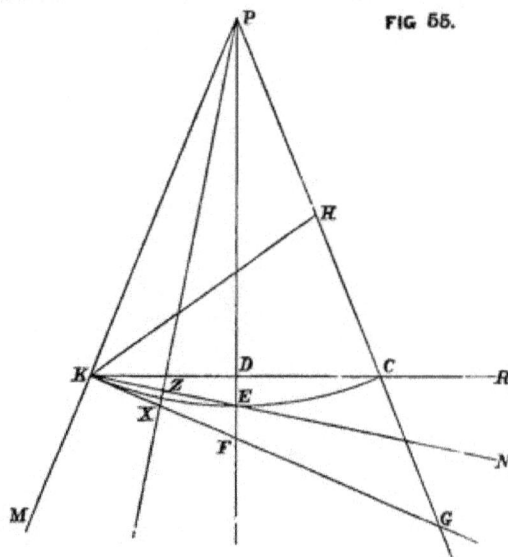

FIG 55.

From K and H, the graduating positions, draw the true bearings, lines K P, P G, which are meridians and will meet at P, the pole of the projection, making the angle K P C, or the Convergency.

Make H C = difference of latitude of H and K. Then P C will equal P K.

Join K C, bisect it in D, and join P D, the central meridian. Lay off K G perpendicular to K P.

Then $\angle$ C K G is the half convergency;

For in $\triangle$ P D K . . . D K P = 90 − D P K,

and as F K P is drawn = 90;

$\therefore$ . . . D K P = 90 − D K F;

$\therefore$ D P K = D K F.

But D P K = $\frac{1}{4}$ K P C the convergency;

$\therefore$ D K F or C K G = $\frac{1}{4}$ convergency.

Q. E. D.

Bisect C K G in K N, making G K N or X K Z = ¼ convergency,

Then E where K N intersects P F is on the parallel K C, or P E = P K.

Bisect K P E in P X.

Now   M K Z = K Z P + K P Z,

but M K Z = 90° + ¼ Convergency (by construction) and K P Z = ¼ Convergency;

∴ 90° + ¼ Conv = K Z P + ¼ Conv ;

∴ K Z P = 90° = P Z E,

and as K P Z = Z P E and P Z is common, the △s K Z P & P Z E are equal and similar ;

∴ P E = P K.

Q. E. D.

C.—*To prove Chord* $= 2 \, rad \left\{ \, Vers \left(90 + \dfrac{\theta}{2} \right) - 1 \right\}.$

FIG 56.

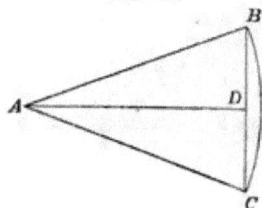

Let $C A B = \theta$, the angle whose chord is required.

At any radius $A C = r$, describe arc $C B$.

Join $C B$, then $C B$ is Chord required.

Bisect $B C$ in D and join $A D$.

Then $D A B = \dfrac{\theta}{2}.$

Now　$D B = A B, \, Sin \, D A B$

$$= r \cdot Sin \frac{\theta}{2},$$

but $B C = 2 \, D B$;

$$\therefore B C = 2 \, r \cdot Sin \frac{\theta}{2} \quad \cdots \quad \cdots \quad (a).$$

But Versine $\dfrac{\theta}{2} = 1 - Cos \dfrac{\theta}{2}$;

$$\therefore Versine \left(90 + \frac{\theta}{2} \right) = 1 - Cos \left(90 + \frac{\theta}{2} \right)$$

$$= 1 - \left(- Sin \frac{\theta}{2} \right)$$

$$= 1 + Sin \frac{\theta}{2}$$

$$\therefore Sin \frac{\theta}{2} = Vers \left(90 + \frac{\theta}{2} \right)$$

Substituting this in (a) we get

$$B C = 2 \, r \left\{ Vers \left(90 + \frac{\theta}{2} \right) - 1 \right\}$$

APPENDIX.

A.— *To prove that Tan Convergency = Tan Dep . Tan Mid Lat.*

FIG. 54.

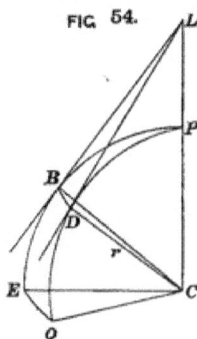

Here C is the centre of the earth, P is the pole, E P, Q P, two meridians a known distance apart. B L, E L, are two tangents to the meridians, at the middle latitude known, in the same plane as the meridian, and meeting one another and the axis of the earth C P, produced, in L.

Then B L D is the Convergency required, and D L C is the middle latitude, and B C D the departure. D C is a radius of the earth = r.

Now as B D is small, it can be taken as a straight line without sensible error.

We can also assume B L D and B C D to be right-angled triangles.

Then B D = D L × Tan B L D.

Similarly B D = r × Tan B C D.

Equating, we have D L × Tan B L D = r × Tan B C D.

But   D L = r × Cot D L C;

∴ r × Cot D L C × Tan B L D = r × Tan B C D,

or Tan B L D = Tan B C D × Tan D L C,

or Tan Convergency = Tan dep × Tan Mid Lat.,

and when Convergency is very small, we can say

Convergency = Dep × Tan Mid Lat.

B.—*In Graduating a Chart on the Gnomonic Projection.*

To show that the angle of half convergency laid off from the rectangle intersects the opposite meridian on the parallel, and also that the further subdivisions of the convergency intersect their respective meridians on the same parallel.

FIG 55.

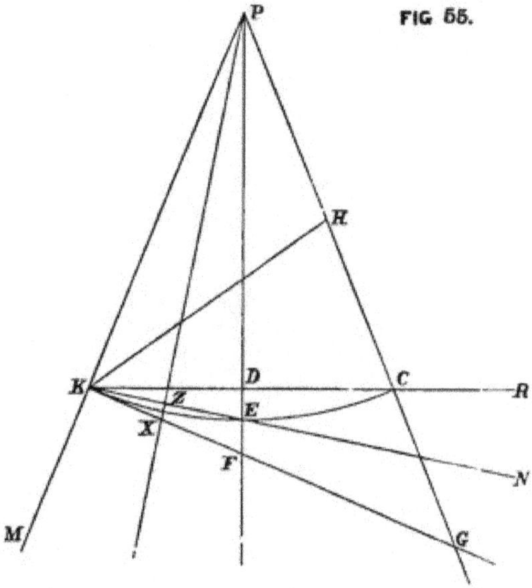

From K and H, the graduating positions, draw the true bearings, lines K P, P G, which are meridians and will meet at P, the pole of the projection, making the angle K P C, or the Convergency.

Make H C = difference of latitude of H and K. Then P C will equal P K.

Join K C, bisect it in D, and join P D, the central meridian. Lay off K G perpendicular to K P.

Then ∠ C K G is the half convergency;
For in △ P D K . . . D K P = 90 − D P K,
and as F K P is drawn = 90;
∴ . . . . D K P = 90 − D K F;
∴ D P K = D K F.
But D P K = ½ K P C the convergency;
∴ . D K F or C K G = ½ convergency.

Q. E. D.

Bisect C K G in K N, making G K N or X K Z = ¼ convergency,

Then E where K N intersects P F is on the parallel K C, or P E = P K.

Bisect K P E in P X.

Now   M K Z = K Z P + K P Z,

but M K Z = $90°$ + ¼ Convergency (by construction) and K P Z = ¼ Convergency;

∴ $90°$ + ¼ Conv = K Z P + ¼ Conv ;

∴ K Z P = $90°$ = P Z E,

and as K P Z = Z P E and P Z is common, the △s K Z P & P Z E are equal and similar;

∴ P E = P K.

Q. E. D.

C.—*To prove Chord* $= 2\ rad\ \left\{\ Vers\left(90 + \dfrac{\theta}{2}\right) - 1\right\}.$

FIG 56.

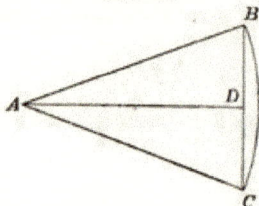

Let $C\,A\,B = \theta$, the angle whose chord is required.

At any radius $A\,C = r$, describe arc $C\,B$.

Join $C\,B$, then $C\,B$ is Chord required.

Bisect $B\,C$ in D and join $A\,D$.

Then $D\,A\,B = \dfrac{\theta}{2}.$

Now $D\,B = A\,B,\ Sin\ D\,A\,B$

$$= r\ .\ Sin\ \frac{\theta}{2},$$

but $B\,C = 2\,D\,B;$

$$\therefore B\,C = 2\,r\ .\ Sin\ \frac{\theta}{2}\ .\ .\ .\ .\ .\ .\ .\ .\ (a).$$

But Versine $\dfrac{\theta}{2} = 1 - Cos\ \dfrac{\theta}{2}\ ;$

$$\therefore Versine\left(90 + \frac{\theta}{2}\right) = 1 - Cos\left(90 + \frac{\theta}{2}\right)$$

$$= 1 - \left(-\ Sin\ \frac{\theta}{2}\right)$$

$$= 1 + Sin\ \frac{\theta}{2}$$

$$\therefore Sin\ \frac{\theta}{2} = Vers\left(90 + \frac{\theta}{2}\right)$$

Substituting this in (*a*) we get

$$B\,C = 2\,r\ \left\{Vers\left(90 + \frac{\theta}{2}\right) - 1\right\}$$

D.—*To prove Reduction to the Meridian* $= \dfrac{\text{Cos } l \,.\, \text{Cos } d}{\text{Sin } z} \,.\, \dfrac{\text{Vers } h}{\text{Sin } 1''}$

FIG. 57.

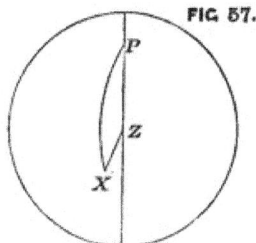

Let X be a heavenly body near the Meridian, P the pole, Z the Zenith. Let Hour Angle $Z P X = h$, Latitude $= 90 - P Z = l$, Zenith distance $X Z = z$, Declination $= 90 - P X = d$.

Then $\text{Cos } Z P X = \dfrac{\text{Cos } X Z - \text{Cos } P X \,.\, \text{Cos } P Z}{\text{Sin } P X \,.\, \text{Sin } P Z}$,

  or $\text{Cos } h = \dfrac{\text{Cos } z - \text{Sin } l \,.\, \text{Sin } d}{\text{Cos } l \,.\, \text{Cos } d}$;

$\therefore \text{Cos } z - \text{Sin } l \,.\, \text{Sin } d = \text{Cos } l \,.\, \text{Cos } d \,.\, \text{Cos } h$
$= \text{Cos } l \,.\, \text{Cos } d \,.\, (1 - \text{Vers } h)$
$= \text{Cos } l \,.\, \text{Cos } d - \text{Cos } l \,.\, \text{Cos } d \,.\, \text{Vers } h$;

$\therefore \text{Cos } z + \text{Cos } l \,.\, \text{Cos } d \,.\, \text{Vers } h = \text{Cos } d + \text{Sin } l \,.\, \text{Sin } d$
$= \text{Cos } (l \backsim d)$
$= 1 - \text{Vers } (l \backsim d)$;

$\therefore$ $\text{Vers } (l \backsim d) = 1 - \text{Cos } z - \text{Cos } l \,.\, \text{Cos } d \,.\, \text{Vers } h$
$= \text{Vers } z - \text{Cos } l \,.\, \text{Cos } d \,.\, \text{Vers } h.$

Working with Declination $= 90 + P X$, we shall get
$= \text{Vers } z - \text{Cos } l \,.\, \text{Cos } d \,.\, \text{Vers } h.$

But $l \backsim d$ or $l + d$ is the Meridian Zenith Distance $= Z$.
Then $\text{Vers } Z = \text{Vers } z - \text{Cos } l \,.\, \text{Cos } d \,.\, \text{Vers } h.$
$- \text{Cos } l \,.\, \text{Cos } d \,.\, \text{Vers } h = \text{Vers } Z - \text{Vers } z$
$= 1 - \text{Cos } Z - 1 + \text{Cos } z$
$= \text{Cos } z - \text{Cos } Z$
$= - 2 \text{ Sin } \dfrac{z + Z}{2} \,.\, \text{Sin } \dfrac{z - Z}{2}$;

but z and Z are nearly alike, so $\dfrac{z + Z}{2}$ may be taken $= z$

and $z - Z$ is very small $\therefore 2 \text{ Sin } \dfrac{z - Z}{2}$ may be taken $= (z - Z) \text{ Sin } 1''$:

$\therefore \text{Cos } l \,.\, \text{Cos } d \,.\, \text{Vers } h = \text{Sin } z \,.\, (z - Z) \text{ Sin } 1''$,

  or $z - Z = \dfrac{\text{Cos } l \,.\, \text{Cos } d}{\text{Sin } z} \,.\, \dfrac{\text{Vers } h}{\text{Sin } 1''}$

but $z - Z$ is the Reduction to the Meridian;

$\therefore$ Reduction to Mer. $= \dfrac{\text{Cos } l \,.\, \text{Cos } d}{\text{Sin } z} \,.\, \dfrac{\text{Vers } h}{\text{Sin } 1''}$

E.—*To show that the Distance of Horizon in English Miles*

$$= \sqrt{\frac{3}{2} \text{ height in feet.}}$$

FIG 58.

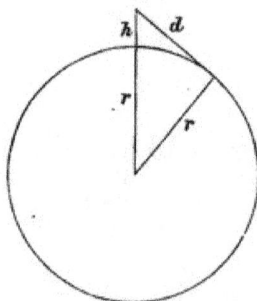

Let *r* be radius of earth.
h   height of observer in feet.
d   distance of horizon.

$$d^2 + r^2 = (h + r)^2$$
$$= h^2 + r^2 + 2\,h\,r,$$

h^2 being small may be omitted.

$$d^2 = 2\,h\,r;$$

but h in Eng. miles is $\dfrac{h}{5280}$

and $2r$ „   is 7910;

$$\therefore d^2 = \frac{7910}{5280}\,h$$

$$= \frac{3}{2}\,h \text{ very nearly};$$

$$\therefore d = \sqrt{\frac{3}{2}\,h.}$$

This is the distance disregarding refraction, which has the effect of increasing the distance of the visible horizon. If having found d as above, we subtract $\frac{1}{18}$ of itself from it, the remainder will be the true distance in sea miles very nearly, with the effects of refraction taken into consideration.

F.—*Base by Sound.*

To prove that $T = \dfrac{2\,t\,t^1}{t + t^1}$

Let d be distance in feet between stations,

v the velocity of sound, | in feet per

x ,,   of the wind, | second,

t the observed interval in seconds with the wind,

t_1 ,,   against the wind,

T the mean interval required.

Then $d = v\,T$.

By observation $d = v\,t + x\,t$

and $\qquad\quad d = v\,t_1 - x\,t_1$.

Dividing by t and t_1 we get

$$\frac{d}{t} = v + x,$$

$$\frac{d}{t_1} = v - x.$$

Adding, we have

$$\frac{d}{t} + \frac{d}{t^1} = 2\,v,$$

$$d\left(\frac{1}{t} + \frac{1}{t^1}\right) = 2\,v,$$

$$d\,\frac{t^1 + t}{t\,t^1} = 2\,v,$$

$$d = v\,\frac{2\,t\,t^1}{t + t^1}\;;$$

but we have also

$$d = v\,T,$$

and by equating

$$T = \frac{2\,t\,t^1}{t + t_1}.$$

Z

FORM G.

Date

Time.	Object.	Angle.	Object.	Angle.	Object.	Sounding at Fix.	Soundings after.	Remarks.

Form H.

Chronometer Comparison Book.

Max.

Ther.

Min.

Chrons.	Time.	Check.	Slow on A.	2nd Diff.	
A					
B					
A					
C					
A					
D					
A					
E					
A					
F					
A					
G					
A					
H					
A					
I					
A					
K					
A					
L					

TABLE J.— *Table of Chords of Arcs from 0° to 60°, to facilitate the Projection of Angles. Radius = 10. By* T. H. TIZARD, *Navigating Lieut. R.N.*

Min.	0°	Parts for "	1°	Parts for "	2°	Parts for "	3°	Parts for "	4°	Parts for "
0	0·00000	0	0·17453	0	0·34905	0	0·52354	0	0·69799	0
1	·00291	5	·17744	5	·35195	5	·52644	5	·70080	5
2	·00582	10	·18035	10	·35486	10	·52935	10	·70380	10
3	·00873	15	·18326	15	·35777	15	·53226	15	·70671	15
4	·01164	20	·18617	20	·36068	20	·53517	20	·70962	20
5	·01455	25	·18907	25	·36359	25	·53808	25	·71252	25
6	·01746	30	·19198	30	·36650	30	·54098	30	·71543	30
7	·02037	35	·19489	35	·36941	35	·54389	35	·71834	35
8	·02328	39	·19780	39	·57231	39	·54680	39	·72129	39
9	·02618	44	·20070	44	·37522	44	·54970	44	·72415	44
10	·02908	49	·20361	49	·37813	49	·55261	49	·72706	49
11	0·03199	53	0·20652	53	0·38104	53	0·55552	53	0·72996	53
12	·03490	58	·20943	58	·38395	58	·55843	58	·73287	58
13	·03781	63	·21234	63	·38685	63	·56134	63	·73578	63
14	·04072	68	·21525	68	·38976	68	·56425	68	·73869	68
15	·04363	73	·21816	73	·39267	73	·56715	73	·74159	73
16	·04654	78	·22107	78	·39558	78	·57006	78	·74450	78
17	·04945	82	·22398	82	·39849	82	·57297	82	·74741	82
18	·05236	87	·22689	87	.40140	87	·57588	87	·75032	87
19	·05527	92	·22979	92	·40430	92	·57878	92	·75322	92
20	·05818	97	·23270	97	·40721	97	·58169	97	·75613	97
21	0·06109	102	0·23561	102	0·41012	102	0·58460	102	0·75903	102
22	·06400	107	·23852	107	·41303	107	·58751	107	·76194	107
23	·06691	112	·24143	112	·41594	112	·59042	112	·76485	112
24	·06982	117	·24434	117	·41884	117	·59333	117	·76775	117
25	·07272	122	·24725	122	·42175	122	·59623	122	·77066	122
26	·07563	127	·25016	127	·42466	127	·59914	127	·77356	127
27	·07854	132	·25306	132	·42757	132	·60205	132	·77647	132
28	·08145	137	·25597	137	·43048	137	·60495	137	·77938	137
29	·08436	141	·25888	141	·43339	141	·60786	141	·78229	141
30	·08727	145	·26179	145	·43629	145	·61077	145	·78519	145
31	0·09017	150	0·26470	150	0·43920	150	0·61368	150	0·78810	150
32	·09308	155	·26761	155	·44211	155	·61658	155	·79107	155
33	·09599	160	·27052	160	·44502	160	·61949	160	·79392	160
34	·09808	165	·27342	165	·44793	165	·62240	165	·79682	165
35	·10181	170	·27633	170	·45084	170	·62531	170	·79973	170
36	·10472	175	·27924	175	·45374	175	·62821	175	·80264	175
37	·10763	180	·28215	180	·45665	180	·63112	180	·80554	180
38	·11054	185	·28506	185	·45956	185	·63403	185	.80845	185
39	·11344	190	·28797	190	·46247	190	·63694	190	·81135	190
40	·11635	194	·29088	194	·46538	194	·63984	194	·81426	194
41	0·11926	199	0·29378	199	0·46828	199	0·64275	199	0·81717	199
42	·12217	204	·29669	204	·47119	204	·64566	204	·82007	204
43	·12508	209	·29960	209	·41410	209	·64857	209	·82297	209
44	·12799	214	·30251	214	·47701	214	·65147	214	·82588	214
45	·13090	219	·30542	219	·47992	219	·65438	219	·82879	219
46	·13381	223	·30833	223	·48283	223	·65728	223	·83170	223
47	·13672	228	·31113	228	·48574	228	·66019	228	·83461	228
48	·13963	232	·31414	232	·48864	232	·66310	232	·83751	232
49	·14153	237	·31705	237	·49155	237	·66601	237	·84042	237
50	·14544	242	·31996	242	·49446	242	·66892	242	·84332	242
51	0·14835	247	0·32287	247	0·49737	247	0·67182	247	0·84623	247
52	·15126	252	·32278	252	·50027	252	·67473	252	·84913	252
53	·15417	257	·32869	257	·50318	257	·67764	257	·85204	257
54	·15708	262	·33160	262	·50609	262	·68055	262	·85495	262
55	·15999	266	·33450	266	·50900	266	·68345	266	·85785	266
56	·16290	271	·33741	271	·51191	271	·68636	271	·86076	271
57	·16580	276	·34032	276	·51481	276	·68927	276	·86367	276
58	·16871	281	·34323	281	·51772	281	·69217	281	·86657	281
59	·17162	286	·34614	286	·52063	286	·69508	286	·86948	226
60	·17453	291	·34905	291	·52354	291	·69799	291	·87239	291

Min.	5°	Parts for "	6°	Parts for "	7°	Parts for "	8°	Parts for "	9°	Parts for "
0	0·87239	0	1·04672	0	1·22097	0	1·39513	0	1·56918	0
1	·87529	5	·04862	5	·22387	5	·39803	5	·57208	5
2	·87820	10	·05252	10	·22677	10	·40093	10	·57498	10
3	·88110	15	·05543	15	·22968	15	·40383	15	·57788	15
4	·88401	20	·05833	20	·23258	20	·40673	20	·58078	20
5	·88691	25	·06124	25	·23548	25	·40964	25	·58368	25
6	·88982	30	·06414	30	·23839	30	·41254	30	·58658	30
7	·89273	35	·06705	35	·24129	35	·41544	35	·58948	35
8	·89563	39	·06995	39	·24419	39	·41834	39	·59238	39
9	·89854	44	·07286	44	·24710	44	·42124	44	·59528	44
10	·90144	49	·07576	48	·25000	48	·42415	48	·59818	48
11	0·90435	53	1.07867	53	1·25292	53	1·42705	53	1·60108	53
12	·90726	58	·08157	58	·25581	58	·42995	58	·60398	58
13	·91016	63	·08448	63	·25871	63	·43285	63	·60688	63
14	·91307	68	·08738	68	·26161	68	·43575	68	·60978	68
15	·91597	73	·09029	73	·26452	73	·43866	73	·61267	73
16	·91888	78	·09319	78	·26742	78	·44156	78	·61557	78
17	·92178	82	·09610	82	·27032	82	·44446	82	·61847	82
18	·92469	87	·09900	87	·27323	87	·44736	87	·62137	87
19	·92759	92	·10190	92	·27613	92	·45026	92	·62427	92
20	·93050	97	·10481	97	·27903	97	·45316	97	·62717	97
21	0·93341	102	1·10771	102	1·28194	102	1·45607	102	1·63007	102
22	·93631	107	·11062	107	·28484	107	·45897	107	·63297	107
23	·93922	112	·11352	112	·28774	112	·46187	112	·63587	112
24	·94212	117	·11643	117	·29064	117	·46477	117	·63876	117
25	·94503	122	·11933	122	·29355	122	·46767	122	·64166	122
26	·94794	127	·12223	127	·29645	127	·47057	127	·64456	127
27	·95084	132	·12514	132	·29935	132	·47347	132	·64746	132
28	·95375	137	·12804	137	·30225	137	·47637	137	·65036	137
29	·95665	141	·13095	141	·30516	141	·47927	141	·65326	141
30	·95956	145	·13385	145	·30806	145	·48217	145	·65616	145
31	0·96246	150	1·13676	149	1·31096	149	1·48507	149	1·65906	149
32	·96537	155	·13966	154	·31387	154	·48797	154	·66196	154
33	·96827	160	·14257	159	·31677	159	·49088	159	·66486	159
34	·97118	165	·14547	164	·31967	164	·49378	164	·66776	164
35	·97409	170	·14837	169	·32257	169	·49668	169	·67065	169
36	·97699	175	·15128	174	·32547	174	·49958	174	·67355	174
37	·97990	180	·15418	179	·32838	179	·50248	179	·67645	179
38	·98280	185	·15709	184	·33128	184	·50538	184	·67935	184
39	·98571	190	·15999	189	·33418	189	·50828	189	·68225	189
40	·98861	194	·16289	193	·33709	193	·51118	193	·68515	193
41	0·99152	199	1·16580	198	1·33999	198	1·51408	198	1·68805	198
42	·99442	204	·16870	203	·34289	203	·51698	203	·69095	203
43	·99733	209	·17160	208	·34579	208	·51988	208	·69384	208
44	1·00023	214	·17451	213	·34869	213	·52278	213	·69674	213
45	·00314	219	·17441	218	·35160	218	·52568	218	·69964	218
46	·00605	223	·18031	222	·35450	222	·52858	222	·70254	222
47	·00895	228	·18322	227	·35740	227	·53148	227	·70544	227
48	·01185	232	·18612	231	·36030	231	·53438	231	·70833	231
49	·01476	237	·18903	236	·36321	236	·53728	236	·71123	236
50	·01766	242	·19193	241	·36611	241	·54018	241	·71413	241
51	1·02057	247	1·19483	246	1.36901	246	1·54308	246	1·71703	246
52	·02348	252	·19774	251	·37191	251	·54598	251	·71993	251
53	·02638	257	·20064	256	·37481	256	·54888	256	·72283	256
54	·02929	262	·20354	261	·37771	261	·51178	261	·72572	261
55	·03219	266	·20645	265	·38062	265	·55468	265	·72862	265
56	·03510	271	·20935	270	·38352	270	·55758	270	·73152	270
57	·03800	276	·21226	275	·38642	275	·56048	275	·73442	275
58	·04090	281	·21516	280	·38932	280	·56338	280	·73732	280
59	·04381	286	·21806	285	·39222	285	·56628	285	·74021	285
60	·04672	291	·22097	290	·39513	290	·56918	290	·74311	290

Min.	10°	Parts for ''	11°	Parts for ''	12°	Parts for ''	13°	Parts for ''	14°	Parts for ''
0	1·74311	0	1·91691	0	2·09057	0	2·26407	0	2·43738	0
1	·74601	5	·91980	5	09346	5	·26696	5	·44026	5
2	·74891	10	·92270	10	·09635	10	·26985	10	·44315	10
3	·75181	15	·92560	15	·09924	15	·27274	15	·44604	15
4	·75470	20	·92849	20	·10214	20	·27563	20	·44892	20
5	·75760	25	·93139	25	·10503	25	·27852	25	·45181	25
6	·76050	30	·93428	30	·10792	30	·28141	30	·45470	30
7	·76340	35	·93718	35	·11082	35	·28430	35	·45758	35
8	·76630	39	·94008	39	·11371	39	·28719	39	·46047	39
9	·76919	44	·94297	44	·11660	44	·29008	44	·46336	44
10	·77209	48	·94587	48	·11950	48	·29297	48	·46625	48
11	1·77499	53	1·94876	53	2·12239	53	2·29586	53	2·46913	53
12	·77789	58	·95166	58	·12528	58	·29875	58	·47202	58
13	·78078	62	·95455	62	·12817	62	·30164	62	·47491	62
14	·78368	67	·95745	67	·13106	67	·30453	67	·47779	67
15	·78658	72	·96034	72	·13396	72	·30742	72	·48068	72
16	·78947	77	·96324	77	·13685	77	·31031	77	·48357	77
17	·79237	81	·96613	81	·13974	81	·31320	81	·48645	81
18	·79527	86	·96903	86	·14263	86	·31609	86	·48934	86
19	·79816	91	·97112	91	·14552	91	·31898	91	·49223	91
20	·80106	96	·97482	96	·14842	96	·32187	96	·49512	96
21	1·80396	101	1·97771	101	2·15131	101	2·32475	101	2·49800	101
22	·80686	106	·98060	106	·15420	106	·32764	106	·50089	106
23	·80975	111	·98350	111	·15709	111	·33053	111	·50377	111
24	·81265	116	·98669	116	·15998	116	·33342	116	·50666	116
25	·81555	121	·98929	121	·16288	121	·33631	121	·50955	121
26	·81844	126	·99218	126	·16577	126	·33919	126	·51243	126
27	·82134	131	·99507	131	·16866	131	·34208	131	·51532	131
28	·82424	136	·99797	136	·17155	136	·34497	136	·51821	136
29	·82713	140	2·00086	140	·17444	140	·34786	140	·52110	140
30	·83003	144	·00376	144	·17734	144	·35075	144	·52399	144
31	1·83292	148	2·00665	148	2·18023	148	2·35363	148	2·52687	148
32	·83582	153	·00954	153	·18312	153	·35652	153	·52976	153
33	·83872	158	·01244	158	·18601	158	·35941	158	·53264	158
34	·84161	163	·01533	163	·18890	163	·36230	163	·53553	163
35	·84451	168	·01823	168	·19179	168	·36519	168	·53841	168
36	·84740	173	·02112	173	·19468	173	·36807	173	·54130	173
37	·85030	178	·02401	178	·19757	178	·37096	178	·54418	178
38	·85320	183	·02691	183	·20046	183	·37385	183	·54707	183
39	·85609	188	·02980	188	·20335	188	·37674	188	·54995	188
40	·85899	192	·03270	192	·20625	192	·37963	192	·55284	192
41	1·86188	197	2·03559	197	2·20914	197	2·38251	197	2·55572	197
42	·86478	202	·03848	202	·21203	202	·38540	202	·55861	202
43	·86768	207	·04138	207	·21492	207	·38829	207	·56149	207
44	·87056	212	·04427	212	·21781	212	·39118	212	·56438	212
45	·87347	217	·04717	217	·22070	217	·39407	217	·56726	217
46	·87637	221	·05006	221	·22359	221	·39695	221	·57015	221
47	·87926	226	·05295	226	·22648	226	·39984	226	·57303	226
48	·88216	230	·05585	230	·22937	230	·40273	230	·57592	230
49	·88506	235	·05874	235	·23226	235	·40562	235	·57880	235
50	·88796	240	·06164	240	·23516	240	·40851	240	·58169	240
51	1·89085	245	2·06453	245	2·23805	245	2·41139	245	2·58457	245
52	·89375	250	·06742	250	·24094	250	·41428	250	·58745	250
53	·89664	255	·07031	255	·24383	255	·41717	255	·59034	255
54	·89954	260	·07321	260	·24672	260	·42005	260	·59322	260
55	·90243	264	·07610	264	·24961	264	·42294	264	·59611	264
56	·90533	269	·07899	269	·25250	269	·42583	269	·59899	269
57	·90822	274	·08189	274	·25539	274	·42871	274	·60187	274
58	·91112	279	·08478	279	·25828	279	·43160	279	·60476	279
59	·91401	284	·08767	284	·26117	284	·43449	284	·60764	284
60	·91691	289	·09057	289	·26407	289	·43738	289	·61053	288

Min.	15°	Parts for ″	16°	Parts for ″	17°	Parts for ″	18°	Parts for ″	19°	Parts for ″
0	2·61053	0	2·78346	0	2·95619	0	3·12868	0	3·30095	0
1	·61341	5	·78634	5	·95906	5	·13155	5	·30382	5
2	·61629	10	·78922	10	·96194	10	·13442	10	·30669	10
3	·61917	15	·79210	15	·96482	15	·13729	15	·30955	15
4	·62206	20	·79498	20	·96769	20	·14017	20	·31242	20
5	·62494	25	·79786	25	·97057	25	·14304	25	·31529	25
6	·62782	30	·80074	30	·97345	30	·14591	30	·31816	30
7	·63071	35	·80362	35	·97632	35	·14879	35	·32103	35
8	·63359	39	·80651	39	·97920	39	·15166	39	·32390	39
9	·63647	44	·80939	44	·98208	44	·15453	44	·32677	44
10	·63936	48	·81227	48	·98496	48	·15741	48	·32964	48
11	2·64224	53	2·81515	53	2·98783	53	3·16028	53	3·33251	53
12	·64512	58	·81803	58	·99071	58	·16315	58	·33537	58
13	·64800	62	·82091	62	·99358	62	·16602	62	·33824	62
14	·65089	67	·82379	67	·99646	67	·16889	67	·34111	67
15	·65377	72	·82667	72	·99934	72	·17177	72	·34398	72
16	·65665	77	·82955	77	3·00221	77	·17464	77	·34685	77
17	·65954	81	·83243	81	·00509	81	·17751	81	·34971	81
18	·66242	86	·83531	86	·00796	86	·18038	86	·35258	86
19	·66530	91	·83819	91	·01084	91	·18325	91	·35545	91
20	·66819	96	·84107	96	·01372	96	·18613	96	·35832	95
21	2·67107	101	2·84394	101	3·01659	101	3·18900	101	3·36118	100
22	·67395	106	·84682	106	·01947	106	·19187	106	·36405	105
23	·67683	111	·84970	111	·02234	111	·19474	111	·36692	110
24	·67971	116	·85258	116	·02522	116	·19762	116	·36979	115
25	·68260	121	·85546	121	·02809	121	·20049	121	·37265	120
26	·68548	126	·85834	126	·03097	126	·20336	126	·37552	125
27	·68836	131	·86122	131	·03384	131	·20623	131	·37839	130
28	·69124	136	·86410	136	·03672	136	·20910	136	·38125	135
29	·69412	140	·86698	140	·03959	140	·21198	140	·38412	139
30	·69701	144	·86986	144	·04247	144	·21485	144	·38699	143
31	2·69989	148	2·87273	148	3·04534	148	3·21772	148	3·38985	147
32	·70277	153	·87561	153	·04821	153	·22059	153	·39272	152
33	·70565	158	·87849	158	·05109	158	·22346	158	·39559	157
34	·70853	163	·88137	163	·05396	163	·22633	163	·39845	162
35	·71142	168	·88425	168	·05684	168	·22920	168	·40132	167
36	·71430	173	·88712	173	·05971	173	·23207	173	·40418	172
37	·71718	178	·89000	178	·06258	178	·23494	178	·40705	177
38	·72006	183	·89288	183	·06546	183	·23782	183	·40992	182
39	·72294	188	·89576	188	·06833	188	·24069	188	·41278	187
40	·72583	192	·89864	192	·07121	192	·24356	192	·41565	191
41	2·72871	197	2·90151	197	3·07408	197	3·24643	197	3·41851	196
42	·73159	202	·90439	202	·07695	201	·24930	201	·42138	200
43	·73447	207	·90727	207	·07983	206	·25217	206	·42425	205
44	·73735	212	·91015	212	·08270	211	·25504	211	·42711	210
45	·74024	217	·91303	217	·08558	216	·25791	216	·42998	215
46	·74312	221	·91590	221	·08845	220	·26078	220	·43284	219
47	·74600	226	·91878	226	·09132	225	·26365	225	·43571	224
48	·74888	230	·92166	230	·09420	229	·26652	229	·43858	228
49	·75176	235	·92454	235	·09707	234	·26939	234	·44144	233
50	·75465	240	·92742	240	·09995	239	·27226	239	·44431	238
51	2·75753	245	2·93029	245	3·10282	244	3·27513	244	3·44717	243
52	·76041	250	·93317	250	·10569	249	·27800	249	·45004	248
53	·76329	255	·93605	255	·10856	254	·28086	254	·45290	253
54	·76617	260	·93892	260	·11144	259	·28373	259	·45577	258
55	·76905	264	·94180	264	·11431	263	·28660	263	·45863	262
56	·77193	269	·94468	269	·11718	268	·28947	268	·46150	267
57	·77481	274	·94755	274	·12006	273	·29234	273	·46436	272
58	·77769	279	·95043	279	·12293	278	·29521	278	·46723	277
59	·78057	284	·95331	284	·12580	283	·29808	283	·47009	282
60	·78346	288	·95619	288	·12868	287	·30095	287	·47296	286

Min.	20°	Parts for ″	21°	Parts for ″	22°	Parts for ″	23°	Parts for ″	24°	Parts for ″
0	3·47296	0	3·64471	0	3·81618	0	3·98735	0	4·15824	0
1	·47582	5	·64757	5	·81904	5	·99020	5	·16109	5
2	·47869	10	·65043	10	·82189	10	·99305	10	·16393	9
3	·48155	15	·65329	15	·82475	15	·99590	15	·16678	14
4	·48441	20	·65615	20	·82760	20	·99875	20	·16962	19
5	·48728	25	·65901	25	·83046	25	4·00160	25	·17247	24
6	·49014	30	·66187	30	·83331	30	·00445	30	·17531	29
7	·49301	35	·66473	35	·83617	35	·00730	35	·17816	34
8	·49587	39	·66759	39	·83902	39	·01015	39	·18100	38
9	·49873	44	·67045	44	·84188	44	·01300	44	·18385	43
10	·50160	48	·67331	48	·84473	48	·01585	48	·18669	47
11	3·50446	53	3·67617	53	3·84758	53	4·01870	53	4·18953	52
12	·50732	58	·67903	58	·85044	58	·02155	58	·19238	57
13	·51019	62	·68189	62	·85329	62	·02440	62	·19522	61
14	·51305	67	·68475	67	·85615	67	·02725	67	·19807	66
15	·51591	72	·68761	72	·85900	72	·03010	72	·20091	71
16	·51878	77	·69046	77	·86185	77	·03294	77	·20375	76
17	·52164	81	·69332	81	·86471	81	·03579	81	·20660	80
18	·52450	86	·69718	86	·86756	86	·03864	86	·20944	85
19	·52736	91	·70004	91	·87042	91	·04149	91	·21229	90
20	·53023	95	·70190	95	·87327	95	·04434	95	·21513	94
21	3·53309	100	3·70476	100	3·87612	100	4·04719	100	4·21797	99
22	·53595	105	·70762	105	·87898	105	·05004	105	·22082	104
23	·53882	110	·71047	110	·88183	110	·05289	110	·22366	109
24	·54168	115	·71333	115	·88468	115	·05574	115	·22650	114
25	·54454	120	·71619	120	·88754	120	·05859	120	·22935	119
26	·54741	125	·71905	125	·89039	125	·06143	125	·23219	124
27	·55027	130	·72191	130	·89324	130	·06428	130	·23503	129
28	·55313	135	·72476	135	·89609	135	·06713	135	·23787	134
29	·55600	139	·72762	139	·89895	139	·06998	139	·24072	138
30	·55886	143	·73048	143	·90180	143	·07283	143	·24356	142
31	3·56172	147	3·73334	147	3·90465	147	4·07568	147	4·24640	146
32	·56458	152	·73619	152	·90750	152	·07853	152	·24924	151
33	·56745	157	·73905	157	·91036	157	·08137	157	·25209	156
34	·57031	162	·74191	162	·91321	161	·08422	161	·25493	160
35	·57317	167	·74477	167	·91606	166	·08707	166	·25777	165
36	·57603	172	·74762	172	·91891	171	·08992	171	·26061	170
37	·57889	177	·75048	177	·92176	176	·09277	176	·26345	175
38	·58176	182	·75334	182	·92462	181	·09561	181	·26630	180
39	·58462	187	·75619	187	·92747	186	·09846	186	·26914	185
40	·58748	191	·75905	191	·93032	191	·10131	190	·27198	189
41	3·59034	196	3·76191	196	3·93317	195	4·10416	195	4·27482	194
42	·59320	200	·76476	200	·93602	199	·10700	199	·27766	198
43	·59607	205	·76762	205	·93888	204	·10985	204	·28050	203
44	·59893	210	·77048	210	·94173	209	·11270	209	·28334	208
45	·60179	215	·77334	215	·94458	214	·11555	214	·28519	213
46	·60465	219	·77619	219	·94743	218	·11839	218	·28903	217
47	·60752	224	·77905	224	·95028	223	·12124	223	·29187	222
48	·61038	228	·78191	228	·95314	227	·12409	227	·29471	226
49	·61324	233	·78476	233	·95599	232	·12693	232	·29755	231
50	·61610	238	·78762	238	·95884	237	·12978	237	·30039	236
51	3·61896	243	3·79048	243	3·96169	242	4·13263	242	4·30323	241
52	·62182	248	·79333	248	·96454	247	·13547	247	·30607	246
53	·62468	253	·79619	253	·96739	252	·13832	252	·30891	251
54	·62754	258	·79904	258	·97024	257	·14116	257	·31175	256
55	·63041	262	·80190	262	·97310	261	·14401	261	·31459	260
56	·63327	267	·80476	267	·97595	266	·14686	266	·31743	265
57	·63613	272	·80761	272	·97880	271	·14970	271	·32027	270
58	·63889	277	·81047	277	·98165	276	·15255	276	·32311	275
59	·64185	282	·81332	282	·98450	281	·15539	281	·32595	280
60	·64471	286	·81618	286	·98735	285	·15824	285	·32879	284

Min.	25°	Parts for "	26°	Parts for "	27°	Parts for "	28°	Parts for "	29°	Parts for "
0	4·32879	0	4·49901	0	4·66890	0	4·83843	0	5·00760	0
1	·33163	5	·50184	5	·67173	5	·84125	5	·01042	5
2	·33447	9	·50468	9	·67456	9	·84407	9	·01323	9
3	·33730	14	·50751	14	·67738	14	·84690	14	·01605	14
4	·34015	19	·51035	19	·68021	19	·84972	19	·01886	19
5	·34298	24	·51318	24	·68304	24	·85254	24	·02168	24
6	·34582	29	·51601	29	·68587	29	·85536	29	·02450	29
7	·34866	34	·51885	34	·68870	34	·85818	34	·02731	34
8	·35150	38	·52168	38	·69152	38	·86101	38	·03013	38
9	·35434	43	·52452	43	·69435	43	·86383	43	·03294	43
10	·35718	47	·52735	47	·69718	47	·86665	47	·03576	47
11	4·36002	52	4·53018	52	4·70001	52	4·86947	51	5·03857	51
12	·36286	57	·53302	57	·70283	57	·87229	56	·04139	56
13	·36569	61	·53585	61	·70566	61	·87511	61	·04420	61
14	·36853	66	·53868	66	·70849	66	·87793	66	·04702	66
15	·37137	71	·54152	71	·71113	71	·88075	71	·04983	71
16	·37421	76	·54435	76	·71414	76	·88358	76	·05265	76
17	·37705	80	·54718	80	·71697	80	·88640	80	·05546	80
18	·37988	85	·55001	85	·71980	85	·88922	85	·05828	85
19	·38272	90	·55285	90	·72262	90	·89204	90	·06109	90
20	·38556	94	·55568	94	·72545	94	·89486	94	·06391	94
21	4·38840	99	4·55851	98	4·72828	98	4·89768	98	5·06672	98
22	·39123	104	·56134	103	·73110	103	·90050	103	·06954	103
23	·39407	109	·56418	108	·73393	108	·90332	108	·07235	108
24	·39691	114	·56701	113	·73675	113	·90614	113	·07516	113
25	·39975	119	·56984	118	·73958	118	·90896	118	·07797	118
26	·40258	124	·57267	123	·74241	123	·91178	123	·08079	123
27	·40542	129	·57550	128	·74523	128	·91460	128	·08360	128
28	·40826	134	·57834	133	·74806	133	·91742	133	·08641	133
29	·41109	138	·58117	137	·75088	137	·92024	137	·08923	137
30	·41393	142	·58400	141	·75371	141	·92306	141	·09204	141
31	4·41677	146	4·58683	145	4·75653	145	4·92588	145	5·09485	145
32	·41960	151	·58966	150	·75936	150	·92870	150	·09766	150
33	·42244	156	·59249	155	·76218	155	·93152	155	·10048	155
34	·42528	160	·59532	159	·76501	159	·93434	159	·10329	159
35	·42812	165	·59816	164	·76783	164	·93615	164	·10610	164
36	·43095	170	·60099	169	·77066	169	·93997	169	·10891	169
37	·43379	175	·60382	174	·77348	174	·94279	174	·11172	174
38	·43663	180	·60665	179	·77631	179	·94561	179	·11454	179
39	·43946	185	·60948	184	·77913	184	·94843	184	·11735	184
40	·44230	189	·61231	188	·78196	188	·95125	188	·12016	188
41	4·44514	194	4·61514	193	4·78478	193	4·95407	193	5·12297	193
42	·44797	198	·61797	197	·78761	197	·95689	197	·12578	197
43	·45080	203	·62080	202	·79043	202	·95970	202	·12859	202
44	·45363	208	·62363	207	·79326	207	·96252	207	·13140	207
45	·45647	213	·62646	212	·79608	212	·96534	212	·13421	212
46	·45931	217	·62929	216	·79890	216	·96816	216	·13703	216
47	·46214	222	·63212	221	·80173	221	·97098	221	·13984	221
48	·46498	226	·63495	225	·80455	225	·97379	225	·14265	225
49	·46781	231	·63778	230	·80738	230	·97661	230	·14546	230
50	·47065	236	·64061	235	·81020	235	·97943	235	·14827	235
51	4·47349	241	4·64344	240	4·81302	240	4·98225	239	5·15108	239
52	·47632	246	·64627	245	·81585	245	·98506	244	·15389	244
53	·47916	251	·64910	250	·81867	250	·98788	249	·15670	249
54	·48199	256	·65193	255	·82149	255	·99070	254	·15951	254
55	·48483	260	·65475	259	·82431	259	·99351	258	·16232	258
56	·48767	265	·65758	264	·82714	264	·90633	263	·16513	263
57	·49050	270	·66041	269	·82996	269	·99915	269	·16794	268
58	·49334	275	·66324	274	·83278	274	5·00197	273	·17075	273
59	·49617	280	·66607	279	·83561	279	·00478	278	·17356	278
60	·49901	284	·66890	283	·83843	283	·00760	282	·17637	282

Min.	30°	Parts for "	31°	Parts for "	32°	Parts for "	33°	Parts for "	34°	Parts for "
0	5·17638	0	5·34476	0	5·51274	0	5·68030	0	5·84744	0
1	·17919	5	·34757	5	·51654	5	·68309	5	·85022	5
2	·18200	9	·35038	9	·51834	9	·68588	9	·85300	9
3	·18481	14	·35318	14	·52114	14	·68867	14	·85578	14
4	·18762	19	·35598	19	·52394	19	·69146	19	·85856	19
5	·19043	23	·35878	23	·52673	23	·69425	23	·86134	23
6	·19324	28	·36158	28	·52952	28	·69704	28	·86412	28
7	·19605	33	·36438	33	·53232	32	·69983	32	·86690	32
8	·19886	37	·36718	37	·53512	36	·70262	36	·86968	36
9	·20167	42	·36999	42	·53791	41	·70541	41	·87246	41
10	·20448	47	·37280	47	·54070	46	·70820	46	·87524	46
11	5·20729	51	5·37560	51	5·54350	50	5·71098	50	5·87802	50
12	·21010	56	·37840	56	·54630	55	·71376	55	·88080	55
13	·21290	61	·38120	61	·54909	60	·71655	60	·88358	60
14	·21570	65	·38400	65	·55188	65	·71934	65	·88636	65
15	·21851	70	·38680	70	·55468	70	·72213	70	·88914	70
16	·22132	75	·38960	75	·55748	75	·72492	75	·89192	75
17	·22413	79	·39240	79	·56027	79	·72771	79	·89470	79
18	·22694	84	·39520	84	·56306	84	·73050	84	·89748	84
19	·22975	89	·39800	89	·56585	89	·73328	89	·90026	89
20	·23256	93	·40080	93	·56864	93	·73606	93	·90304	93
21	5·23537	98	5·40360	98	5·57144	97	5·73885	97	5·90582	97
22	·23818	103	·40640	103	·57424	102	·74164	102	·90860	102
23	·24098	108	·40920	108	·57703	106	·74443	106	·91138	106
24	·24378	112	·41200	112	·57982	111	·74722	111	·91416	111
25	·24659	117	·41481	117	·58261	116	·75000	116	·91694	116
26	·24940	122	·41762	122	·58540	121	·75278	121	·91972	121
27	·25221	126	·42042	126	·58820	125	·75557	125	·92250	125
28	·25502	131	·42322	131	·59100	130	·75836	130	·92528	130
29	·25782	136	·42601	136	·59379	135	·76114	135	·92806	135
30	·26062	140	·42880	140	·59658	139	·76392	139	·93084	139
31	5·26343	145	5·43160	145	5·59937	144	5·76671	144	5·93361	144
32	·26624	150	·43440	150	·60216	149	·76950	148	·93638	148
33	·26905	154	·43720	154	·60496	153	·77228	152	·93916	152
34	·27186	159	·44000	159	·60776	158	·77506	157	·94194	157
35	·27466	164	·44280	164	·61055	163	·77885	162	·94472	162
36	·27746	168	·44560	168	·61334	167	·78064	166	·94750	166
37	·28027	173	·44840	173	·61613	172	·78342	171	·95028	171
38	·28308	178	·45120	178	·61892	177	·78620	176	·95306	176
39	·28588	183	·45400	183	·62171	182	·78899	181	·95583	181
40	·28868	187	·45680	187	·62450	186	·79178	185	·98560	185
41	5·29149	192	5·45960	192	5·62729	191	5·79456	190	5·96138	190
42	·29430	196	·46240	196	·63008	195	·79734	194	·96416	194
43	·29710	201	·46520	201	·63287	200	·80013	199	·96694	199
44	·29990	206	·46800	206	·63566	205	·80292	204	·96972	204
45	·30271	210	·47079	210	·63845	209	·80570	208	·97249	208
46	·30552	215	·47358	215	·64124	214	·80848	213	·97526	213
47	·30832	220	·47638	219	·64404	218	·81126	217	·97804	217
48	·31112	224	·47918	223	·64684	222	·81404	221	·98082	221
49	·31393	229	·48198	228	·64963	227	·81683	226	·98359	226
50	·31674	234	·48478	233	·65242	232	·81962	231	·98636	231
51	5·31954	238	5·48758	237	5·65521	236	5·82240	235	5·98914	235
52	·32234	243	·49038	242	·65800	241	·82518	240	·99192	240
53	·32514	248	·49317	247	·66079	246	·82796	245	·99469	245
54	·32794	252	·49596	251	·66358	250	·83074	249	·99746	249
55	·33075	257	·49876	256	·66637	255	·83352	254	6·00024	254
56	·33356	262	·50156	261	·66916	260	·83630	259	·00302	259
57	·33636	266	·50436	265	·67194	264	·83909	263	·00579	263
58	·35916	271	·50716	270	·67472	269	·84188	268	·00856	268
59	·34196	276	·50995	275	·67751	274	·84466	273	·01134	273
60	·34476	281	·51274	280	·68030	279	·84744	278	·01412	278

Min.	35°	Parts for ″	36°	Parts for ″	37°	Parts for ″	38°	Parts for ″	39°	Parts for ″
0	6·01412	0	6·18034	0	6·34610	0	6·51136	0	6·67614	0
1	·01689	5	·18311	5	·34886	5	·51411	5	·67888	5
2	·01966	9	·18588	9	·35162	9	·51686	9	·68162	9
3	·02244	14	·18864	14	·35437	14	·51961	14	·68436	14
4	·02522	18	·19140	18	·35712	18	·52236	18	·68710	18
5	·02799	23	·19417	23	·35988	23	·52511	23	·68984	23
6	·03076	28	·19694	28	·36264	28	·52786	28	·69258	27
7	·03353	32	·19970	32	·36540	32	·53061	32	·69532	31
8	·03630	37	·20246	37	·36816	37	·53336	37	·69806	36
9	·03908	41	·20523	41	·37092	41	·53611	41	·70081	40
10	·04186	46	·20800	46	·37368	46	·53886	46	·70356	45
11	6·04463	50	6·21076	50	6·37643	50	6·54161	50	6·70630	50
12	·04740	55	·21352	55	·37918	55	·54436	55	·70904	55
13	·05017	59	·21629	59	·38194	59	·54711	59	·71178	59
14	·05294	64	·21906	64	·38470	64	·54986	64	·71452	64
15	·05571	69	·22182	69	·38746	69	·55261	68	·71726	68
16	·05848	74	·22458	74	·39022	73	·55536	73	·72000	73
17	·06126	78	·22735	78	·39297	77	·55810	77	·72274	77
18	·06404	83	·23012	83	·39572	82	·56084	82	·72548	82
19	·06681	88	·23288	88	·39848	87	·56359	87	·72822	87
20	·06958	92	·23564	92	·40124	91	·56634	91	·73096	91
21	6·07235	96	6·23841	96	6·40399	95	6·56909	95	6·73369	95
22	·07512	101	·24118	101	·40674	100	·57184	100	·73642	100
23	·07789	105	·24394	105	·40950	104	·57459	104	·73916	104
24	·08066	110	·24670	110	·41226	109	·57734	109	·74190	109
25	·08343	115	·24946	115	·41502	114	·58008	114	·74464	114
26	·08620	120	·25222	120	·41778	119	·58282	119	·74738	119
27	·08897	124	·25499	124	·42053	123	·58557	123	·75012	123
28	·09174	129	·25776	129	·42328	128	·58832	128	·75286	128
29	·09451	134	·26052	134	·42604	133	·59107	133	·75560	133
30	·09728	138	·26328	138	·42880	137	·59382	137	·75834	137
31	6·10005	143	6·26604	143	6·43155	142	6·59656	142	6·76108	142
32	·10282	147	·26880	147	·43430	146	·59930	146	·76382	146
33	·10559	151	·27156	151	·43705	150	·60205	150	·76655	150
34	·10836	156	·27432	156	·43980	155	·60480	155	·76928	155
35	·11113	161	·27709	161	·44256	160	·60754	160	·77202	160
36	·11390	165	·27986	165	·44532	164	·61028	164	·77476	164
37	·11667	170	·28262	170	·44807	169	·61303	169	·77750	169
38	·11944	175	·28538	175	·45082	174	·61578	174	·78024	173
39	·12221	180	·28814	180	·45357	179	·61852	179	·78297	178
40	·12498	184	·29090	184	·45632	183	·62126	183	·78570	182
41	6·12775	189	6·29366	189	6·45908	188	6·62401	188	6·78844	187
42	·13052	193	·29642	193	·46184	192	·62676	192	·79118	191
43	·13329	198	·29918	198	·46459	197	·62950	197	·79392	196
44	·13606	203	·30194	203	·46734	202	·63224	202	·79666	201
45	·13883	207	·30470	207	·47009	206	·63499	206	·79939	205
46	·14160	212	·30746	212	·47284	211	·63774	211	·80212	210
47	·14437	216	·31022	216	·47559	215	·64048	215	·80486	214
48	·14714	220	·31298	220	·47834	219	·64322	219	·80760	218
49	·14990	225	·31574	225	·48110	224	·64597	224	·81033	223
50	·15266	230	·31850	230	·48386	229	·64872	229	·81306	228
51	6·15543	234	6·32126	234	6·48661	233	6·65146	233	6·81580	232
52	·15820	239	·32402	239	·48936	238	·65420	238	·81854	237
53	·16097	244	·32678	244	·49211	243	·65694	243	·82127	242
54	·16374	248	·32954	248	·49486	247	·65968	247	·82400	246
55	·16651	253	·33230	253	·49761	252	·66242	252	·82673	251
56	·16928	258	·33506	257	·50036	256	·66516	256	·82946	255
57	·17204	262	·33782	261	·50311	260	·66791	260	·83220	259
58	·17480	267	·34058	266	·50586	265	·67066	265	·83494	264
59	·17757	272	·34334	271	·50861	270	·67340	270	·83767	269
60	·18034	277	·34610	276	·51136	275	·67614	275	·84040	274

Min.	40°	Parts for "	41°	Parts for "	42°	Parts for "	43°	Parts for "	44°	Parts for "
0	6·84040	0	7·00414	0	7·16736	0	7·33002	0	7·49214	0
1	·84314	5	·00687	5	·17008	5	·33273	5	·49483	5
2	·84588	9	·00960	9	·17280	9	·33544	9	·49752	9
3	·84861	14	·01232	14	·17551	14	·33815	14	·50022	14
4	·85134	18	·01504	18	·17822	18	·34086	18	·50292	18
5	·85407	23	·01777	23	·18094	23	·34356	23	·50562	23
6	·85680	27	·02050	27	·18366	27	·34626	27	·50832	27
7	·85953	31	·02322	31	·18637	31	·34897	31	·51101	31
8	·86226	36	·02594	36	·18908	36	·35168	36	·51370	36
9	·86500	40	·02866	40	·19179	40	·35438	40	·51640	40
10	·86774	45	·03138	45	·19450	45	·35708	45	·51910	45
11	6·87047	50	7·03411	50	7·19722	50	7·35979	50	7·52179	50
12	·87320	55	·03684	55	·19994	55	·36250	54	·52448	54
13	·87593	59	·03956	59	·20265	59	·36520	59	·52718	59
14	·87866	64	·04228	64	·20536	64	·36790	63	·52988	63
15	·88139	68	·04500	68	·20808	68	·37060	68	·53257	68
16	·88412	73	·04772	73	·21080	73	·37330	72	·53526	72
17	·88685	77	·05044	77	·21351	77	·37601	77	·53796	77
18	·88958	82	·05316	82	·21622	81	·37872	81	·54066	81
19	·89231	87	·05589	87	·21893	86	·38142	86	·54335	86
20	·89504	91	·05862	91	·22164	90	·38412	90	·54604	90
21	6·89777	95	7·06134	95	7·22435	94	7·38683	95	7·54873	95
22	·90050	100	·06406	100	·22706	99	·38954	99	·55142	99
23	·90323	104	·06678	104	·22978	103	·39224	104	·55412	104
24	·90596	109	·06950	109	·23250	108	·39494	108	·55682	108
25	·90869	113	·07222	113	·23521	112	·39764	113	·55951	113
26	·91142	118	·07494	118	·23792	117	·40034	117	·56220	117
27	·91415	122	·07766	122	·24063	121	·40304	122	·56489	122
28	·91688	127	·08038	127	·24334	126	·40574	126	·56758	126
29	·91961	132	·08310	132	·24605	131	·40844	131	·57028	131
30	·92234	136	·08582	136	·24876	135	·41114	135	·57298	135
31	6·92507	141	7·08854	141	7·25147	140	7·41385	140	7·57567	140
32	·92780	146	·09126	146	·25418	145	·41656	144	·57836	144
33	·93053	150	·09398	150	·25689	149	·41926	149	·58105	149
34	·93326	155	·09670	155	·25960	154	·42196	153	·58374	153
35	·93599	160	·09942	159	·26231	158	·42466	158	·58643	158
36	·93872	164	·10214	163	·26502	162	·42736	162	·58912	162
37	·94145	169	·10486	168	·26773	167	·43006	167	·59181	167
38	·94418	173	·10758	172	·27044	171	·43276	171	·59450	171
39	·94690	178	·11030	177	·27315	176	·43546	176	·59719	175
40	·94962	182	·11302	181	·27586	180	·43816	180	·59988	179
41	6·95235	187	7·11574	186	7·27857	185	7·44086	185	7·60257	184
42	·95508	191	·11846	190	·28128	189	·44356	189	·60526	188
43	·95781	196	·12117	195	·28399	194	·44626	194	·60795	193
44	·96054	200	·12388	199	·28670	198	·44896	198	·61064	197
45	·96326	204	·12660	203	·28941	202	·45166	203	·61333	202
46	·96598	209	·12932	208	·29212	207	·45436	207	·61602	206
47	·96871	213	·13204	212	·29483	211	·45706	212	·61871	211
48	·97144	217	·13476	216	·29754	215	·45976	216	·62140	215
49	·97417	222	·13748	221	·30025	220	·46246	221	·62409	220
50	·97690	227	·14020	226	·30296	225	·46516	225	·62678	224
51	6·97962	231	7·14291	230	7·30566	229	7·46786	230	7·62947	229
52	·98234	236	·14562	235	·30836	234	·47056	234	·63216	233
53	·98507	241	·14834	240	·31107	239	·47325	239	·63485	238
54	·98780	245	·15106	244	·31378	243	·47594	243	·63754	242
55	·99052	250	·15378	249	·31649	248	·47864	248	·64023	247
56	·99324	254	·15650	253	·31920	252	·48134	252	·64292	251
57	·99597	258	·15921	257	·32191	256	·48404	257	·64561	256
58	·99870	263	·16192	262	·32462	261	·48674	261	·64830	260
59	7·00142	268	·16464	267	·32732	266	·48944	266	·65098	265
60	·00414	273	·16736	272	·33002	271	·49214	270	·65366	269

Min.	45°	Parts for "	46°	Parts for "	47°	Parts for "	48°	Parts for "	49°	Parts for "
0	7·65366	0	7·81462	0	7·97498	0	8·13474	0	8·29386	0
1	·65635	5	·81730	5	·97765	4	·13739	4	·29651	4
2	·65904	9	·81998	9	·98032	8	·14004	8	·29916	8
3	·66173	14	·82266	14	·98299	13	·14270	13	·30181	13
4	·66442	18	·82534	18	·98566	17	·14536	17	·30446	17
5	·66711	23	·82801	23	·98832	22	·14802	22	·30710	22
6	·66980	27	·83068	27	·99098	26	·15068	26	·30974	26
7	·67248	31	·83336	31	·99365	30	·15333	30	·31239	30
8	·67516	36	·83604	36	·99632	35	·15598	35	·31504	35
9	·67785	40	·83872	40	·99899	39	·15864	39	·31768	39
10	·68054	45	·84140	45	8·00166	44	·16130	44	·32032	44
11	7·68322	50	7·84407	50	8·00432	49	8·16396	49	8·32297	49
12	·68590	54	·84674	54	·00698	53	·16661	53	·32562	53
13	·68859	59	·84942	59	·00965	58	·16927	58	·32826	58
14	·69128	63	·85210	63	·01232	62	·17192	62	·33090	62
15	·69396	67	·85477	67	·01498	66	·17458	66	·33355	66
16	·69664	71	·85744	71	·01764	70	·17724	70	·33620	70
17	·69933	76	·86012	76	·02031	75	·17989	75	·33884	75
18	·70202	80	·86280	80	·02298	79	·18254	79	·34148	79
19	·70470	85	·86547	85	·02564	84	·18519	84	·34413	84
20	·70738	89	·86814	89	·02830	88	·18784	88	·34678	88
21	7·71007	94	7·87082	94	8·03096	93	8·19050	93	8·34942	93
22	·71276	98	·87350	98	·03362	97	·19316	97	·35206	97
23	·71544	103	·87617	103	·03629	102	·19581	102	·35470	102
24	·71812	107	·87884	107	·03896	106	·19846	106	·35734	106
25	·72080	112	·88151	112	·04162	111	·20111	111	·35998	111
26	·72348	116	·88418	116	·04428	115	·20376	115	·36262	115
27	·72617	121	·88686	121	·04694	120	·20642	120	·36527	120
28	·72886	125	·88954	125	·04960	124	·20908	124	·36792	124
29	·73154	130	·89221	130	·05227	129	·21173	128	·37056	128
30	·73422	134	·89488	134	·05494	133	·21438	132	·37320	132
31	7·73690	139	7·89755	139	8·05760	138	8·21703	137	8·37584	137
32	·73958	143	·90022	143	·06026	142	·21968	141	·37848	141
33	·74226	148	·90289	148	·06292	147	·22233	146	·38112	146
34	·74494	152	·90556	152	·06558	151	·22498	150	·38376	150
35	·74763	157	·90824	157	·06824	156	·22763	155	·38640	155
36	·75032	161	·91092	161	·07090	160	·23028	159	·38904	159
37	·75300	166	·91359	166	·07359	165	·23294	164	·39168	164
38	·75568	170	·91626	170	·07622	169	·23560	168	·39432	168
39	·75836	175	·91893	174	·07889	173	·23825	172	·39696	172
40	·76104	179	·92160	178	·08156	177	·24090	176	·39960	176
41	7·76372	184	7·92427	183	8·08422	182	8·24355	181	8·40224	181
42	·76640	188	·92694	187	·08688	186	·24620	185	·40488	185
43	·76908	193	·92961	192	·08954	191	·24885	190	·40752	190
44	·77176	197	·93228	196	·09220	195	·25150	194	·41016	194
45	·77444	202	·93495	201	·09486	200	·25415	199	·41280	199
46	·77712	206	·93762	205	·09752	204	·25680	203	·41544	203
47	·77980	211	·94029	210	·10018	209	·25944	208	·41808	208
48	·78248	215	·94296	214	·10284	213	·26208	212	·42072	212
49	·78516	219	·94563	218	·10550	217	·26473	216	·42336	216
50	·78784	223	·94830	222	·10816	221	·26738	220	·42600	220
51	7·79052	228	7·95097	227	8·11081	226	8·27003	225	8·42863	224
52	·79320	232	·95364	231	·11346	230	·27268	229	·43126	228
53	·79588	237	·95631	236	·11612	235	·27533	234	·43390	233
54	·79856	241	·95898	240	·11878	239	·27798	238	·43654	237
55	·80124	246	·96165	245	·12144	244	·28063	243	·43918	242
56	·80392	250	·96432	249	·12410	248	·28328	247	·44182	246
57	·80659	255	·96698	254	·12676	253	·28593	252	·44446	251
58	·80926	259	·96964	258	·12942	257	·28858	257	·44710	256
59	·81194	264	·97231	263	·13208	262	·29122	261	·44973	260
60	·81462	268	·97498	267	·13474	266	·29386	265	·45236	264

Min.	60°	Parts for "	51°	Parts for "	52°	Parts for "	53°	Parts for "	54°	Parts for "
0	8·45256	0	8·61022	0	8·76742	0	8·92396	0	9·07982	0
1	·45500	4	·61285	4	·77004	4	·92656	4	·08241	4
2	·45764	8	·61548	8	·77266	8	·92916	8	·08500	8
3	·46028	13	·61810	13	·77527	13	·93176	12	·08759	12
4	·46292	17	·62072	17	·77788	17	·93436	16	·09018	16
5	·46555	22	·62335	22	·78049	22	·93697	21	·09277	21
6	·46818	26	·62598	26	·78310	26	·93958	25	·09536	25
7	·47082	30	·62860	30	·78572	30	·94218	29	·09795	29
8	·47346	35	·63122	35	·78834	35	·94478	34	·10054	34
9	·47609	39	·63384	39	·79095	39	·94738	38	·10313	38
10	·47872	44	·63646	44	·79356	44	·94998	43	·10572	43
11	8·48135	49	8·63909	48	8·79617	48	8·95258	47	9·10831	47
12	·48398	53	·64172	52	·79878	52	·95518	51	·11090	51
13	·48662	58	·64434	57	·80139	57	·95778	56	·11349	56
14	·48926	62	·64696	61	·80400	61	·96038	60	·11608	60
15	·49189	66	·64958	65	·80662	65	·96298	64	·11867	64
16	·49452	70	·65220	69	·80924	69	·96558	68	·12126	68
17	·49716	75	·65483	74	·81185	74	·96818	73	·12385	73
18	·49980	79	·65746	78	·81446	78	·97078	77	·12644	77
19	·50243	84	·66008	83	·81707	83	·97338	82	·12902	82
20	·50506	88	·66270	87	·81968	87	·97598	86	·13160	86
21	8·50769	93	8·66532	92	8·82229	92	8·97858	91	9·13419	91
22	·51032	97	·66794	96	·82490	96	·98118	95	·13678	95
23	·51295	101	·67056	101	·82751	101	·98378	100	·13937	100
24	·51558	106	·67318	105	·83012	105	·98638	104	·14196	104
25	·51822	111	·67580	110	·83273	110	·98898	109	·14455	109
26	·52086	115	·67842	114	·83534	114	·99158	114	·14714	113
27	·52349	120	·68104	119	·83795	119	·99418	118	·14972	118
28	·52612	124	·68366	123	·84056	123	·99678	122	·15230	122
29	·52875	128	·68628	127	·84317	127	·99937	126	·15489	126
30	·53138	132	·68890	131	·84578	131	9·00196	130	·15748	130
31	8·53401	136	8·69152	135	8·84839	135	9·00456	134	9·16007	134
32	·53664	140	·69414	139	·85100	139	·00716	138	·16266	138
33	·53927	145	·69676	144	·85360	144	·00976	143	·16524	143
34	·54190	149	·69938	148	·85620	148	·01236	147	·16782	147
35	·54453	154	·70200	153	·85881	153	·01496	152	·17041	152
36	·54716	158	·70462	157	·86142	157	·01756	156	·17300	156
37	·54979	163	·70724	162	·86403	162	·02015	161	·17558	161
38	·55242	167	·70986	166	·86664	166	·02274	165	·17816	165
39	·55505	171	·71248	170	·86925	170	·02534	169	·18075	169
40	·55768	175	·71510	174	·87186	174	·02794	173	·18334	173
41	8·56031	180	8·71772	179	8·87446	178	9·03053	177	9·18592	177
42	·56294	184	·72034	183	·87706	182	·03312	181	·18850	181
43	·56557	189	·72295	188	·87967	187	·03572	186	·19108	186
44	·56820	193	·72556	192	·88228	191	·03832	190	·19366	190
45	·57082	198	·72818	197	·88489	196	·04091	195	·19625	195
46	·57344	202	·73080	201	·88750	200	·04350	199	·19884	199
47	·57607	207	·73342	206	·89010	205	·04610	204	·20142	204
48	·57870	211	·73604	210	·89270	209	·04870	208	·20400	208
49	·58133	215	·73865	214	·89531	213	·05129	212	·20658	212
50	·58396	219	·74126	218	·89792	217	·05388	216	·20916	216
51	8·58659	223	8·74388	222	8·90052	221	9·05648	220	9·21174	220
52	·58922	227	·74650	226	·90312	225	·05908	224	·21432	224
53	·59184	232	·74912	231	·90573	230	·06167	229	·21690	229
54	·59446	236	·75174	235	·90834	234	·06426	233	·21948	233
55	·59709	241	·75435	240	·91094	239	·06685	238	·22207	238
56	·59972	245	·75696	244	·91354	243	·06944	242	·22466	242
57	·60235	250	·75958	249	·91615	248	·07203	247	·22724	247
58	·60498	255	·76220	254	·91876	253	·07462	252	·22982	251
59	·60760	259	·76481	258	·92136	257	·07722	256	·23240	255
60	·61022	263	·76742	262	·92396	261	·07982	260	·23498	259

Min.	55°	Parts for "	56°	Parts for "	57°	Parts for "	58°	Parts for "	59°	Parts for "
0	9·23498	0	9·38944	0	9·54318	0	9·69620	0	9·84848	0
1	·23756	4	·39200	4	·54573	4	·69874	4	·85101	4
2	·24014	8	·39456	8	·54828	8	·70128	8	·85354	8
3	·24272	12	·39713	12	·55084	12	·70382	12	·85607	12
4	·24530	16	·39970	16	·55340	16	·70636	16	·85860	16
5	·24788	21	·40227	21	·55596	21	·70891	21	·86113	21
6	·25046	25	·40484	25	·55852	25	·71146	25	·86366	25
7	·25303	29	·40741	29	·56107	29	·71400	29	·86619	29
8	·25560	34	·40998	34	·56362	34	·71654	34	·86872	34
9	·25818	38	·41254	38	·56617	38	·71908	38	·87125	38
10	·26076	43	·41511	43	·56872	42	·72162	42	·87378	42
11	9·26334	47	9·41767	47	9·57128	47	9·72416	47	9·87631	46
12	·26592	51	·42024	51	·57384	51	·72670	51	·87884	50
13	·26850	56	·42281	55	·57639	55	·72925	55	·88137	54
14	·27108	60	·42538	59	·57894	59	·73180	59	·88390	58
15	·27366	64	·42794	63	·58150	63	·73434	63	·88643	62
16	·27624	68	·43050	67	·58406	67	·73688	67	·88896	66
17	·27881	73	·43307	72	·58661	72	·73942	72	·89149	71
18	·28138	77	·43564	76	·58916	76	·74196	76	·89402	75
19	·28396	82	·43820	81	·59171	81	·74450	81	·89654	80
20	·28654	86	·44076	85	·59426	85	·74704	85	·89906	84
21	9·28912	90	9·44332	89	9·59681	89	9·74958	89	9·90159	89
22	·29170	94	·44588	93	·59936	93	·75212	93	·90412	92
23	·29427	99	·44845	98	·60191	97	·75466	97	·90665	96
24	·29684	103	·45102	102	·60448	101	·75720	101	·90918	100
25	·29942	108	·45358	107	·60703	106	·75974	106	·91170	105
26	·30200	112	·45614	111	·60958	110	·76228	110	·91422	109
27	·30457	117	·45870	116	·61213	115	·76481	115	·91675	114
28	·30714	121	·46126	120	·61468	119	·76734	119	·91928	118
29	·30972	125	·46383	124	·61723	123	·76988	123	·92181	122
30	·31230	129	·46640	128	·61978	127	·77242	127	·92434	126
31	9·31487	133	9·46896	132	9·62233	131	9·77496	131	9·92686	130
32	·31744	137	·47152	136	·62488	135	·77750	135	·92938	134
33	·32001	142	·47408	141	·62743	140	·78004	139	·93191	138
34	·32258	146	·47664	145	·62998	144	·78258	143	·93444	142
35	·32516	151	·47920	150	·63253	149	·78512	148	·93696	147
36	·32774	155	·48176	154	·63508	153	·78766	152	·93948	151
37	·33031	160	·48432	159	·63763	158	·79019	157	·94200	156
38	·33288	164	·48688	163	·64018	162	·79272	161	·94452	160
39	·33545	168	·48944	167	·64272	166	·79526	165	·94705	164
40	·33802	172	·49200	171	·64526	170	·79780	169	·94958	168
41	9·34059	176	9·49456	175	9·64781	174	9·80033	173	9·95210	172
42	·34316	180	·49712	179	·65036	178	·80286	177	·95462	176
43	·34574	185	·49968	184	·65291	183	·80540	182	·95714	180
44	·34832	189	·50224	188	·65546	187	·80794	186	·95966	184
45	·35089	194	·50480	193	·65801	192	·81047	191	·96219	189
46	·35346	198	·50736	197	·66056	196	·81300	195	·96472	193
47	·35603	203	·50992	202	·66310	201	·81554	200	·96724	198
48	·35860	207	·51248	206	·66564	205	·81808	204	·96976	202
49	·36117	211	·51504	210	·66819	209	·82061	208	·97228	206
50	·36374	215	·51760	214	·67074	213	·82314	212	·97480	210
51	9·36631	219	9·52016	218	9·67329	217	9·82568	216	9·97732	214
52	·36888	223	·52272	222	·67584	221	·82822	220	·97984	218
53	·37145	227	·52528	226	·67838	225	·83075	224	·98236	222
54	·37402	231	·52784	230	·68092	229	·83328	228	·98488	226
55	·37659	236	·53039	235	·68347	234	·83581	233	·98740	231
56	·37916	240	·53294	239	·68602	238	·83834	237	·98992	235
57	·38173	245	·53550	244	·68856	243	·84087	242	·99244	240
58	·38430	249	·53806	248	·69110	247	·84340	246	·99496	244
59	·38687	253	·54062	252	·69365	251	·84594	250	·99748	248
60	·38944	257	·54318	256	·69620	255	·84848	254	10.00000	252

TABLE L.—*Tables showing the length in feet of a degree, minute, and second of latitude and longitude, for every ten minutes of the quadrant. Based on the Ordnance Geodetical Tables, compression* $\frac{1}{284}$. *By Robert C. Carrington, F.R.G.S., F.A.S.L.*

LATITUDE.				LONGITUDE.			
Latitude.	Length in Feet of a			Latitude	Length in Feet of a		
	Degree.	Minute.	Second.		Degree.	Minute.	Second.
0° 0′	362755·6	6045·93	100·77	0° 0′	365233·7	6087·23	101·454
10	362755·6	6045·93	100·77	10	365232·1	6087·20	101·453
20	362755·7	6045·93	100·77	20	365227·5	6087·13	101·452
30	362755·9	6045·93	100·77	30	365219·9	6087·00	101·450
40	362756·1	6045·93	100·77	40	365209·1	6086·82	101·447
50	362756·4	6045·94	100·77	50	365195·3	6086·59	101·443
1° 0′	362756·7	6045·94	100·77	1° 0′	365178·4	6086·31	101·438
10	362757·1	6045·95	100·77	10	365158·5	6085·98	101·433
20	362757·6	6045·96	100·77	20	365135·5	6085·59	101·427
30	362758·1	6045·97	100·77	30	365109·4	6085·16	101·419
40	362758·7	6045·98	100·77	40	365080·2	6084·67	101·411
50	362759·4	6045·99	100·77	50	365048·0	6084·13	101·402
2° 0′	362760·1	6046·00	100·77	2° 0′	365012·7	6083·54	101·392
10	362760·9	6046·01	100·77	10	364974·3	6082·91	101·381
20	362761·7	6046·03	100·77	20	364932·9	6082·22	101·370
30	362762·6	6046·04	100·77	30	364888·4	6081·47	101·358
40	362763·6	6046·06	100·77	40	364840·8	6080·68	101·345
50	362764·6	6046·08	100.77	50	364790·2	6079·84	101·331
3° 0′	362765·7	6046·09	100·77	3° 0′	364736·5	6078·94	101·316
10	362766·9	6046·11	100·77	10	364679·8	6078·00	101·300
20	362768·1	6046·13	100·77	20	364619·9	6077·00	101·283
30	362769·4	6046·16	100·77	30	364557·0	6075·95	101·266
40	362770·7	6046·18	100·77	40	364491·1	6074·85	101·248
50	362772·1	6046·20	100·77	50	364422·1	6073·70	101·228
4° 0′	362773·6	6046·23	100·77	4° 0′	364350·0	6072·50	101·208
10	362775·1	6046·25	100·77	10	364274·9	6071·25	101·187
20	362776·7	6046·28	100·77	20	364196·7	6069·95	101·166
30	362778·3	6046·30	100·77	30	364115·4	6068·59	101·143
40	362780·0	6046·33	100·77	40	364031·1	6067·19	101·120
50	362781·8	6046·36	100·77	50	363943·7	6065·73	101·096

	LATITUDE.				LONGITUDE.		
Latitude.	Length in Feet of a			Latitude.	Length in Feet of a		
	Degree.	Minute.	Second.		Degree.	Minute.	Second.
5° 0'	362783·6	6046·39	100·77	5° 0'	363853·2	6064·22	101·070
10	362785·5	6046·42	100·77	10	363759·7	6062·66	101·044
20	362787·5	6046·46	100·77	20	363663·2	6061·05	101·018
30	362789·5	6046·49	100·77	30	363563·5	6059·39	100·990
40	362791·6	6046·53	100·78	40	363460·9	6057·68	100·961
50	362793·7	6046·56	100·78	50	363355·1	6055·92	100·932
6° 0'	362795·9	6046·60	100·78	6° 0'	363246·3	6054·11	100·902
10	362798·2	6046·64	100·78	10	363134·5	6052·24	100·871
20	362800·5	6046·68	100·78	20	363019·6	6050·33	100·839
30	362802·9	6046·72	100·78	30	362901·7	6048·36	100·806
40	362805·4	6046·76	100·78	40	362780·7	6046·35	100·772
50	362807·9	6046·80	100·78	50	362656·6	6044·28	100·738
7° 0'	362810·4	6046·84	100·78	7° 0'	362529·5	6042·16	100·703
10	362813·1	6046·89	100·78	10	362399·4	6039·99	100·667
20	362815·2	6046·93	100·78	20	362266·2	6037·77	100·630
30	362818·5	6046·98	100·78	30	362130·0	6035·50	100·592
40	362821·3	6047·02	100·78	40	361990·7	6033·18	100·553
50	362824·2	6047·07	100·78	50	361848·4	6030·81	100·513
8° 0'	362827·1	6047·12	100·79	8° 0'	361703·0	6028·38	100·473
10	362830·1	6047·17	100·79	10	361554·6	6025·91	100·432
20	362833·2	6047·22	100·79	20	361403·2	6023·39	100·390
30	362836·3	6047·27	100·79	30	361248·7	6020·81	100·347
40	362839·4	6047·32	200·79	40	361091·2	6018·19	100·303
50	362842·7	6047·38	100·79	50	360930·6	6015·51	100·258
9° 0'	362846·0	6047·43	100·79	9° 0'	360767·0	6012·78	100·213
10	362849·3	6047·49	100·79	10	360600·4	6010·01	100·167
20	362852·7	6047·55	100·79	20	360430·7	6007·18	100·120
30	362856·2	6047·60	100·79	30	360258·0	6004·30	100·072
40	362859·7	6047·66	100·79	40	360082·3	6001·37	100·023
50	362863·3	6047·72	100·80	50	359903·5	5998·39	99·973
10° 0'	362866·9	6047·78	100·80	10° 0'	359721·7	5995·36	99·923
10	362870·7	6047·85	100·80	10	359536·7	5992·28	99·871
20	362874·4	6047·91	100·80	20	359349·1	5989·15	99·819
30	362878·2	6047·97	100·80	30	359158·3	5985·97	99·766
40	362882·1	6048·04	100·80	40	358964·4	5982·74	99·712
50	362886·1	6048·10	100·80	50	358767·5	5979·46	99·658

	LATITUDE				LONGITUDE		
	Length in Feet of a				Length in Feet of a		
Latitude.	Degree.	Minute.	Second.	Latitude.	Degree.	Minute.	Second.
11° 0'	362890·1	6048·17	100·80	11° 0'	358567·6	5976·13	99·602
10	362894·1	6048·23	100·80	10	358364·7	5972·75	99·546
20	362898·2	6048·30	100·80	20	358158·7	5969·31	99·489
30	362902·4	6048·37	100·81	30	357949·8	5965·83	99·431
40	362906·6	6048·44	100·81	40	357737·8	5962·30	99·372
50	362910·9	6048·52	100·81	50	357522·8	5958·71	99·312
12° 0'	362915·2	6048·59	100·81	12° 0'	357304·8	5955·08	99·251
10	362919 6	6048·66	100·81	10	357083·9	5951·40	99·190
20	362924·1	6048·74	100·81	20	356859·9	5947·67	99·128
30	362928·6	6048·81	100·81	30	356632·9	5943·88	99·065
40	362933·2	6048·89	100·81	40	356402·9	5940·05	99·001
50	362937·8	6048·96	100·82	50	356169·9	5936·17	98·936
13° 0'	362942·5	6049·04	100·82	13° 0'	355933·9	5932·23	98·871
10	362947·2	6049·12	100·82	10	355694·9	5928·25	98·804
20	362952·0	6049·20	100·82	20	355452·9	5924·22	98·737
30	362956·9	6049·28	100·82	30	355207·9	5920·13	98·669
40	362961·8	6049·36	100·82	40	354959·9	5916·00	98·600
50	362966·8	6049·45	100·82	50	354709·0	5911·82	98·530
14° 0'	362971·8	6049·53	100·83	14° 0'	354455·1	5907·59	98·460
10	362976·9	6049·62	100·83	10	354198·1	5903·30	98·388
20	362982·0	6049·70	100·83	20	353938·2	5898·97	98·316
30	362987·2	6049·79	100·83	30	353675·3	5894·59	98·243
40	362992·4	6049·87	100·83	40	353409·4	5890·16	98·169
50	362997·7	6049·96	100·83	50	353140·6	5885·68	98·095
15° 0'	363003·1	6050·05	100·83	15° 0'	352868·8	5881·15	98·019
10	363008·5	6050·14	100·84	10	352594·1	5876·57	97·943
20	363013·9	6050·23	100·84	20	352316·3	5871·94	97·866
30	363019·4	6050·32	100·84	30	352035·6	5867·26	97·788
40	363025·0	6050·42	100·84	40	351751·9	5862·53	97·709
50	363030·6	6050·51	100·84	50	351465·3	5857·76	97·629
16° 0'	363036·3	6050·61	100·84	16° 0'	351175·7	5852·93	97·549
10	363042·0	6050·70	100·84	10	350883·1	5848·05	97·468
20	363047·8	6050·80	100·85	20	350587·6	5843·13	97·386
30	363053·6	6050·89	100·85	30	350289·1	5838·15	97·303
40	363059·3	6050·99	100·85	40	349987·7	5833·13	97·219
50	363065·4	6051·09	100·85	50	349683·4	5828·06	97·134

Latitude	LATITUDE Length in Feet of a			Latitude	LONGITUDE Length in Feet of a		
	Degree.	Minute.	Second.		Degree.	Minute.	Second.
17° 0′	363071·4	6051·19	100·85	17° 0′	349376·0	5822·93	97·049
10	363077·4	6051·29	100·85	10	349065·8	5817·76	96·963
20	363083·5	6051·39	100·86	20	348752·6	5812·54	96·876
30	363089·7	6051·50	100·86	30	348436·5	5807·28	96·788
40	363095·9	6051·60	100·86	40	348117·4	5801·96	96·699
50	363102·1	6051·70	100·86	50	347795·4	5796·59	96·610
18° 0′	363108·4	6051·81	100·86	18° 0′	347470·5	5791·18	96·520
10	363114·8	6051·91	100·87	10	347142·6	5785·71	96·429
20	363121·2	6052·02	100·87	20	346811·8	5780·20	96·337
30	363127·6	6052·13	100·87	30	346478·1	5774·64	96·244
40	363134·1	6052·24	100·87	40	346141·5	5769·03	96·150
50	363140·7	6052·35	100·87	50	345801·9	5763·37	96·056
19° 0′	363147·3	6052·46	100·87	19° 0′	345459·5	5757·66	95·961
10	363153·9	6052·57	100·88	10	345114·1	5751·90	95·865
20	363160·6	6052·68	100·88	20	344765·8	5746·10	95·768
30	363167·4	6052·79	100·88	30	344414·6	5740·24	95·671
40	363174·2	6052·90	100·88	40	344060·6	5734·34	95·572
50	363181·0	6053·02	100·88	50	343703·6	5728·39	95·473
20° 0′	363187·9	6053·13	100·89	20° 0′	343343·7	5722·40	95·373
10	363194·8	6053·25	100·89	10	342980·9	5716·35	95·272
20	363201·8	6053·36	100·89	20	342615·2	5710·25	95·171
30	363208·8	6053·48	100·89	30	342246·7	5704·11	95·069
40	363215·9	6053·60	100·89	40	341875·2	5697·92	94·965
50	363223·1	6053·72	100·90	50	341500·9	5691·68	94·861
21° 0′	363230·2	6053·84	100·90	21° 0′	341123·7	5685·40	94·756
10	363237·5	6053·96	100·90	10	340743·6	5679·06	94·651
20	363244·7	6054·08	100·90	20	340360·6	5672·68	94·545
30	363252·1	6054·20	100·90	30	339974·8	5666·25	94·438
40	363259·4	6054·32	100·91	40	339586·1	5659·77	94·330
50	363266·8	6054·45	100·91	50	339194·5	5653·24	94·221
22° 0′	363274·3	6054·57	100·91	22° 0′	338800·1	5646·67	94·111
10	363281·8	6054·70	100·91	10	338402·8	5640·05	94·001
20	363289·3	6054·82	100·91	20	338002·7	5633·38	93·890
30	363296·9	6054·95	100·92	30	337599·7	5626·66	93·778
40	363304·6	6055·08	100·92	40	337193·9	5619·90	93·665
50	363312·2	6055·20	100·92	50	336785·2	5613·09	93·551

	LATITUDE.				LONGITUDE.		
	Length in Feet of a				Length in Feet of a		
Latitude.	Degree.	Minute.	Second.	Latitude.	Degree.	Minute.	Second.
23° 0'	363320·0	6055·33	100·92	23° 0'	336373·6	5606·23	93·437
10	363327·7	6055·46	100·92	10	335959·3	5599·32	93·322
20	363335·5	6055·59	100·93	20	335542·1	5592·37	93·206
30	363343·4	6055·72	100·93	30	335122·0	5585·37	93·089
40	363351·3	6055 86	100·93	40	334699·2	5578·32	92·972
50	363359·2	6055·99	100·93	50	334273·5	5571·23	92·854
24° 0'	363367·2	6056·12	100·94	24° 0'	333845·0	5564·08	92·735
10	363375·2	6056·25	100·94	10	333413·7	5556·89	92·615
20	363383·3	6056·39	100·94	20	332979·5	5549·66	92·494
30	363391·4	6056·52	100·94	30	332542·6	5542·38	92·373
40	363399·6	6056·66	100·94	40	332102·8	5535·05	92·251
50	363407·8	6056·80	100·95	50	331660·3	5527·67	92·128
25° 0'	363416·0	6056·93	100·95	25° 0'	331214·9	5520·25	92·004
10	363424·3	6057·07	100·95	10	330766·7	5512·78	91·879
20	363432·6	6057·21	100·95	20	330315·8	5505·26	91·754
30	363440·9	6057·35	100·96	30	329862·0	5497·70	91·628
40	363449·3	6057·49	100·96	40	329405·5	5490·09	91·502
50	363457·7	6057·63	100·96	50	328946·2	5482·44	91·374
26° 0'	363466·2	6057·77	100·96	26° 0'	328484·1	5474·74	91·245
10	363474·7	6057·91	100·97	10	328019·2	5466·99	91·116
20	363483·3	6058·06	100·97	20	327551·6	5459·19	90·987
30	363491·9	6058·20	100·97	30	327081·2	5451·35	90·856
40	363500·5	6058·34	100·97	40	326608·0	5443·47	90·724
50	363509·2	6058·49	100·97	50	326132·1	5435·54	90·592
27° 0'	363517·9	6058·63	100·98	27° 0'	325653·4	5427·56	90·459
10	363526·6	6058·78	100·98	10	325171·9	5419·53	90·326
20	363535·4	6058·92	100·98	20	324687·7	5411·46	90·191
30	363544·2	6059·07	100·98	30	324200·8	5403·35	90·056
40	363553·0	6059·22	100·99	40	323711·2	5395·19	89·920
50	363561·9	6059·37	100·99	50	323218·8	5386·98	89·783
28° 0'	363570·8	6059·51	100·99	28° 0'	322723.6	5378·73	89·645
10	363579·8	6059·66	100·99	10	322225.7	5370·43	89·507
20	363588·8	6059·81	101·00	20	321725·1	5362·09	89·368
30	363597·8	6059·96	101·00	30	321221·8	5353·70	89·228
40	363606·8	6060·11	101·00	40	320715·8	5345·26	89.088
50	363615·9	6060·27	101·00	50	320207·1	5336·78	88·946

	LATITUDE				LONGITUDE		
Latitude.	Length in Feet of a			Latitude.	Length in Feet of a		
	Degree.	Minute.	Second.		Degree.	Minute.	Second.
29° 0'	363625·0	6060·42	101·01	29° 0'	319695·6	5328·26	88·804
10	363634·2	6060·57	101·01	10	319181·5	5319·69	88·661
20	363643·4	6060·72	101·01	20	318664·6	5311·08	88·518
30	363652·6	6060·88	101·01	30	318145·1	5302·42	88·374
40	363661·9	6061·03	101·02	40	317622·8	5293·71	88·229
50	363671·2	6061·19	101·02	50	317097·9	5284·97	88·083
30° 0'	363680·5	6061·34	101·02	30° 0'	316570·3	5276·17	87·936
10	363689·9	6061·50	101·03	10	316040·0	5267·33	87·789
20	363699·3	6061·66	101·03	20	315507·0	5258·45	87·641
30	363708·7	6061·81	101·03	30	314971·4	5249·52	87·492
40	363718·1	6061·97	101·03	40	314433·1	5240·55	87·343
50	363727·6	6062·13	101·04	50	313892·1	5231·54	87·192
31° 0'	363737·1	6062·29	101·04	31° 0'	313348·5	5222·48	87·041
10	363746·7	6062·45	101·04	10	312802·2	5213·57	86·889
20	363756·2	6062·60	101·04	20	312253·3	5204·22	86·737
30	363765·8	6062·76	101·05	30	311701·7	5195·03	86·584
40	363775·4	6062·92	101·05	40	311147·5	5185·79	86·430
50	363785·1	6063·09	101·05	50	310590·7	5176·51	86·275
32° 0'	363794·8	6063·25	101·05	32° 0'	310031·2	5167·19	86·119
10	363804·5	6063·41	101·06	10	309469·1	5157·82	85·963
20	363814·2	6063·57	101·c6	20	308904·4	5148·41	85·807
30	363824·0	6063·73	101·06	30	308337·1	5138·95	85·649
40	363833·8	6063·90	101·07	40	307767·2	5129·45	85·491
50	363843·6	6064·06	101·07	50	307194·6	5119·91	85·332
33° 0'	363853·5	6064·23	101·07	33° 0'	306619·5	5110·33	85·172
10	363863·4	6064·39	101·07	10	306041·7	5100·70	85·011
20	363873·3	6064·56	101·08	20	305461·4	5091·02	84·850
30	363883·2	6064·72	101·08	30	304878·5	5081·31	84·688
40	363893·1	6064·89	101·08	40	304293·0	5071·55	84·526
50	363903·1	6065·05	101·08	50	303704·9	5061·75	84·362
34° 0'	363913·1	6065·22	101·09	34° 0'	303114·2	5051·90	84·198
10	363923·1	6065·39	101·09	10	302521·0	5042·02	84·034
20	363933·2	6065·55	101·09	20	301925·2	5032·09	83·868
30	363943·2	6065·72	101·10	30	301326·8	5022·11	83·702
40	363953·3	6065·89	101·10	40	300725·9	5012·10	83·535
50	363963·4	6066·06	101·10	50	300122·4	5002·04	83·367

	LATITUDE				LONGITUDE		
Latitude.	Length in Feet of a			Latitude.	Length in Feet of a		
	Degree.	Minute.	Second.		Degree.	Minute.	Second.
35° 0'	363973·6	6066·23	101·10	35° 0'	299516·4	4991·94	83·199
10	363983·7	6066·40	101·11	10	298907·8	4981·80	83·030
20	363993·9	6066·57	101·11	20	298296·8	4971·61	82·860
30	364004·1	6066·74	101·11	30	297683·1	4961·38	82·690
40	364014·3	6066·91	101·12	40	297067·0	4951·12	82·519
50	364024·6	6067·08	101·12	50	296448·4	4940·81	82·347
36° 0'	364034·9	6067·25	101·12	36° 0'	295827·2	4930·45	82·174
10	364045·1	6067·42	101·12	10	295203·5	4920·06	82·001
20	364055·4	6067·59	101·13	20	294577·3	4909·62	81·827
30	364065·8	6067·76	101·13	30	293948·7	4899·15	81·652
40	364076·1	6067·94	101·13	40	293317·5	4888·63	81·477
50	364086·4	6068·11	101·14	50	292683·8	4878·06	81·301
37° 0'	364096·8	6068·28	101·14	37° 0'	292047·7	4867·46	81·124
10	364107·2	6068·45	101·14	10	291409·0	4856·82	80·947
20	364117·6	6068·63	101·14	20	290767·9	4846·13	80·769
30	364128·1	6068·80	101·15	30	290124·4	4835·41	80·590
40	364138·5	6068·98	101·15	40	289418·3	4824·64	80·411
50	364149·0	6069·15	101·15	50	288829·8	4813·83	80·231
38° 0'	364159·5	6069·33	101·16	38° 0'	288178·9	4802·98	80·050
10	364170·0	6069·50	101·16	10	287525·5	4792·09	79·868
20	364180·5	6069·68	101·16	20	286869·7	4781·16	79·686
30	364191·0	6069·85	101·16	30	286211·4	4770·19	79·503
40	364201·5	6070·03	101·17	40	285550·7	4759·18	79·320
50	364212·1	6070·20	101·17	50	284887·0	4748·13	79·136
39° 0'	364222·6	6070·38	101·17	39° 0'	284222·0	4737·03	78·951
10	364233·2	6070·55	101·18	10	283554·0	4725·90	78·765
20	364243·8	6070·73	101·18	20	242883·7	4714·73	78·579
30	364254·4	6070·91	101·18	30	282210·9	4703·52	78·392
40	364265·1	6071·09	101·18	40	281535·8	4692·26	78·204
50	364275·7	6071·27	101·19	50	280858·2	4680·97	78·016
40° 0'	364286·3	6071·44	101·19	40° 0'	280178·2	4669·64	77·827
10	364297·0	6071·62	101·19	10	279495·9	4658·27	77·638
20	364307·7	6071·80	101·20	20	278811·2	4646·85	77·448
30	364318·3	6071·97	101·20	30	278124·1	4635·40	77·257
40	364329·0	6072·15	101·20	40	277434·7	4623·91	77·065
50	364339·7	6072·33	101·21	50	276742·9	4612·38	76·873

LATITUDE.				LONGITUDE.			
	Length in Feet of a				Length in Feet of a		
Latitude.	Degree.	Minute.	Second.	Latitude.	Degree.	Minute.	Second.
41° 0'	364350·4	6072·51	101·21	41° 0'	276048·7	4600·81	76·680
10	364361·1	6072·69	101·21	10	275352·2	4589·20	76·480
20	364371·9	6072·87	101·21	20	274653·4	4577·56	76·293
30	364382·6	6073·04	101·22	30	273952·2	4565·87	76·098
40	364393·4	6073·22	101·22	40	273248·7	4554·75	75·902
50	364404·1	6073·40	101·22	50	272542·9	4542·38	75·706
42° 0'	364414·9	6073·58	101·23	42° 0'	271834·7	4530·58	75·509
10	364425·6	6073·76	101·23	10	271124·3	4518·74	75·312
20	364436·4	6073·94	101·23	20	270411·5	4506·86	75·114
30	364447·2	6074·12	101·24	30	269696·4	4494·94	74·916
40	364458·0	6074·30	101·24	40	268979·1	4482·99	74·717
50	364468·8	6074·48	101·24	50	268259·5	4470·99	74·517
43° 0'	364479·6	6074·66	101·24	43° 0'	267537·5	4458·96	74·316
10	364490·4	6074·84	101·25	10	266813·3	4446·89	74·115
20	364501·2	6075·02	101·25	20	266086·8	4434·78	73·913
30	364512·0	6075·20	101·25	30	265358·1	4422·64	73·711
40	364522·8	6075·38	101·26	40	264627·1	4410·45	73·508
50	364533·6	6075·56	101·26	50	263893·8	4398·23	73·304
44° 0'	364544·4	6075·74	101·26	44° 0'	263158·3	4385·97	73·100
10	364555·2	6075·92	101·27	10	262420·5	4373·68	72·895
20	364566·1	6076·10	101·27	20	261680·6	4361·34	72·689
30	364576·9	6076·28	101·27	30	260938·4	4348·97	72·483
40	364587·7	6076·46	101·27	40	260193·9	4336·57	72·276
50	364598·5	6076·64	101·28	50	259447·3	4324·12	72·069
45° 0'	364609·4	6076·82	101·28	45° 0'	258698·4	4311·64	71·861
10	364620·2	6077·00	101·28	10	257947·3	4299·12	71·652
20	364631·0	6077·18	101·29	20	257194·1	4286·57	71·443
30	364641·9	6077·37	101·29	30	256438·6	4273·98	71·233
40	364652·7	6077·55	101·29	40	255681·0	4261·35	71·022
50	364663·5	6077·73	101·30	50	254921·2	4248·69	70·811
46° 0'	364674·4	6077·91	101·30	46° 0'	254159·2	4235·99	70·600
10	364685·2	6078·09	101·30	10	253395·0	4223·25	70·388
20	364696·0	6078·27	101·30	20	252628·7	4210·48	70·175
30	364706·8	6078·45	101·31	30	251860·2	4197·67	69·961
40	364717·7	6078·63	101·31	40	251089·6	4184·83	69·747
50	364728·5	6078·81	101·31	50	250316·8	4171·95	69·532

| | LATITUDE. | | | | LONGITUDE | | |
| | Length in Feet of a | | | | Length in Feet of a | | |
Latitude.	Degree.	Minute.	Second.	Latitude.	Degree.	Minute.	Second.
47° 0'	364739·3	6078·99	101·32	47° 0'	249541·9	4159·03	69·317
10	364750·1	6079·17	101·32	10	248764·9	4146·08	69·101
20	364760·9	6079·35	101·32	20	247985·8	4133·10	68·885
30	364771·7	6079·53	101·33	30	247204·5	4120·08	68·668
40	364782·5	6079·71	101·33	40	246421·2	4107·02	68·450
50	364793·3	6079·89	101·33	50	245635·8	4093·93	68·232
48° 0'	364804·1	6080·07	101·33	48° 0'	244848·2	4080·80	68·013
10	364814·9	6080·25	101·34	10	244058·5	4067·64	67·794
20	364825·6	6080·43	101·34	20	243266·8	4054·45	67·574
30	364836·4	6080·61	101·34	30	242473·0	4041·22	67·353
40	364847·1	6080·79	101·35	40	241677·1	4027·95	67·132
50	364857·9	6080·97	101·35	50	240879·2	4014·65	66·911
49° 0'	364868·6	6081·14	101·35	49° 0'	240079·2	4001·32	66·689
10	364879·4	6081·32	101·36	10	239277·1	3987·95	66·466
20	364890·1	6081·50	101·36	20	238473·1	3974·55	66·242
30	364900·8	6081·68	101·36	30	237667·0	3961·12	66·018
40	364911·5	6081·86	101·36	40	236858·9	3947·65	65·794
50	364922·2	6082·04	101·37	50	236048·7	3934·15	65·569
50° 0'	364932·9	6082·22	101·37	50° 0'	235236·5	3920·61	65·343
10	364943·6	6082·39	101·37	10	234422·3	3907·04	65·117
20	364954·2	6082·57	101·38	20	233606·1	3893·44	64·890
30	364964·9	6082·75	101·38	30	232787·9	3879·80	64·663
40	364975·5	6082·93	101·38	40	231967·8	3866·13	64·435
50	364986·1	6083·10	101·38	50	231145·7	3852·43	64·207
51° 0'	364996·8	6083·28	101·39	51° 0'	230321·4	3838·69	63·978
10	365007·4	6083·46	101·39	10	229495·3	3824·92	63·749
20	365018·0	6083·63	101·39	20	228667·2	3811·12	63·519
30	365028·6	6083·81	101·40	30	227837·2	3797·29	63·288
40	365039·1	6083·99	101·40	40	227005·3	3783·42	63·057
50	365049·7	6084·16	101·40	50	226171·4	3769·52	62·825
52° 0'	365060·2	6084·34	101·41	52° 0'	225335·5	3755·59	62·593
10	365070·7	6084·51	101·41	10	224497·7	3741·63	62·360
20	365081·2	6084·69	101·41	20	223658·1	3727·64	62·127
30	365091·7	6084·86	101·41	30	222816·5	3713·61	61·893
40	365102·2	6085·04	101·42	40	221973·0	3699·55	61·659
50	365112·7	6085·21	101·42	50	221127·6	3685·46	61·424

	LATITUDE.				LONGITUDE.		
		Length in Feet of a				Length in Feet of a	
Latitude.	Degree.	Minute.	Second.	Latitude.	Degree.	Minute.	Second.
53° 0'	365123·1	6085·39	101·42	53° 0'	220280·3	3671·34	61·189
10	365133·6	6085·56	101·43	10	219431·1	3657·19	60·953
20	365144·0	6085·73	101·43	20	218580·0	3643·00	60·717
30	365154·4	6085·91	101·43	30	217727·1	3628·79	60·480
40	365164·7	6086·08	101·43	40	216872·3	3614·54	60·242
50	365175·1	6086·25	101·44	50	216015·7	3600·26	60·004
54° 0'	365185·4	6086·42	101·44	54° 0'	215157·2	3585·95	59·766
10	365195·7	6086·60	101·44	10	214296·9	3571·62	59·527
20	365206·1	6086·77	101·45	20	213434·7	3557·25	59·287
30	365216·3	6086·94	101·45	30	212570·7	3542·85	59·047
40	365226·6	6087·11	101·45	40	211704·9	3528·42	58·807
50	365236·8	6087·28	101·45	50	210837·3	3513·96	58·566
55° 0'	365247·0	6087·45	101·46	55° 0'	209968·0	3499·47	58·324
10	365257·2	6087·62	101·46	10	209096·8	3484·95	58·082
20	365267·4	6087·79	101·46	20	208223·8	3470·40	57·840
30	365277·6	6088·96	101·47	30	207349·0	3455·82	57·597
40	365287·7	6088·13	101·47	40	206472·5	3441·21	57·353
50	365297·8	6088·30	101·47	50	205594·2	3426·57	57·109
56° 0'	365307·9	6088·47	101·47	56° 0'	204714·0	3411·90	56·865
10	365318·0	6088·63	101·48	10	203832·2	3397·20	56·620
20	365328·0	6088·80	101·48	20	202948·6	3382·48	56·375
30	365338·0	6088·97	101·48	30	202063·3	3367·72	56·129
40	365348·0	6089·13	101·49	40	201176·2	3352·94	55·882
50	365358·0	6089·30	101·49	50	200287·4	3338·12	55·635
57° 0'	565367·9	6089·47	101·49	57° 0'	199396·9	3323·28	55·388
10	365377·8	6089·63	101·49	10	198504·7	3308·41	55·140
20	365387·7	6089·80	101·50	20	197610·8	3293·51	54·892
30	365397·6	6089·96	101·50	30	196715·2	3278·59	54·643
40	365407·4	6090·12	101·50	40	195817·9	3263·63	54·394
50	365417·2	6090·29	101·50	50	194919·0	3248·65	54·144
58° 0'	365427·0	6090·45	101·51	58° 0'	194018·3	3233·64	53·643
10	365436·8	6090·61	101·51	10	193116·0	3218·60	53·643
20	365446·5	6090·78	101·51	20	192212·1	3203·54	53·392
30	365456·2	6090·94	101·52	30	191306·5	3188·44	53·141
40	365465·9	6091·11	101·52	40	190399·3	3173·32	52·889
50	365475·5	6091·26	101·52	50	189490·4	3158·17	52·636

	LATITUDE.				LONGITUDE.		
Latitude.	Length in Feet of a			Latitude.	Length in Feet of a		
	Degree.	Minute.	Second.		Degree.	Minute.	Second.
59° 0'	365485·1	6091·42	101·52	59° 0'	188579·9	3143·00	52·383
10	365494·7	6091·58	101·53	10	187667·8	3127·80	52·130
20	365504·3	6091·74	101·53	20	186754·1	3112·57	51·876
30	365513·8	6091·90	101·53	30	185838·8	3097·31	51·622
40	365523·3	6092·06	101·53	40	184921·9	3082·03	51·367
50	365532·8	6092·21	101·54	50	184003·4	3066·72	51·112
60° 0'	365542·2	6092·37	101·54	60° 0'	183083·3	3051·59	50·856
10	365551·6	6092·53	101·54	10	182161·6	3036·03	50·600
20	365561·0	6092·68	101·54	20	181238·4	3020·64	50·344
30	365570·3	6092·84	101·55	30	180313·7	3005·23	50·087
40	365579·6	6092·99	101·55	40	179387·4	2989·79	49·830
50	365588·9	6093·15	101·55	50	178459·5	2974·33	49·572
61° 0'	365598·1	6093·30	101·56	61° 0'	177530·1	2958·84	49·314
10	365607·3	6093·46	101·56	10	176599·2	2943·32	49·055
20	365616·5	6093·61	101·56	20	175666·8	2927·78	48·796
30	365625·7	6093·76	101·56	30	174732·8	2912·21	48·537
40	365634·8	6093·91	101·57	40	173797·4	2896·62	48·277
50	365643·9	6094·07	101·57	50	172860·5	2881·01	48·017
62° 0'	365652·9	6094·22	101·57	62° 0'	171922·1	2865·37	47·750
10	365661·9	6094·37	101·57	10	170982·2	2849·70	47·495
20	365670·9	6094·52	101·58	20	170040·9	2834·02	47·234
30	365679·8	6094·66	101·58	30	169098·1	2818·30	46·972
40	365688·7	6094·81	101·58	40	168153·8	2802·56	46·709
50	365697·6	6094·96	101·58	50	167208·1	2786·80	46·447
63° 0	365706·4	6095·11	101·59	63° 0'	166261·0	2771·01	45·184
10	365715·2	6095·25	101·59	10	165312·4	2755·21	45·920
20	365723·9	6095·40	101·59	20	164362·5	2739·38	45·656
30	365732·6	6095·54	101·59	30	163411·1	2723·52	45·392
40	365741·3	6095·69	101·59	40	162458·4	2707·64	45·127
50	365749·9	6095·83	101·60	50	161504·2	2691·74	44·862
64° 0	365758·5	6095·98	101·60	64° 0'	160548·6	2675·81	44·587
10	365767·1	6096·12	101·60	10	159591·6	2659·86	44·331
20	365775·6	6096·26	101·60	20	158633·2	2643·89	44·065
30	365784·1	6096·40	101·61	30	157673·5	2627·90	43·798
40	365792·6	6096·54	101·61	40	156712·5	2611·88	43·531
50	365801·0	6096·68	101·61	50	155750·1	2595·84	43·264

	LATITUDE.				LONGITUDE.		
Latitude.	Length in Feet of a			Latitude.	Length in Feet of a		
	Degree.	Minute.	Second.		Degree.	Minute.	Second.
65° 0′	365809·3	6096·82	101·61	65° 0′	154786·3	2579·77	42·996
10	365817·6	6096·96	101·62	10	153821·2	2563·69	42·728
20	365825·9	6097·10	101·62	20	152854·8	2547·58	42·460
30	365834·2	6097·24	101·62	30	151887·2	2531·45	42·191
40	365842·4	6097·37	101·62	40	150918·2	2515·30	41·922
50	365850·5	6097·51	101·63	50	149947·9	2499·13	41·652
66° 0′	365858·6	6097·64	101·63	66° 0′	148976·3	2482·94	41·382
10	365866·7	6097·78	101·63	10	148003·4	2466·72	41·112
20	365874·7	6097·91	101·63	20	147029·3	2450·49	40·841
30	365882·7	6098·05	101·63	30	146053·9	2434·23	40·570
40	365890·7	6098·18	101·64	40	145977·3	2417·96	40·299
50	365898·6	6098·31	101·64	50	144099·3	2401·66	40·028
67° 0′	365906·4	6098·44	101·64	67° 0′	143120·2	2385·34	39·756
10	365914·3	6098·57	101·64	10	142139·8	2369·00	39·483
20	365922·0	6098·70	101·65	20	141158·2	2352·64	39·211
30	365929·8	6098·83	101·65	30	140175·4	2336·26	38·938
40	365937·4	6098·96	101·65	40	139191·4	2319·86	38·664
50	365945·1	6099·09	101·65	50	138206·1	2303·44	38·390
68° 0′	365952·7	6099·21	101·65	68° 0′	137219·7	2287·00	38·116
10	365960·2	6099·34	101·66	10	136232·1	2270·54	37·842
20	365967·7	6099·46	101·66	20	135243·3	2254·06	37·568
30	365975·2	6099·59	101·66	30	134253·4	2237·56	37·293
40	365982·6	6099·71	101·66	40	133262·3	2221·04	37·017
50	365989·9	6099·83	101·66	50	132270·1	2204·50	36·742
69° 0′	365997·3	6099·96	101·67	69° 0′	131276·7	2187·95	36·466
10	366004·5	6100·08	101·67	10	130282·2	2171·37	36·190
20	366011·7	6100·20	101·67	20	129286·6	2154·78	35·913
30	366018·9	6100·32	101·67	30	128289·9	2138·17	35·636
40	366026·1	6100·44	101·67	40	127292·1	2121·54	35·359
50	366033·1	6100·55	101·68	50	126293·2	2104·89	35·082
70° 0′	366040·2	6100·67	101·68	70° 0′	125293·2	2088·22	34·804
10	366047·2	6100·79	101·68	10	124292·1	2071·54	34·526
20	366054·1	6100·90	101·68	20	123289·9	2054·83	34·247
30	366061·0	6101·02	101·68	30	122286·7	2038·11	33·968
40	366067·8	6101·13	101·69	40	121282·4	2021·37	33·690
50	366074·6	6101·24	101·69	50	120277·1	2004·62	33·410

	LATITUDE				LONGITUDE		
	Length in Feet of a				Length in Feet of a		
Latitude.	Degree.	Minute.	Second.	Latitude.	Degree.	Minute.	Second.
71° 0'	366081·3	6101·36	101·69	71° 0'	119270·7	1987·85	33·131
10	366088·0	6101·47	101·69	10	118263·3	1971·06	32·851
20	366094·6	6101·58	101·69	20	117254·9	1954·25	32·571
30	366101·2	6101·69	101·69	30	116245·6	1937·43	32·290
40	366107·8	6101·80	101·70	40	115235·2	1920·59	32·009
50	366114·3	6101·91	101·70	50	114223·8	1903·73	31·729
72° 0'	366120·7	6102·01	101·70	72° 0'	113211·4	1886·86	31·448
10	366127·1	6102·12	101·70	10	112198·0	1869·97	31·166
20	366133·4	6102·22	101·70	20	111183·7	1853·06	30·884
30	366139·7	6102·33	101·71	30	110168·4	1836·14	30·602
40	366145·9	6102·43	101·71	40	109132·2	1819·20	30·320
50	366153·1	6102·54	101·71	50	108135·0	1802·25	30·038
73° 0'	366158·2	6102·64	101·71	73° 0'	107116·9	1785·28	29·755
10	366164·3	6102·74	101·71	10	106098·0	1768·30	29·472
20	366170·3	6102·84	101·71	20	105077·9	1751·30	29·189
30	366176·3	6102·94	101·72	30	104057·0	1734·28	28·905
40	366182·2	6103·04	101·72	40	103035·3	1717·26	28·621
50	366188·1	6103·14	101·72	50	102012·8	1700·21	28·337
74° 0'	366193·9	6103·23	101·73	74° 0'	100989·1	1683·15	28·053
10	366199·6	6103·33	101.73	10	99964·7	1666·08	27·768
20	366205·3	6103·42	101·73	20	98939·5	1648·99	27·483
30	366211·0	6103·52	101·73	30	97913·4	1631·89	27·198
40	366216·6	6103·61	101·73	40	96886·5	1614·78	26·913
50	366222·1	6103·70	101·73	50	95858·7	1597·65	26·627
75° 0'	366227·6	6103·79	101·73	75° 0'	94830·1	1580·50	26·342
10	366233·0	6103·88	101·73	10	93800·6	1563·34	26·056
20	366238·4	6103·97	101·73	20	92730·4	1546·17	25·770
30	366243·7	6104·06	101·73	30	91739·4	1528·99	25·483
40	366249·0	6104·15	101·74	40	90707·6	1511·79	25·196
50	366254·2	6104·24	101·74	50	89675·0	1494·58	24·901
76° 0'	366259·6	6104·32	101·74	76° 0'	88641·6	1477·36	24·623
10	366264·4	6104·41	101·74	10	87607·4	1460·12	24·335
20	366269·5	6104·49	101·74	20	86572·5	1442·88	24·048
30	366274·5	6104·58	101·74	30	85536·9	1425·62	23·760
40	366279·4	6104·66	101·74	40	84500·5	1408·34	23·472
50	366284·3	6104·74	101·75	50	83463·4	1391·06	23·184

	LATITUDE				LONGITUDE		
	Length in Feet of a				Length in Feet of a		
Latitude.	Degree.	Minute.	Second.	Latitude.	Degree.	Minute.	Second.
77° 0'	366289·1	6104·82	101·75	77° 0'	82425·6	1373·76	22·896
10	366293·8	6104·90	101·75	10	81387·0	1356·45	22·108
20	366298·5	6104·98	101·75	20	80347·8	1339·13	22·319
30	366303·1	6105·05	101·75	30	79307·9	1321·80	22·030
40	366307·7	6105·13	101·75	40	78267·3	1304·46	21·741
50	366312·3	6105·21	101·75	50	77226·0	1287·10	21·452
78° 0'	366316·7	6105·28	101·75	78° 0'	76184·0	1269·73	21·162
79° 0'	366342·3	6105·71	101·76	79° 0'	69918·8	1165·31	19·422
80° 0'	366365·8	6106·10	101·77	80° 0'	63631·8	1060·53	17·676
81° 0'	366387·1	6106·45	101·77	81° 0'	57325·2	955·42	15·924
82° 0'	366406·3	6106·77	101·78	82° 0'	51000·6	850·01	14·167
83° 0'	366423·2	6107·05	101·78	83° 0'	44660·3	744·34	12·406
84° 0'	366438·0	6107·30	101·79	84° 0'	38306·1	638·44	10·641
85° 0'	366450·5	6107·51	101·79	85° 0'	31939·9	532·33	8·872
86° 0'	366460·7	6107·68	101·79	86° 0'	25563·9	426·07	7·101
87° 0'	366468·7	6107·81	101·80	87° 0'	19179·8	319·66	5·328
88° 0'	366474·4	6107·91	101·80	88° 0'	12789·9	213·17	3·553
89° 0'	366477·9	6107·97	101·80	89° 0'	6395·9	106·60	1·777
90° 0'	366479·0	6107·98	101·80	90° 0'	0·0	0·0	0·0

Table M.—*For computing the reduction to the meridian in seconds, or* $\dfrac{\text{Vers H.A.}}{\text{Sin }1''}$.

0 Hours.

s.	0ᵐ	1ᵐ	2ᵐ	3ᵐ	4ᵐ	5ᵐ	6ᵐ	7ᵐ	8ᵐ	9ᵐ	10ᵐ	11ᵐ	12ᵐ	13ᵐ	14ᵐ	15ᵐ	16ᵐ	17ᵐ	18ᵐ	19ᵐ	20ᵐ	21ᵐ	22ᵐ	23ᵐ	24ᵐ	

s.	35ᵐ	36ᵐ	37ᵐ	38ᵐ	39ᵐ	40ᵐ	41ᵐ	42ᵐ	43ᵐ	44ᵐ	45ᵐ	46ᵐ	47ᵐ	48ᵐ	49ᵐ	50ᵐ	51ᵐ	52ᵐ	53ᵐ	54ᵐ	55ᵐ	56ᵐ	57ᵐ	58ᵐ	59ᵐ

TABLE N.—*Dip Table for calculation of Heights to 30 miles.*

Dip in feet $= 0·8815\ d^2$ (in miles).

Dist.	Dip in feet.	Dist.	Dip in feet.	Dist.	Dip in feet.	Dist.	Dip in feet.	Dist.	Dip in feet.
1	0·9	10½	97	16½	240	21¼	398	26	596
1½	2·0	11	107	16¾	247	21½	407	26¼	607
2	3·5	11½	117	17	255	21¾	417	26½	619
2½	5·5	12	127	17¼	262	22	427	26¾	631
3	7·9	12½	138	17½	270	22¼	436	27	643
3½	10·8	13	149	17¾	278	22½	446	27¼	655
4	14·1	13¼	155	18	286	22¾	456	27½	667
4½	17·8	13½	161	18¼	294	23	466	27¾	679
5	22·0	13¾	167	18½	302	23¼	476	28	691
5½	26·6	14	173	18¾	310	23½	486	28¼	703
6	31·7	14¼	179	19	318	23¾	497	28½	716
6½	37·2	14½	185	19¼	327	24	507	28¾	729
7	43·1	14¾	192	19½	335	24¼	518	29	741
7½	49·6	15	198	19¾	344	24½	529	29¼	754
8	56·4	15¼	205	20	353	24¾	540	29½	767
8½	63·7	15½	212	20¼	361	25	551	29¾	780
9	71	15¾	219	20½	370	25¼	562	30	793
9½	79	16	226	20¾	379	25½	573		
10	88	16¼	233	21	388	25¾	584		

TABLE O.—*Angles subtended by various lengths at different distances.*

Distance in Miles Nautical.

Feet	8	7	6	5½	5	4½	4	3½	3	2½	2	1¾	1½	1¼	1	¾	½	¼
1	0·04	0·05	0·06	0·6	0·7	0·8	0·8	0·10	0·11	0·14	0·17	0·20	0·22	0·28	0·34	0·44	1·08	2·16
2	0·08	0·10	0·11	0·12	0·13	0·15	0·17	0·20	0·23	0·26	0·34	0·40	0·46	0·56	1·08	1·32	2·16	4·32
3	0·12	0·15	0·17	0·19	0·20	0·22	0·25	0·30	0·34	0·40	0·51	1·00	1·08	1·20	1·42	2·16	3·24	6·48
4	0·17	0·20	0·23	0·25	0·27	0·30	0·34	0·40	0·45	0·54	1·08	1·20	1·31	1·48	2·16	3·02	4·32	9·04
5	0·21	0·25	0·28	0·31	0·34	0·38	0·42	0·49	0·57	1·08	1·25	1·39	1·53	2·16	2·50	3·46	5·40	11·20
6	0·26	0·30	0·34	0·37	0·41	0·46	0·51	0·59	1·08	1·22	1·42	1·59	2·16	2·50	3·24	4·32	6·48	13·36
7	0·30	0·34	0·40	0·43	0·48	0·53	0·59	1·08	1·19	1·36	1·59	2·16	2·38	3·12	3·57	5·16	7·54	15·48
8	0·34	0·39	0·45	0·49	0·55	1·00	1·08	1·18	1·31	1·50	2·16	2·36	3·02	3·40	4·32	6·04	9·04	18·08
9	0·38	0·44	0·51	0·55	1·01	1·08	1·16	1·28	1·42	2·02	2·33	2·56	3·24	4·04	5·06	6·48	10·12	20·24
10	0·42	0·49	0·56	1·01	1·08	1·17	1·24	1·37	1·52	2·16	2·49	3·14	3·44	4·32	5·39	7·28	11·18	22·36
11	0·46	0·54	1·02	1·08	1·15	1·25	1·32	1·47	2·03	2·30	3·06	3·34	4·06	5·00	6·13	8·12	12·26	24·52
12	0·51	0·59	1·08	1·14	1·22	1·31	1·41	1·57	2·16	2·44	3·23	3·54	4·32	5·28	6·47	9·04	13·34	27·08
13	0·55	1·03	1·14	1·20	1·29	1·38	1·50	2·06	2·27	2·58	3·40	4·12	4·54	5·56	7·21	9·48	14·42	29·24
14	0·59	1·08	1·19	1·26	1·36	1·45	1·58	2·16	2·37	3·12	3·57	4·32	5·14	6·24	7·55	10·28	15·50	31·40
15	1·04	1·13	1·24	1·33	1·42	1·52	2·07	2·26	2·48	3·24	4·14	4·54	5·36	6·47	8·29	11·12	16·58	33·56

TABLE P.—*Table of Distances at which Objects can be Seen at Sea, according to their respective Elevations and the Elevation of the Eye of the Observer.*

Height in Feet.	Distance in English or Statute Miles.	Distance in Geographical or Nautical Miles.	Height in Feet.	Distance in English or Statute Miles.	Distance in Geographical or Nautical Miles.
5	2·958	2·565	100	13·228	11·47
10	4·184	3·628	110	13·874	12·03
15	5·123	4·443	120	14·490	12·56
20	5·916	5·130	130	15·083	13·08
25	6·614	5·736	140	15 652	13·57
30	7·245	6·283	150	16·201	14·22
35	7·826	6·787	200	18·708	16·22
40	8·366	7·255	250	20·916	18·14
45	8·874	7·696	300	22·912	19·87
50	9·354	8·112	350	24·748	21·46
55	9·811	8·509	400	26·457	22·94
60	10·246	8·886	450	28·062	24·33
65	10·665	9·249	500	29·580	25·65
70	11·067	9·598	550	36·024	26·90
75	11·456	9·935	600	32·403	28·10
80	11·832	10·26	650	33·736	29·25
85	12·196	10·57	700	35·000	30·28
90	12·549	10·88	800	37·416	32·45
95	12·893	11·18	900	39·836	34·54
			1000	41·833	36·28

Example.—A tower 150 feet high will be visible to an observer whose eye is elevated 15 feet above the water 19 nautical miles; thus, from the Table:

15 feet elevation distance visible 4·44 nautical miles.

150 ⋅⋅ ⋅⋅ 14·22 ,,
————
18·66

From Admiralty Tables.

TABLE Q.—*True Depression or Distance of the Sea Horizon.*

Height.	Dep.	Square.	Height.	Dep.	Square.	Height.	Dep.	Square.	Dep.	Square.
ft.	′		ft.	′		ft.	′		′	
1·1	1	1	3293	61	3721	12966	121	14641	181	32761
3·5	2	4	3403	62	3844	13183	122	14884	182	33124
8·0	3	9	3513	63	3969	13397	123	15129	183	33489
14·2	4	16	3924	64	4096	13615	124	15376	184	33856
22·1	5	25	3740	65	4225	13836	125	15625	185	34225
31·9	6	36	3855	66	4356	14061	126	15876	186	34596
43·3	7	49	3974	67	4489	14282	127	16129	187	34969
56·6	8	64	4093	68	4624	14502	128	16384	188	35344
71·7	9	81	4213	69	4761	14737	129	16641	189	35721
88·5	10	100	4337	70	4900	14970	130	16900	190	36100
107	11	121	4461	71	5041	15197	131	17161	191	36481
127	12	144	4587	72	5184	15429	132	17424	192	36864
149	13	169	4716	73	5329	15664	133	17689	193	37249
174	14	196	4846	74	5476	15901	134	17956	194	37636
199	15	225	4976	75	5625	16139	135	18225	195	38025
226	16	256	5112	76	5776	16380	136	18496	196	38416
256	17	289	5249	77	5929	16622	137	18769	197	38809
287	18	324	5385	78	6084	16866	138	19044	198	39204
319	19	361	5524	79	6241	17111	139	19321	199	39601
354	20	400	5665	80	6400	17362	140	19600	200	40000
390	21	441	5808	81	6561	17608	141	19881	201	40401
428	22	484	5952	82	6724	17860	142	20164	202	40804
468	23	529	6098	83	6889	18111	143	20449	203	41209
510	24	576	6246	84	7056	18366	144	20736	204	41616
550	25	625	6394	85	7225	18622	145	21025	205	42025
598	26	676	6547	86	7396	18878	146	21316	206	42436
645	27	729	6700	87	7569	19140	147	21609	207	42849
694	28	784	6855	88	7744	19401	148	21904	208	43264
744	29	841	7012	89	7921	19664	149	22201	209	43681
797	30	900	7172	90	8100	19930	150	22500	210	44100
850	31	961	7332	91	8281	20197	151	22801	211	44521
906	32	1024	7492	92	8464	20465	152	23104	212	44944
964	33	1089	7656	93	8649	20736	153	23409	213	45369
1023	34	1156	7824	94	8836	21008	154	23716	214	45796
1084	35	1225	7987	95	9025	21282	155	24025	215	46225
1147	36	1296	8158	96	9216	21558	156	24336	216	46656
1211	37	1369	8330	97	9409	21836	157	24649	217	47089
1278	38	1444	8504	98	9604	22115	158	24964	218	47524
1346	39	1521	8678	99	9801	22397	159	25281	219	47961
1416	40	1600	8852	100	10000	22680	160	25600	220	48400
1487	41	1681	9032	101	10201	22964	161	25921	221	48841
1561	42	1764	9210	102	10404	23251	162	26244	222	49284
1636	43	1849	9393	103	10609	23540	163	26569	223	49729
1713	44	1936	9577	104	10816	23830	164	26896	224	50176
1792	45	2025	9760	105	11025	24121	165	27225	225	50625
1872	46	2116	9951	106	11236	24415	166	27556	226	51076
1954	47	2209	10135	107	11449	24711	167	27889	227	51529
2039	48	2304	10325	108	11664	25008	168	28224	228	51984
2124	49	2401	10518	109	11881	25307	169	28561	229	52441
2212	50	2500	10712	110	12100	25608	170	28900	230	52900
2301	51	2601	10908	111	12321	25911	171	29241	231	53361
2393	52	2704	11105	112	12544	26215	172	29584	232	53824
2485	53	2809	11304	113	12769	26521	173	29929	233	54289
2581	54	2916	11506	114	12996	26829	174	30276	234	54756
2677	55	3025	11700	115	13225	27139	175	30625	235	55225
2775	56	3136	11913	116	13465	27451	176	30976	236	55696
2875	57	3246	12120	117	13689	27764	177	31329	237	56169
2977	58	3364	12328	118	13924	28079	178	31684	238	56644
3081	59	3481	12538	119	14161	28396	179	32041	239	57121
3186	60	3600	12749	120	14400	28715	180	32400	240	57600

From Raper.

TABLE R.—*Angles subtended at different distances by a pole ten feet in length.*

Yds.	Angle. °	′	″	Yds.	Angle. °	′	″	Yds.	Angle. °	′	″	Yds.	Angle. °	′	″
17	11	25	20	92	2	05	00	167	1	08	46	242	0	47	26
18	10	23	20	93	2	02	50	168	1	08	00	243	0	47	08
20	9	00	00	95	2	00	40	170	1	07	24	245	0	46	48
22	8	47	48	97	1	58	30	172	1	06	44	247	0	46	28
23	8	10	18	98	1	56	30	173	1	06	04	248	0	46	08
25	7	37	54	101	1	54	30	175	1	05	30	250	0	45	48
27	7	09	10	102	1	52	40	177	1	04	40	252	0	45	30
28	6	45	48	103	1	50	54	178	1	04	16	253	0	45	12
30	6	21	36	105	1	49	00	180	1	03	40	255	0	44	54
32	6	01	32	107	1	47	20	182	1	03	04	257	0	44	38
33	5	43	28	108	1	45	44	183	1	02	30	258	0	44	20
35	5	27	10	110	1	44	10	185	1	01	56	260	0	44	02
37	5	12	20	112	1	42	20	187	1	01	22	262	0	43	50
38	4	58	44	113	1	41	00	188	1	00	50	263	0	43	36
40	4	46	16	115	1	39	40	190	1	00	18	265	0	43	14
42	4	34	50	117	1	38	14	192	0	59	46	267	0	43	00
43	4	24	20	118	1	36	50	193	0	59	16	268	0	42	42
45	4	14	00	120	1	35	30	195	0	58	46	270	0	42	26
47	4	05	30	122	1	34	10	197	0	58	16	272	0	42	11
48	3	57	00	123	1	32	50	198	0	57	48	273	0	41	55
50	3	49	06	125	1	31	40	200	0	57	18	275	0	41	40
52	3	41	42	127	1	30	30	202	0	56	48	277	0	41	25
53	3	34	50	128	1	29	16	203	0	56	20	278	0	41	10
55	3	28	20	130	1	28	10	205	0	55	54	280	0	40	55
57	3	22	10	132	1	27	00	207	0	55	26	282	0	40	41
58	3	16	24	133	1	26	00	208	0	55	00	283	0	40	26
60	3	10	50	135	1	24	50	210	0	54	34	285	0	40	12
62	3	05	40	137	1	23	50	212	0	54	16	287	0	39	58
63	3	01	00	138	1	22	50	213	0	54	00	288	0	39	44
65	2	56	14	140	1	21	50	215	0	53	26	290	0	39	30
67	2	51	40	142	1	20	50	217	0	52	54	292	0	39	17
68	2	47	40	143	1	20	00	218	0	52	28	293	0	39	04
70	2	43	40	145	1	19	00	220	0	52	04	295	0	38	50
72	2	40	00	147	1	18	08	222	0	51	42	297	0	38	37
73	2	36	14	148	1	17	16	223	0	51	20	298	0	38	24
75	2	32	44	150	1	16	22	225	0	50	56	300	0	38	11
77	2	29	30	152	1	15	30	227	0	50	34	302	0	37	59
78	2	26	20	153	1	14	44	228	0	50	12	303	0	37	47
80	2	23	14	155	1	14	00	230	0	49	50	305	0	37	34
82	2	20	16	157	1	13	08	232	0	49	28	307	0	37	22
83	2	17	30	158	1	12	20	233	0	49	06	308	0	37	10
85	2	14	50	160	1	11	40	235	0	48	44	310	0	36	58
87	2	12	10	162	1	11	00	237	0	48	22	311	0	36	46
88	2	09	44	163	1	10	10	238	0	48	02	313	0	36	34
90	2	07	20	165	1	09	30	240	0	47	44	315	0	36	23

TABLE S.—*For converting Intervals of Time or Longitude into Decimals of a Day.*

Long.	Time.	Decimals of a Day.	Long.	Time.	Decimals of a Day.	Long.	Time.	Decimals of a Day.
°	h		° '	m		° '	m	
15	1	·0417	0 15	1	·0007	7 45	31	·0215
30	2	·0833	0 30	2	·0014	8 0	32	·0222
45	3	·1250	0 45	3	·0021	8 15	33	·0229
60	4	·1667	1 0	4	·0028	8 30	34	·0236
75	5	2083	1 15	5	·0035	8 45	35	·0243
90	6	·2500	1 30	6	·0042	9 0	36	·0250
105	7	·2917	1 45	7	·0049	9 15	37	·0257
120	8	·3333	2 0	8	·0056	9 30	38	·0264
135	9	·3750	2 15	9	·0062	9 45	39	·0271
150	10	·4167	2 30	10	·0069	10 0	40	·0278
165	11	·4583	2 45	11	·0076	10 15	41	·0285
180	12	·5000	3 0	12	·0083	10 30	42	·0292
195	13	·5417	3 15	13	·0090	10 45	43	·0299
210	14	·5833	3 30	14	·0097	11 0	44	·0306
225	15	·6250	3 45	15	·0104	11 15	45	·0311
240	16	·6667	4 0	16	·0111	11 30	46	·0319
255	17	·7083	4 15	17	·0118	11 45	47	·0326
270	18	·7500	4 30	18	·0125	12 0	48	·0333
285	19	·7917	4 45	19	·0132	12 15	49	·0340
300	20	8333	5 0	20	·0139	12 30	50	·0347
315	21	·8750	5 15	21	·0146	12 45	51	·0354
330	22	·9167	5 30	22	·0153	13 0	52	·0361
345	23	·9583	5 45	23	·0160	13 15	53	·0368
360	24	1·0000	6 0	24	·0167	13 30	54	·0375
			6 15	25	·0174	13 45	55	·0382
			6 30	26	·0181	14 0	56	·0389
			6 45	27	·0187	14 15	57	·0396
			7 0	28	·0194	14 30	58	·0403
			7 15	29	·0201	14 45	59	·0410
			7 30	30	·0208	15 0	60	·0417

From Shadwell's " Chronometers."

TABLE T.—*Metrical and English Barometers.*

Barometer Scales.		Barometer Scales.		Barometer Scales.	
Fr. Mill.	Eng. In.	Fr. Mill.	Eng. In.	Fr. Mill.	Eng. In.
640	25·2	691	27·2	742	29·2
643	25·3	693	27·3	744	29·3
645	25·4	696	27·4	747	29·4
648	25·5	698	27·5	749	29·5
650	25·6	701	27·6	752	29·6
653	25·7	704	27·7	754	29·7
655	25·8	706	27·8	757	29·8
658	25·9	709	27·9	759	29·9
660	26·0	711	28·0	762	30·0
663	26·1	714	28·1	765	30·1
665	26·2	716	28·2	767	30·2
668	26·3	719	28·3	770	30·3
670	26·4	721	28·4	772	30·4
673	26·5	724	28·5	775	30·5
676	26·6	726	28·6	777	30·6
678	26·7	729	28·7	780	30·7
681	26·8	732	28·8	782	30·8
683	26·9	734	28·9	785	30·9
686	27·0	737	29·0	787	31·0
688	27·1	739	29·1		

TABLE U. — *Corresponding Thermometers, Fahrenheit, Centigrade, Réaumur.*

F.	C.	R.	F.	C.	R.	F.	C.	R.
°	°	°	°	°	°	°	°	°
0	-17.8	-14.2	41	5.0	4.0	81	27.2	21.8
1	-17.2	-13.8	42	5.6	4.4	82	27.8	22.2
2	-16.7	-13.3	43	6.1	4.9	83	28.3	22.7
3	-16.1	-12.9	44	6.7	5.3	84	28.9	23.1
4	-15.6	-12.4	45	7.2	5.8	85	29.4	23.6
5	-15.0	-12.0	46	7.8	6.2	86	30.0	24.0
6	-14.4	-11.6	47	8.3	6.7	87	30.6	24.4
7	-13.9	-11.1	48	8.9	7.1	88	31.1	24.9
8	-13.3	-10.7	49	9.4	7.5	89	31.7	25.3
9	-12.8	-10.2	50	10.0	8.0	90	32.2	25.8
10	-12.2	-9.8	51	10.6	8.4	91	32.8	26.2
11	-11.7	-9.3	52	11.1	8.9	92	33.3	26.7
12	-11.1	-8.9	53	11.7	9.3	93	33.9	27.1
13	-10.6	-8.4	54	12.2	9.8	94	34.4	27.6
14	-10.0	-8.0	55	12.8	10.2	95	35.0	28.0
15	-9.4	-7.5	56	13.3	10.7	96	35.6	28.4
16	-8.9	-7.1	57	13.9	11.1	97	36.1	28.9
17	-8.3	-6.7	58	14.4	11.6	98	36.7	29.3
18	-7.8	-6.2	59	15.0	12.0	99	37.2	29.8
19	-7.2	-5.8	60	15.6	12.4	100	37.8	30.2
20	-6.7	-5.3	61	16.1	12.9	101	38.3	30.7
21	-6.1	-4.9	62	16.7	13.3	102	38.9	31.1
22	-5.6	-4.4	63	17.2	13.8	103	39.4	31.6
23	-5.0	-4.0	64	17.8	14.2	104	40.0	32.0
24	-4.4	-3.6	65	18.3	14.7	105	40.6	32.4
25	-3.9	-3.1	66	18.9	15.1	106	41.1	32.9
26	-3.3	-2.7	67	19.4	15.6	107	41.7	33.3
27	-2.8	-2.2	68	20.0	16.0	108	42.2	33.8
28	-2.2	-1.8	69	20.6	16.4	109	42.8	34.2
29	-1.7	-1.3	70	21.1	16.9	110	43.3	34.7
30	-1.1	-0.9	71	21.7	17.3	111	43.9	35.1
31	-0.6	-0.4	72	22.2	17.8	112	44.4	35.5
32	0	0	73	22.8	18.2	113	45.0	36.0
33	0.6	0.4	74	23.3	18.7	114	45.6	36.4
34	1.1	0.9	75	23.9	19.1	115	46.1	36.9
35	1.7	1.3	76	24.4	19.6	116	46.7	37.3
36	2.2	1.8	77	25.0	20.0	117	47.2	37.8
37	2.8	2.2	78	25.6	20.5	118	47.8	38.2
38	3.3	2.7	79	26.1	20.9	119	48.3	38.7
39	3.9	3.1	80	26.7	21.3	120	48.9	39.1
40	4.4	3.6						

TABLE V.—*Measures used to express depths in Foreign Charts.*

National Measure.		Eng. Feet.	Eng. Fathoms.
French	Metre	3·281	0·5468
	Brasse	5·329	0·8881
Spanish	Braza	5·492	0·9153
Swedish	Fonn	5·843	0·974
Danish	Favn	6·175	1·0292
Norwegian	,,	6·175	1·0292
German	Faden	5·906	0·984
Dutch	Vaden	5·575	0·929
Russian	Marine Sashine	6·000	1·000
Portuguese	Braca	6·004	1·000

INDEX.

—◆◇◆—

A.

	PAGE
Accuracy, remarks on	60
Adjusting theodolite	14
Altitudes, circummeridian	198
„ „ at sea	297
„ equal	221
„ „ short	294
Aneroids, pocket	33
„ use of in heights	193
Angles, observing main	76
„ plotting	102
„ repeating theodolite	77
„ subtended by different lengths	App. O.
Artificial horizon	13
Astronomical observations for scale	197, 301
„ „ when taken	62
„ positions, correcting triangulation to..	97

B.

	PAGE
Bank, sounding a	149
„ searching for	156
Barometer, aneroid	33
„ metrical and English compared	App. T.
Bases	63
„ by angle of short measured length	68
„ by chain..	63
„ by difference of latitude	66
„ by masthead angle	67
„ by patent log	132
„ by rope	69
„ by sound	69
„ „ formula	App. F.

2 c

		PAGE
Beacon, fixed	.. ..	56
Beacons, floating	.. ..	54
„ use of..	.. ..	149
Bearing, mercatorial	.. ..	92, 95
„ true ..	.. ..	272
Boats' fittings ..	.. ..	48
„ gear ..	.. ..	50
Boards, drawing	.. ..	39
„ field ..	.. ..	43, 181, 296
Books, blank ..	.. ..	44
„ deck ..	.. ..	155
„ form for height	.. ..	188
„ sight, form of ..	.. ..	230
Bore, tidal ..	.. ..	176
Buoy, beacon ..	.. ..	54
„ small, for boats..	.. ..	148

C.

Catalogues, star	.. ..	200
„ „ obtaining apparent place from	.. .. ..	214
Centring error ..	.. ..	7, 210
Chains, measuring	.. ..	32
Chart, colouring on	.. ..	300
„ completed	.. ..	295
„ delineation of ..	.. ..	298
„ distortion of printed	.. ..	321
„ fair ..	.. ..	295
„ graduation of ..	.. ..	302
„ names on	.. ..	300
„ original ..	.. ..	295
„ soundings on original ..	.. ..	302
„ transferring to Mercator	.. ..	307
„ transmission home	.. ..	295
Chords, calculating ..	.. ..	103
„ plotting by ..	.. ..	102
„ proof of formula	.. ..	App. C.
„ table of	.. ..	App. J.
Chronometers ..	.. ..	44
„ comparing	.. ..	225
„ comparison book	.. ..	46
„ defects in pocket	.. ..	226
„ effect of temperature on	.. ..	257
„ observations for error of	.. ..	218
„ rejecting results of	.. ..	254

PAGE

Chronometers, stowage of 45, 260
" variation in rates of 256
" winding 45
Circummeridian altitudes of stars 198
" " sun 214
" " " at sea 297
Coast-line of island 139
Coast-lining 133
Collimation of theodolite, adjustment for 15
" " in heights.. 183
Colouring 300
Comparing watches 225
Compass, deviation of.. 327
" variation of 282
Contouring 180
Convergency of meridians 87, 94
" by spherical triangle 92, 304
" formulæ 92
" neglect of 112
" proof of formula .. .. App. A.
" proof of rule in graduating .. .. .. App. B.
Current, ascertaining rate of 171
" drag 322
" log 171
" under, observations on 322
" meters.. 325

D.

Datum for reduction 161
" " approximating a 164
Decimals of a day, time in .. .. App. S.
Deck book 155
" form .. .. App. G.
Deep sea soundings 308
Definitions, tidal 160
Degrees, lengths of .. .. App. L.
Delineation of charts 298
Deviation by distant object 327
" reciprocal bearings 328
" swinging.. 327
Dip in heights 186
" Table of .. .. App. N.
Distortion of printed charts 321
Diurnal inequality 161

		PAGE
Double altitude at sea		288
Drawing boards		39
„ rectangular lines		120
Dredging		318
Dry proofs		321

E.

Elevations		138, 162 _et seq._
Equal altitudes		221-2
„ „ at inferior transit		222, 235
„ „ elimination of errors by		221
„ „ meaning		231
„ „ short—at sea		294
„ „ of the stars		238
„ „ equation of		223, 233, 235
„ „ example		236
„ „ principle of		223
Equation of equal altitudes		223, 233, 235
Error centring		7, 210
„ collimation of theodolite		183
„ index of sextant		229
„ level, of theodolite		184
„ of chronometer, by stars		223
„ „ observations for		218
„ by equal altitude of two stars		238
„ personal		232
Errors of observation, eliminating		197
Establishment, tidal		166
Eyepieces, dark		228

F.

False station		80
Feet, number in degree and minutes		App. L.
Field boards		43, 181
Fittings for boats		48
Fix, by tracing paper		109
Fixing, by calculation from angles at position		118
„ care in choosing objects for		23
„ marks		115
„ marks from ship		116
„ soundings		142, 145
Foreign measures of depth		App. V.
Form for comparison book		App. H.

PAGE

Form for deck book .. .. App. G.
  ,, height book 188

G.

Galton Sun signal 35
Gauge, automatic tide 163
Gnomonic projection 88
  ,,   ,, graduating on 302
Graduation of chart 302
  ,, before plotting 114, 122

H.

Height book form 188
  ,, problems 191
Heights, absolute 192
  ,, by aneroid 193
  ,, by sextant 184
  ,, by theodolite 183
  ,, calculating 190
  ,, dependent 193
  ,, dip in .. .. 186, App. N.
  ,, distance by 193
  ,, formulæ for 191
  ,, obtaining 183
  ,, refraction in obtaining 185
Heliostat 35
  ,, arrangement of 79
  ,, use of 116
High water, obtaining.. 163
Hills, contouring 180
  ,, delineation of 300
Horizon, artificial 13
  ,, artificial precautions 201
  ,, distance of true .. .. App. Q.
  ,,   ,, visible .. .. Apps. E., P.
  ,, stand 13

I.

Inferior transit, equal altitudes at 222, 235
Inequality, diurnal 161
  ,, semi-menstrual 161
Interpolation, meridian distance by 261

	PAGE
Interpolation, meridian distance by, with harbour rates	269
Intervals of time in decimals of a day	App. S.
Irregular methods of plotting	109

L.

	PAGE
Latitude (at sea), by circummeridian altitudes of sun	294
„ by circummeridian altitudes of stars	198
„ „ „ sun	214
„ by pole star	209
„ by stars, example	208
„ by stars, valuing	212
„ observations for	197
Lead-lines	52
„ marking	53
„ measuring	150
Levelling	194
Level, mean water	169, 176
Line, straight ruling	106
Log, current	171
Logs, patent	51
Longitude, absolute	218
„ at sea, double altitude	288
„ by short equal altitude	294
„ differential	218
Lunitidal interval	161

M.

	PAGE
Main station, making	76
„ stations	73
„ triangulation	73
Marks	46
„ fixing	115
„ tripod	47
Measures, foreign—of depth	App. V.
Measuring chains	32
„ lead-lines	150
Meridian distance	219, 248
„ „ by harbour rates	264, 267
„ „ by interpolation	261
„ „ „ by harbour rates	267
„ „ by rockets	269
„ „ by travelling rates	253
„ „ chronometric	220, 253

	PAGE
Meridian distance, return of	271
„ „ telegraphic	248
Meridian, reduction to..	205
„ secondary	219
Mirrors, resilvering	10
Moon's transit	161
Mounting field boards..	43
„ paper	41

N.

Names on chart	300
Natural scale	302
New navigation	293

O.

Observations, astronomical—when to obtain	62
„ calculating time for	229
„ elimination of errors in	197
„ for error of chronometer, set of	224
„ for error of chronometer	218
„ for error, form for	231
„ for error, method of	227
„ for latitude	197
„ for true bearing	272
„ general remarks on	197
„ preparations for	201
„ sea for position	286
„ on undercurrents	322
Observing stars, method of	203
„ tides	162

P.

Paper mounting	41
„ sizes of	43
„ stretching of	108
„ transfer	40
Parallax, adjustment of theodolite for	14
„ in reading sextant	204
Personal error	232
Plans, reducing	297
„ scale in	301
Plotting	101
„ by chords	102
„ by distances	109
„ coast line	133

		PAGE
Plotting, irregular methods of	.. ..	109
„ with tracing paper	.. ..	109
Points, necessity of marking	.. ..	297
„ in transit	.. ..	26
Polaris, latitude by	.. ..	209
„ true bearing by	.. ..	281
Pole, ten-foot	.. ..	38
„ use of	.. ..	135, 139
„ tide	.. ..	162
Position, calculated from angles at it	.. ..	118
„ at sea, by Sumner's method	.. .. ..	289
Projection, gnomonic	.. ..	88
Proofs of rules	.. ..	App. A. to F.
Protractors	.. ..	32

R.

Rate, causes of variation of	.. ..	256
„ epochs for accumulation of	.. ..	267
„ harbour	.. ..	264
„ sea	.. ..	261
„ Tiark's formula for	.. ..	267
„ travelling	.. ..	220, 253
Rectangular lines, drawing	.. ..	120
Reducing plans	.. ..	297
„ soundings	.. ..	151
Reduction to meridian	.. ..	205
„ „ proof of formula	.. .. ..	App. D.
Reef sections	.. ..	150
Refraction in obtaining heights	.. ..	185
„ sea observations	.. ..	286
Resilvering mirrors	.. ..	10
Rivers, exploring	.. ..	325
Rock, sweeping for	.. ..	148
Rockets in meridian distance	.. ..	269
Rods, sounding	.. ..	310
Ruling a straight line	.. ..	106
Running survey	.. ..	122

S.

Scale, natural	.. ..	302
„ of chart	.. ..	62, 197, 301
Scales, brass	.. ..	31
Sections of reef slopes	.. ..	150

		PAGE
Sentry, submarine		158
Sextant angles		75, 182
,, ,, from ship		116
,, elevations by		193
,, Hadley's		5
,, sounding		9
,, stand		1?, 199
,, triangulation by		75
Sheet, graduation of, before plotting		114, 122
Ship sounding		152
Ship, use of, as station		116
Shoals, searching for		156
Sights (see Observations).		
Sketch		83
Sound, base by		69
,, ,, formula		App. F.
,, velocity of		64
Sounding		142
,, a bar		151
,, banks		149
,, book		145
,, deep sea		308
,, ,, management of ship		316
,, ,, preparing for		314
,, ,, routine casts		318
,, ,, signals for		313
,, ,, taking a		314
,, ,, time occupied		317
,, fittings for ship		152
,, importance of		142
,, lines, doubling		148
,, machines		308
,, machine, method of working		311
,, out of sight of land		149
,, rods		310
,, sextant		9
,, ship		152
,, taking a deep-sea		314
,, weights used for		310
Soundings, calling		151
,, direction of lines of		144
,, in a harbour		148
,, instruments for		153
,, on original chart		302
,, recording		145

2 D

	PAGE
Soundings, reducing	151
„ suspicious	147
Spherical excess	86
Spheroid, correction for	99
Squaring in	125, 297
Stand, artificial horizon	13
„ sextant	12, 199
Star atlas	202
„ catalogues	200
„ example of latitude by	208
Stars, calculating apparent places of	214
„ choosing pairs	200
„ daybreak at sea	287
„ latitude by	198
„ „ at sea	287
„ observing	203
„ pairing results	210
Station, false	80
„ main	73
„ „ making	76
„ pointer	21
„ „ circle for testing	29
„ „ caution as to use of	30
„ „ testing a	29
„ secondary	73
Steam cutters	48
Stores for boats	50
Straight-edge	31
Streams, observing tidal	171
Submarine sentry	158
Sumner's method	289
Sun, equal altitudes of	222
„ latitude by circummeridian altitude	214
„ signal, Galton's	35
Survey, detailed	58
„ general description	57
„ general plan of	61
„ modified running	126
„ ordinary	58
„ running	122
„ sketch	57
Suspicious ground	147
Sweeping for a rock	148
Swinging ship	327
Symbols	298

T.

	PAGE
Telegraphic meridian distance	248
„ „ „ example of	251
Temperatures, serial	317
Ten-foot pole	38
„ use of	135, 139
Theodolite	14
„ adjusting	14
„ collimation error	184
„ level error	184
„ repeating angles	77
„ measuring angles with	19
„ webs	18
Thermometers, corresponding	App. U.
Tidal bore	176
„ datum	164
„ definitions	160
„ establishment	166
„ reduction, table of	167
„ streams	171
„ „ time of change of	172
Tide, age of	161
„ effect of atmospheric pressure on	161
„ gauge, automatic	163
„ interpolating height of	166
„ reference mark for	163
„ pole	162
„ range of	160
Tides	159
„ general remarks on	173
„ graphic representation of ..	168
„ inequality of	173
„ interference	174
„ mean level	169
„ theory of	172
Time taking	230
Topography	178
„ specimen of	181
Tracing paper in plotting	109
Transfer paper..	40
Triangles, correcting	86
„ ill-conditioned	98
Triangulation, calculating	93
„ calculated	74

		PAGE
Triangulation, calculated, example of	.. ..	94
„ correcting for error of base	.. ..	97
„ definition of	.. ..	73
„ general	.. ..	59
„ kinds of	.. ..	73
„ main	.. ..	73
„ preparation for calculation of	.. .. ..	85
„ by sextant	.. ..	75
True bearing	.. ..	272
„ by equal altitudes	.. ..	273
„ by polaris	.. ..	155, 281
„ by sextant	.. ..	277
„ by single altitude	.. ..	275
„ for orientation	.. ..	85
„ use of—in plotting	.. ..	110, 113

U.

Undercurrents, observations of	.. .. ..	322

V.

Valuing results of observations	.. .. ..	212
Variation	.. ..	282
„ by sea observations	.. .. ..	283
„ by shore observations	.. .. ..	283
Venus, use of—by day	.. ..	288
Vernier, setting	.. ..	228
Vigias, searching for	.. ..	156

W.

Watches, comparing	.. ..	225
Water level, mean	.. ..	169, 176
„ line, low	.. ..	138
Weights	.. ..	40
Whitewashing	.. ..	46
Wire, splices in	.. ..	310
„ splicing to hemp	.. ..	311
„ strength of	.. ..	310
„ weight of	.. ..	313
„ winding	.. ..	311

Z.

Zero, choice of	.. ..	76
„ verifying	.. ..	79

LONDON : PRINTED BY WILLIAM CLOWES AND SONS, LIMITED,
STAMFORD STREET AND CHARING CROSS.

www.ingramcontent.com/pod-product-compliance
Lightning Source LLC
Chambersburg PA
CBHW032339280326
41935CB00008B/389